Meiosis

CELL BIOLOGY: A Series of Monographs

EDITORS

D. E. BUETOW
*Department of Physiology
and Biophysics
University of Illinois
Urbana, Illinois*

I. L. CAMERON
*Department of Cellular and
Structural Biology
The University of Texas
Health Science Center at San Antonio
San Antonio, Texas*

G. M. PADILLA
*Department of Physiology
Duke University Medical Center
Durham, North Carolina*

A. M. ZIMMERMAN
*Department of Zoology
University of Toronto
Toronto, Ontario, Canada*

Recently published volumes

Gary L. Whitson (editor). NUCLEAR-CYTOPLASMIC INTERACTIONS IN THE CELL CYCLE, 1980
Danton H. O'Day and Paul A. Horgen (editors). SEXUAL INTERACTIONS IN EUKARYOTIC MICROBES, 1981
Ivan L. Cameron and Thomas B. Pool (editors). THE TRANSFORMED CELL, 1981
Arthur M. Zimmerman and Arthur Forer (editors). MITOSIS/CYTOKINESIS, 1981
Ian R. Brown (editor). MOLECULAR APPROACHES TO NEUROBIOLOGY, 1982
Henry C. Aldrich and John W. Daniel (editors). CELL BIOLOGY OF *PHYSARUM* AND *DIDYMIUM*. Volume I: Organisms, Nucleus, and Cell Cycle, 1982; Volume II: Differentiation, Metabolism, and Methodology, 1982
John A. Heddle (editor). MUTAGENICITY: New Horizons in Genetic Toxicology, 1982
Potu N. Rao, Robert T. Johnson, and Karl Sperling (editors). PREMATURE CHROMOSOME CONDENSATION: Application in Basic, Clinical, and Mutation Research, 1982
George M. Padilla and Kenneth S. McCarty, Sr. (editors). GENETIC EXPRESSION IN THE CELL CYCLE, 1982
David S. McDevitt (editor). CELL BIOLOGY OF THE EYE, 1982
P. Michael Conn (editor). CELLULAR REGULATION OF SECRETION AND RELEASE, 1982
Govindjee (editor). PHOTOSYNTHESIS, Volume I: Energy Conversion by Plants and Bacteria, 1982; Volume II: Development, Carbon Metabolism, and Plant Productivity, 1982
John Morrow. EUKARYOTIC CELL GENETICS, 1983
John F. Hartmann (editor). MECHANISM AND CONTROL OF ANIMAL FERTILIZATION, 1983
Gary S. Stein and Janet L. Stein (editors). RECOMBINANT DNA AND CELL PROLIFERATION, 1984
Prasad S. Sunkara (editor). NOVEL APPROACHES TO CANCER CHEMOTHERAPY, 1984
Burr G. Atkinson and David B. Walden (editors). CHANGES IN EUKARYOTIC GENE EXPRESSION IN RESPONSE TO ENVIRONMENTAL STRESS, 1985
Reginald M. Gorczynski (editor). RECEPTORS IN CELLULAR RECOGNITION AND DEVELOPMENTAL PROCESSES, 1986
Govindjee, Jan Amesz, and David Charles Fork (editors). LIGHT EMISSION BY PLANTS AND BACTERIA, 1986
Peter B. Moens (editor). MEIOSIS, 1987

In preparation

Robert A. Schlegel, Margaret S. Halleck, and Potu N. Rao (editors). MOLECULAR REGULATION OF NUCLEAR EVENTS IN MITOSIS AND MEIOSIS, 1987
Monique C. Braude and Arthur M. Zimmerman (editors). GENETIC AND PERINATAL EFFECTS OF ABUSED SUBSTANCES, 1987
E. J. Rauckman and George M. Padilla (editors). THE ISOLATED HEPATOCYTE: USE IN TOXICOLOGY AND XENOBIOTIC BIOTRANSFORMATIONS, 1987

Meiosis

Edited by

Peter B. Moens
Department of Biology
York University
Downsview, Ontario, Canada

1987

ACADEMIC PRESS, INC.
Harcourt Brace Jovanovich, Publishers
Orlando   San Diego   New York   Austin
Boston   London   Sydney   Tokyo   Toronto

COPYRIGHT © 1987 BY ACADEMIC PRESS, INC.
ALL RIGHTS RESERVED.
NO PART OF THIS PUBLICATION MAY BE REPRODUCED OR
TRANSMITTED IN ANY FORM OR BY ANY MEANS, ELECTRONIC
OR MECHANICAL, INCLUDING PHOTOCOPY, RECORDING, OR
ANY INFORMATION STORAGE AND RETRIEVAL SYSTEM, WITHOUT
PERMISSION IN WRITING FROM THE PUBLISHER.

ACADEMIC PRESS, INC.
Orlando, Florida 32887

United Kingdom Edition published by
ACADEMIC PRESS INC. (LONDON) LTD.
24–28 Oval Road, London NW1 7DX

Library of Congress Cataloging in Publication Data

Meiosis.

   (Cell biology)
   Includes index.
   1. Meiosis. I. Moens, Peter B. II. Series.
QH605.3.M45   1987   574.87'62   86-10940
ISBN 0–12–503365–6 (alk. paper)

PRINTED IN THE UNITED STATES OF AMERICA

87 88 89 90   9 8 7 6 5 4 3 2 1

Contents

Preface xi

1 Introduction to Meiosis
Peter B. Moens

   I.   The Basics 1
   II. Commitment to Meiosis 5
   III. Chromosome Pairing 9
   IV. Recombination 13
         References 17

I Evolution

2 Genetic Transmission and the Evolution of Reproduction: The Significance of Parent–Offspring Relatedness to the "Cost of Meiosis"
Marcy K. Uyenoyama

   I.   Introduction 21
   II. The Covariance between Fitness and Genotype ... 22
   III. On the Evolution of Selfing 25
   IV. On the Evolution of Parthenogenesis 31
   V.   Summary 38
         References 38

3 The Evolution of Parthenogenesis: A Historical Perspective

Orlando Cuellar

I.	Introduction	43
II.	Meiosis and Parthenogenesis	44
III.	The Hybridization Hypothesis in Plants	47
IV.	Mutations as the Source of Parthenogenesis in Animals	55
V.	Hybridization as the Source of Parthenogenesis in Insects	58
VI.	The Prime Case of a Hybrid Origin in Insects	63
VII.	The Spontaneous versus Hybridization Controversy in Vertebrates	66
VIII.	The Mechanisms of Meiotic Restitution	80
	References	97

II Recombination

4 Meiotic Recombination Interpreted as Heteroduplex Correction

P. J. Hastings

I.	Introduction	107
II.	The Observations	107
III.	Heteroduplex Repair Models of Conversion	120
IV.	Alternatives to Excision Repair of Heteroduplex	131
	References	135

5 Models of Heteroduplex Formation

P. J. Hastings

I.	Introduction	139
II.	Direction of Travel of the Event	139
III.	Models of Heteroduplex Formation	140
IV.	Control of Crossover Position by an Aviemore Process	148
V.	Conversion and Crossing-Over as Separate Events	151
VI.	Speculation	154
	References	155

Contents

6 Investigating the Genetic Control of Biochemical Events in Meiotic Recombination

Michael A. Resnick

I.	Introduction	157
II.	Genetically Identifiable Factors Affecting Meiotic Recombination	163
III.	Methods of Isolating and Characterizing Mutants	175
IV.	Molecular and Genetic Studies of Meiotic Recombination in Yeast	182
V.	Conclusions	196
	References	198

III Chromosomes

7 Chiasmata

G. H. Jones

I.	Historical and Descriptive	213
II.	The Chiasmatype Theory	216
III.	The Mechanism of Chiasma Formation	217
IV.	Do Chiasmata Terminalize?	220
V.	The Nature of "Terminal" Bivalent Associations	221
VI.	Chiasma Variation: An Overview	223
VII.	Indications of Chiasma Control	225
VIII.	Genetic Control of Chiasma Distribution	230
IX.	Models of Chiasma Control	231
X.	Mechanisms of Chiasma Distribution Control	232
	References	238

8 The Synaptonemal Complex and Meiosis: An Immunocytochemical Approach

Michael E. Dresser

I.	Introduction	245
II.	An Introduction to SC Biology	246
III.	An Immunocytochemical Approach	252
IV.	Preliminary Results	255
V.	Summary	268
	References	269

9 The Rabl Orientation: A Prelude to Synapsis
Catharine P. Fussell

I.	Introduction	275
II.	The Rabl Orientation in Mitotic Cells	276
III.	Meiosis and the Rabl Orientation	284
IV.	Summary	294
	References	295

IV Chemistry of Meiosis

10 The Biochemistry of Meiosis
Herbert Stern and Yasuo Hotta

I.	Introduction	303
II.	The Prezygotene Phase	305
III.	The Zygotene Phase	308
IV.	The Pachytene Phase: DNA	313
V.	Meiosis-Specific Recombinogenic Proteins	318
VI.	Regulation of Meiotic Events	325
	References	329

11 Proteins of the Meiotic Cell Nucleus
Marvin L. Meistrich and William A. Brock

I.	Introduction	333
II.	Methods for Studying Nuclear Proteins in Meiosis	334
III.	Histones and HMG Proteins of Meiotic Cells	336
IV.	Proteins of Structural Subcomponents of the Meiotic Cell Nucleus	347
V.	Conclusions	349
	References	349

12 Transcription during Meiosis
P. T. Magee

I.	Introduction	355
II.	Transcription in Mammalian Meiocytes	356
III.	Transcription in Meiocytes of *Lilium*	358

IV.	Transcription in Meiocytes of *Saccharomyces cerevisiae*	358
V.	Summary and Future Prospects	376
	References	379

Index 383

Preface

This volume celebrates the coming of age of meiosis research. Although meiosis is the essence of sexual reproduction—and thereby deserves a central position in biological research—rarely has a symposium or a monograph been devoted to this topic. Meiosis usually dwells in the wings at genetic, chromosome, and cell biology meetings, while its rival, the mitotic cell cycle, frequently takes center stage.

Modern meiosis research became more consolidated when, at the conclusion of the 1973 annual meeting of the Genetics Society of America, Herbert Stern invited meiosis-oriented researchers to La Jolla for an informal discussion. There the topic was analyzed not so much within the confines of classical cytogenetics, but more in terms of its biochemistry, its recombinational mechanisms, and its ultrastructural phenomena.

Subsequently, meiosis claimed a more prominent position through "A Discussion on the Meiotic Process," organized by R. Riley, M. D. Bennett, and R. B. Flavell for the Royal Society of London on December 10 and 11, 1975 (*Phil. Trans. R. Soc. Lond.* B **277**, 183–376, 1977). On this occasion, new techniques and insights came to the fore, symbolized to an extent by the frontispiece depicting the ingenious Counce–Meyer spermatocyte surface spread, which revolutionized the study of meiotic chromosome behavior and structure for the next decade and beyond. To my own pleasure, the structural analysis of meiotic processes in yeast (*Saccharaomyces cerevisiae*) has since become commonplace. Curiosity about the genetic regulation of meiosis-specific functions is implicit in most papers, and presently the genetic dissection of meiosis has indeed become a field of intense research. The many observations on the synaptonemal complex—the ubiquitous nuclear organelle of the meiotic prophase nucleus—demanded knowledge of its molecular structure. Through the use of antibodies this information is now emerging.

As ten years have passed since the Royal Society discussion, it was a timely decision to organize a monograph on this topic. I gladly accepted the opportunity to bring together a number of researchers in the field to report on progress and, thereby, to show the direction of future research. I am pleased to acknowledge the personal assistance of my wife, Maria, and of my colleague Barbara Spyropoulos in the administration of the task. I am particularly grateful to Dr. A. Zimmerman for the initiative, to the staff of Academic Press for their assistance, and to the twelve outstanding scientists who were willing to contribute to this volume. We all share an interest in the topic and hope this volume will instruct the reader in the processes of meiosis and possibly recruit some into its study.

Peter B. Moens

Meiosis

1

Introduction to Meiosis

PETER B. MOENS

Department of Biology
York University
Downsview, Ontario
Canada M3J 1P3

I. THE BASICS

Chromosome behavior at meiosis differs in detail between species. To avoid unwarranted generalizations, I will illustrate chromosome behavior at meiosis by a specific example, the grasshopper *Chloealtis abdominalis*. To demonstrate the universality of the process of meiosis, the Easter lily *Lilium longiflorum* is shown in comparison. Variations in the regulation of meiosis are illustrated in *Chloealtis conspersa*, a close relative of *C. abdominalis*.

In sexually reproducing organisms, each of two parents contributes one set of chromosomes, a genome, to an offspring. In *C. abdominalis*, one genome consists of nine chromosomes with or without a sex chromosome (X). There are two long chromosomes with the centromere near the middle (Fig. 1j). These two metacentrics look V-shaped. The remainder have the centromere near the end; these telocentrics have a rodlike appearance. Chromosome #3 and the X chromosome are the two larger rods and chromosomes #4 to #9 are successively smaller. Upon fertilization between an egg (one genome + X) and a spermatozoan (one genome), the resulting zygote has two genomes plus an X chromosome per nucleus. The chromosomes of such a nucleus are shown in Fig. 1a, b. There are two of each kind of chromosome, except the X chromosome. The two smallest chromosomes #9 are in the middle of the group and the four V-shaped chromosomes are on the periphery. To emphasize that there are two genomes in Fig. 1b, one set of chromosomes is drawn in solid black and the other is stippled. The positions of the chromosomes relative to each other appear

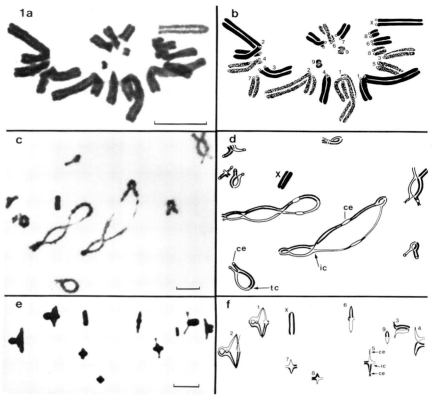

Fig. 1. Meiosis in the male grasshopper *Chloealtis abdominalis*. All scale bars are 10 μm. (a) Mitotic metaphase with 18 autosomes and an X chromosome. (b) Diagram of Fig. 1a to demonstrate that the somatic cells have two sets of chromosomes as a result of fertilization between a male and a female gamete, each carrying a single set of nine chromosomes (9 + X in the oocyte). The chromosomes are numbered according to length with the maternal complement in solid black and the paternal contribution stippled. Assignment of origin is arbitrary. The single genome is illustrated in Fig. 1j. (c) During meiosis, corresponding chromosomes of each of the two genomes become paired. As a result, there are now nine bivalents and an X chromosome per spermatocyte nucleus. At the time of pairing, the bivalents are too long and tangled to recognize singly, but after contraction into the diplotene stage of meiosis, the individual bivalents are evident. (d) Diagram of Fig. 1c to illustrate the pairing arrangement of each maternal chromosome (thick line) with the corresponding paternal homologue (thin line). The homologues have formed non-sister chromatid cross connections at several points. These reciprocal crossovers are visible as an interstitial chiasma (ic) or a terminal chiasma (tc, the size of the exchanged segment is too small to be drawn in). Centromere = ce. (e) Meiotic metaphase I. (f) The diagram of Fig. 1e. The centromeres (ce) of each pair of homologues

1. Introduction to Meiosis

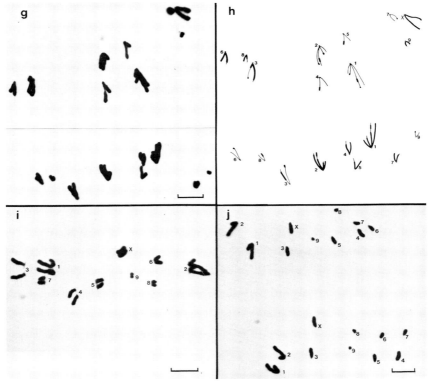

are directed to opposite poles. The chromosomes that originally came from the female parent (thick line) and the male parent (thin line) are directed randomly to one pole or another (assignment of origins is arbitrary in the drawing). The sister chromatids of each chromosome are still joined and, as a result, crossovers prevent precocious separation of the chromosomes. (g) Anaphase I of meiosis. The essential feature of meiosis, the reduction of two genomes per nucleus to one genome per nucleus, is accomplished at this time. (h) Diagram of Fig. 1g shows that only one homologue of each pair goes to the upper pole and the other to the lower. As a result, each new nucleus has only one set of chromosomes. The mix of thick and thin lines demonstrates the distribution of maternally and paternally derived chromosome material as a result of random assortment of centromeres and of reciprocal genetic exchange. One nucleus receives the unpaired X chromosome. The exchange points are taken from the chiasma positions in metaphase I (Fig. 1e, f). (i) Metaphase II of meiosis. Since the chromosomes are duplicated during S phase prior to pairing, the anaphase I of Fig. 1g can immediately enter second metaphase. Half the metaphase II nuclei have an X chromosome. (j) Anaphase II, the final step in the formation of a nucleus with one genome. The nine chromosomes are identified by number and can be compared with chromosomes of a somatic nucleus in Fig. 1a, b.

random in nuclei of this organism but they may be ordered in other organisms (Chapter 9).

In the adult male, the spermatogenic cells enter meiosis and the uniquely meiotic pairing of chromosomes takes place. The chromosomes that came from one parent become paired to the corresponding chromosomes that came from the other parent. The phenomenon becomes evident when the paired chromosomes, now bivalents, have contracted far enough to be recognized individually (Fig. 1c, d). It is clear that there are no longer $9 + 9 + 1 = 19$ bodies in the nucleus. Instead there are 9 bivalents and an unpaired sex chromosome in the *C. abdominalis* spermatocyte nucleus. The oocyte, not shown here, has 10 bivalents, the two paired X chromosomes being one of them. At meiosis, chromosomes become paired after they have duplicated so that there are four chromatids per bivalent. The chromatids of the same chromosome are referred to as sister chromatids and they are genetically identical. The chromatids of two homologues are non-sister chromatids. There can be small or even extensive differences between the two homologues of a bivalent (see Fig. 8). The diagram of Fig. 1d emphasizes that each chromosome which came from one parent (thick black line) is paired to the corresponding chromosome from the other parent (thin black line). The assignments are arbitrary and serve illustrative purposes only.

While the chromosomes are paired, a regulated program of chromosome breakage and repair causes non-sister chromatids to become cross connected (Chapters 4, 5, 6, and 10). These cross connections are visible as chiasmata in Figs. 1c, d, 2b, d, and 8c, e (Chapter 7). At this point, the genetic contributions of the two original parents are no longer distinct. The two genomes have become mixed. The degree of mixing is under genetic control and differs between species. It is an evolutionary adaptation of the organism which can generate or reduce genetic variability between offspring and it can thereby promote similarity or dissimilarity between parents and offspring (Chapter 2).

The synapsis of homologous chromosomes followed by their segregation to separate nuclei is the essential meiotic mechanism whereby sexually reproducing organisms produce cells with a single genome from cells with two genomes. The process of separation starts at the first metaphase of meiosis when the bivalents orient on the equator of the cell (Fig. 1e,f). The microtubules of the spindle become attached to the centromeres, or kinetochores, and appear to pull the centromeres of a bivalent to opposite poles. However, since sister chromatids still adhere to each other, the crossovers between non-sister chromatids prevent the two chromosomes of a bivalent from separating (Fig. 1f). The X chromosome is an exception and may go to either pole. Under abnormal conditions, where the adhesion between sister chromatids or crossing-over is interfered with, precocious separation results in unorderly distribution of chromosomes.

In a microscope preparation of live spermatocytes, metaphase I stage suddenly ends as the separating chromosomes move simultaneously and rapidly (1

1. Introduction to Meiosis

μm/min) to the poles of the cell. This first meiotic anaphase (Fig. 1g, h) is the cardinal moment in the meiocyte when the mixture of two genomes is sorted out into two single genomes. Each genome contains a complete set of chromosomes but the two sets are different in genetic detail. If, for example, the metaphase I bivalents of Fig. 1f divide into the anaphase I of Fig. 1h, then the distribution of thick and thin lines represents the mix of original maternal and paternal contributions in the two separating genomes. No two spermatocytes will have the same mix. The thoroughness of the mix is generated from two sources, the random orientation of bivalents at the metaphase plate in regard to parental origin and the exchange between non-sister chromatids.

Since the chromosomes were duplicated before they paired at meiotic prophase, the anaphase I chromosomes can enter the second metaphase of meiosis immediately, without an intervening duplicating stage (Fig. 1i). Metaphase II is followed by a second meiotic anaphase which reduces the single genomes of duplicated chromosomes to single genomes of single DNA content (Fig. 1j).

The remarkable similarity of the meiotic process, even between biological kingdoms, is evident from a comparison of meiosis in a grasshopper (Fig. 1) with meiosis in the Easter lily, *Lilium longiflorum* (Fig. 2). In the lily two sets of 12 chromosomes pair during meiotic prophase producing the pachytene stage of meiosis with an intractable tangle of long bivalents (Fig. 2a). Much of the biochemistry of meiosis discussed in Chapter 10 was done with these chromosomes. When the 12 bivalents shorten during prophase, they become individually recognizable (Fig. 2b). The diplotene bivalents obviously resemble, in chromatid and chiasma structure, those of the grasshopper in Fig. 1c. Both are in fact similar to most organisms that have genetic recombination at meiosis. At metaphase I, the bivalents orient on the equatorial plate in Fig. 2c and the centromeres are directed to the poles (Fig. 2c, d). At anaphase I (Fig. 2e, f) the undivided chromosomes move to the poles (as in Fig. 1g). The segregating chromosomes, consisting of two chromatids each, are already duplicated and can therefore enter metaphase II right away (Fig. 2g). The four single genomes produced by meiosis are evident in Fig. 2h.

II. COMMITMENT TO MEIOSIS

In complex organisms, meiosis is a genetically programmed step in the life of the organism. Meiosis-specific events may not be recognizable as such because they are embedded in the differentiation of the gonad and the gamete (Chapter 11). In free-living single-celled organisms as well as in relatively simple multicellular organisms, meiosis can occur as a response to environmental conditions. Since the environment can be artificially manipulated, the regulation of meiosis

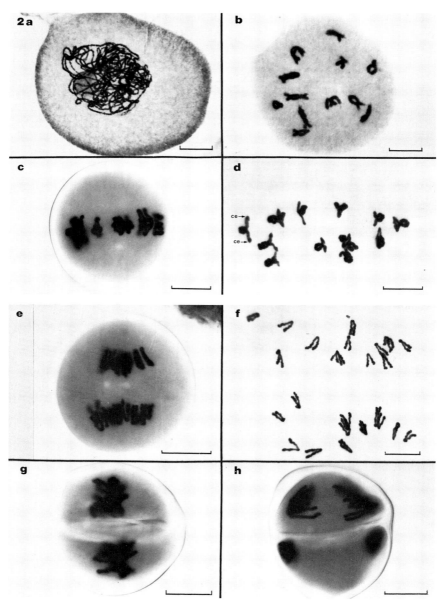

Fig. 2. Meiosis in pollen mother cells of the Easter lily, *Lilium longiflorum*. Bars = 10 μm. (a) During meiotic prophase, after chromosome duplication, the homologous chromosomes synapse and produce 12 long bivalents. This is the "pachytene" (thick strand) stage of meiosis. Because of their great lengths, the individual bivalents are not recognizable in this type of preparation. The grey

1. Introduction to Meiosis

is more accessible for analysis in such organisms (Chapters 6 and 12). Meiosis in the yeast *Saccharomyces cerevisiae* is given as an example here (Fig. 3).

The yeast *S. cerevisiae* can proliferate either as a haploid with one set of 16 chromosomes or as a diploid with two sets. The haploid cells cannot undergo meiosis. After a cell of mating type a fuses with a cell of α mating type, the diploid zygote can continue vegetative reproduction or it can enter meiosis. Under experimental conditions it is possible to induce meiosis in an α cell that has one set of chromosomes plus one extra chromosome #3 which carries the a allele. It follows that diploidy per se is not required for entrance into meiosis. a/α heterozygosity at the mating locus is necessary. Diploid cells of the unusual a/a or α/α mating type are incapable of undergoing meiosis.

Competent yeast cells can be made to switch from vegetative growth—budding—to sexual reproduction by starving the cells of nitrogen sources and sugar (Fig. 3). The cell ceases to bud; it becomes an ascus which will eventually contain four haploid ascospores, the products of two meiotic divisions. As the cells enter the meiotic pathway, they make a number of meiosis-specific biochemical adjustments (Chapters 6 and 12). Chromosome behavior at meiosis can be inferred from genetic experiments but cannot be observed directly because the yeast chromosomes are not clearly visible with the light microscope. The paired chromosomes, however, form synaptonemal complexes (SCs) which are visible with the light microscope (see examples of insect SCs in Figs. 4, 5, and 6, and discussion of SCs in Chapter 8). The 16 bivalents per nucleus are represented by 16 SCs in the meiotic prophase nucleus. In spite of the clearly meiotic characteristics of such a cell, it is apparently not committed to undergo meiosis. Return to an enriched medium causes cells to revert to vegetative growth and budding. Genetic analysis of reverted cells indicates that meiotic levels of recombination had occurred in the cells. In a complex organism, such as the Easter lily, pachytene cells can be reverted to a mitotic division when protein synthesis is inhibited (Fig. 2a) but, unlike yeast, such cells are moribund.

Once the kinetic apparatus of the yeast meiotic prophase cell is in place, the

---

sphere inside the nucleus is the nucleolus. (b) At diplotene, the 12 shortened bivalents are recognizable. Each bivalent has one or more chiasmata. These represent reciprocal exchanges between non-sister chromatids, as in Fig. 1c, d. (c) The 12 bivalents orient on the metaphase I plane of the pollen mother cell with the centromeres pointing to opposite poles. (d) A squashed preparation of a pollen mother cell in meiotic metaphase I, showing 12 bivalents, centromere orientation (ce), and chiasmata. Note the similarity to metaphase I in a grasshopper (Fig. 1e, f). There is no sex chromosome in *Lilium*. (e, f) Anaphase I in an intact cell and a squashed preparation. The duplicated but undivided chromosomes move to the poles. As in Fig. 1h, there is one of each chromosome at the pole so that the reduction from diploid to haploid cell is accomplished at this stage. (g) Metaphase II. A cell plate forms between the two groups of chromosomes. The chromosomes line up on the new equatorial plates. (h) Telophase II. At the end of the meiotic divisions, four haploid nuclei are formed.

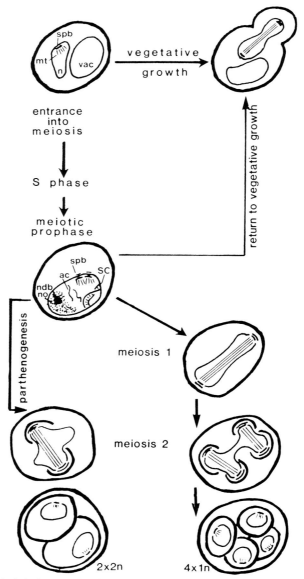

Fig. 3. Meiosis in the yeast *Saccharomyces cerevisiae*. Yeast cells reproduce through budding (vac = vacuole, spb = spindle pole body, mt = microtubule, n = nucleus). Under defined nutritional conditions, competent diploid cells of a/α mating type will enter meiotic development. Budding ceases, and following S phase, chromosomes pair and each pair forms a synaptonemal complex (sc), characteristic of meiotic prophase (ac = axial core, no = nucleolus, ndb = nuclear-dense body). Such cells are not committed to meiosis and can return to vegetative growth if the proper nutrients are added to the medium. Normally, however, a metaphase I spindle forms after about 6 hr in sporulation medium. The spindle elongates at anaphase I and immediately thereafter two anaphase II spindles are formed. The four haploid nuclei produced by meiosis become encapsulated in a cell wall. In the end the ascus contains four ascospores. In some parthenogenic strains, genes are present that suppress the first meiotic division. The second meiotic division than produces two diploid ascospores.

cell is committed to complete the meiotic divisions. The spindle pole bodies (Fig. 3, spb) move apart and come to bracket an intranuclear metaphase I spindle. At anaphase I, the spindle elongates within the confines of the ascus. The spindle pole bodies each duplicate and two spindles are formed. At anaphase II, these elongate and four haploid nuclei are produced. After the new cells are formed, there are four ascospores per ascus. In various mutant yeast strains with genetic or chromosomal defects, the kinetic apparatus follows its commitment to the meiotic division pattern, even to the extreme of forming anucleate spores.

Under conditions where the genetic variability generated by the meiotic process no longer contributes positively to the fitness of parents, offspring, and relatives, natural selection would favor modifications that reduce or abolish genetic differences between parents and offspring. In plants, inbreeding through self-pollination has that effect. In higher animals, parthenogenesis, the production of offspring from female reproductive cells without a genetic contribution from a male parent, also has that effect (Chapter 3). To produce clones of genetically identical offspring, yeast can rely on the vegetative cycle. The appearance of parthenogenesis in yeast suggests that selection has favored not only a reduction in genetic variability, but at the same time the advantages conferred by characteristics of the ascospore. A common form of parthenogenesis in yeast and other fungi is homothallism. An ascospore of a mating type produces a bud of type a and then, through a directed mutation, changes its mating type from a to α. Mother and bud are now capable of conjugation and the resultant zygote is genetically homozygous, except for the mating-type alleles. An alternative form of parthenogenesis is found in some naturally occurring yeast strains that produce two diploid, instead of four haploid, ascospores (Fig. 3). After chromosome pairing at meiotic prophase, the segregation of chromosomes is suppressed and instead the chromosomes unpair and divide in a meiosis II manner. The genetic simplicity of these modifications is remarkable. One to three alternative alleles can restore normal sexual reproduction. Even though the parthenogenic forms make use of the sexual reproductive machinery—meiosis and conjugation—the functions in terms of genetic variability are lost and the process is technically no longer sexual reproduction.

III. CHROMOSOME PAIRING

The meiotic process accomplishes the reduction from two genomes to one genome per nucleus by the juxtaposition of each individual chromosome from one genome with the corresponding chromosome of the other genome and by the subsequent segregation of one member of each pair to one pole and the other member to the opposite pole. The synapsis of homologous chromosomes is a meiosis-specific function not found in somatic cells. Homologous chromosome

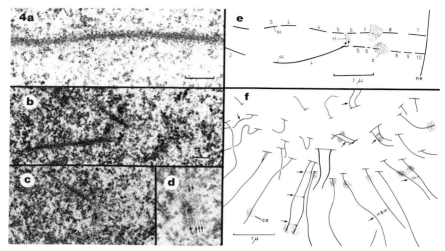

Fig. 4. Meiotic chromosome pairing in a *Locusta migratoria* spermatocyte. (a) At meiotic prophase, each chromosome (two chromatids) has an axial core with which the DNA is associated in a regular pattern (see Fig. 5). The cross-striated substructure of the core varies between species with some of the fungi having elaborate patterns. Bar = 0.2 μm. (b, c, d) The pairing of chromosomes is mediated by filaments which appear to pull the axial cores of homologous chromosomes together. Figures 4b, and c are from consecutive sections through a pairing fork and Fig. 4d is a higher magnification of Fig. 4c. Bar = 0.2 μm. (e) Diagram of the positions of axial cores in sections 2 to 10, through a set of homologous axial cores in the process of pairing. The transverse filaments (tf) have formed at one point. The centromeres (Fig. 4c) are not yet paired and all cores are attached to the nuclear envelope (ne). (f) Reconstruction of part of a meiotic prophase nucleus where chromosome pairing has started (arrows) but a number of cores are still unpaired.

synapsis of giant chromosomes in *Diptera* is a different phenomenon, based on the annealing of multiple DNA copies, rather than chromosomal cores. Also, nonhomologous synapsis, which is not uncommon at meiosis, does not occur in giant chromosomes.

Following meiotic S phase, each duplicated chromosome develops an axial core (Fig. 4a). At the same time, the synapsis of homologous chromosomes begins by the parallel alignment of pairs of axial cores (Fig. 4). The pairing of

---

Fig. 5. Synaptonemal complexes. When chromosome pairing is completed, all bivalents have a synaptonemal complex, formed by the parallel alignment of the axial cores of homologous chromosomes. (a) The nuclear contents of a spermatocyte of the moth *Hyalophora columbia* were spread in a low-salt solution on a plastic film, stained with ammoniacal silver, and photographed with an electron microscope. The technique leaves the nucleosomes intact and the looped DNA arrangement on the axial cores is evident. The dense-staining body is the nucleolus. This particular male was a trisome, having three instead of two homologues of one of the shorter chromosomes (center). Bar = 5 μm. (b) Axial cores with associated chromatin from the spread oocyte of the moth *Ephestia kuehniella* (Weith and Traut, 1980). Bar = 5 μm.

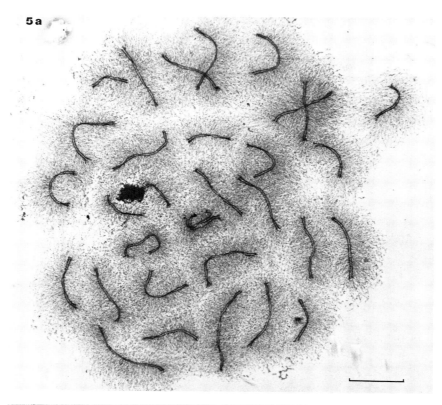
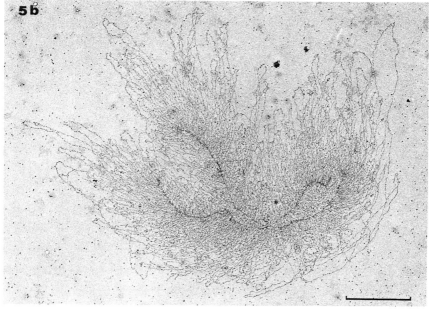

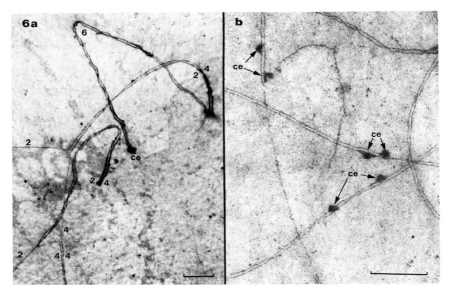

Fig. 6. (a) Translocation fine structure in a *Chloealtis conspersa* spermatocyte. Axial cores of chromosomes #4 are correctly paired and have a synaptonemal complex at 4-4. The ends of chromosome #4, however, are homologous to the ends of a pair of metacentric chromosomes 2-2. As a result, the ends of axial cores #4 are paired with the ends of axial cores #2, 2-4. Normal SC #6 with terminal centromere is also shown. Bar = 2 μm. (b) Meiotic mispairing in a *Chloealtis conspersa* spermatocyte. The centromeres (ce) of a pair of homologues are usually exactly opposite each other (as in Fig. 6a). Where they are not (arrows) it means that the homologous chromosomes are not accurately paired. Bar = 2 μm.

axial cores appears to be mediated by filaments extended between the cores (Fig. 4b, c, d; Chapter 8). When synapsis is completed, there is a haploid set of long bivalents per nucleus (Fig. 2a), each with two parallel cores, now together referred to as a *synaptonemal complex* (Fig. 5). At least some of this pairing results in the juxtaposition of homologous DNA with molecular accuracy. Genetic evidence shows that meiotic genetic exchange between non-sister chromatids is mostly exact and reciprocal. Surface spread preparations of synaptonemal complexes (Fig. 5) show that the chromatin (DNA + nucleosomes) loops back and forth to the axial core, now the lateral element of the synaptonemal complex. Evidently, only a small fraction of the total chromosomal DNA is associated with the SC, and of that fraction only a few locations become involved in exchange.

Where there are chromosomal rearrangements, pairing can become confounded by opposing demands on the pairing mechanism. For example, the spermatocyte nucleus in Fig. 5a came from a moth, *Hyalophora columbia*, which had three copies instead of two for one of the chromosomes. Pairing was

1. Introduction to Meiosis

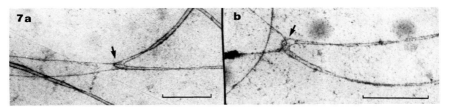

Fig. 7. (a, b) Interlocking. Two examples of chromosome interlocks (arrows) at meiotic prophase. Usually, few occur even though many might be expected, considering the length, the number, and the apparent tangle of single strands that pair during meiotic prophase. Bar = 2 μm.

initiated at one end between copies #1 and #2 and at the other end between copies #2 and #3. As pairing extends toward the middle, a conflict arises between the forces that bring the cores together and the mechanism designed to align two rather than three cores. The complex consequences have been reviewed elsewhere (von Wettstein *et al.*, 1984). Another example from a *C. conspersa* spermatocyte in Fig. 6a illustrates a case where chromosomes #2 have ends in common with chromosomes #4. If pairing initiation occurs within the common ends as well as in the remainder of the homologues, the configuration in Fig. 6a results. If pairing is not initiated in the ends, normal synaptonemal complexes form in the ends, even though, in the case of a translocation heterozygote, the paired ends are nonhomologous. Further evidence that synaptonemal complex formation can be broadly homologous but inaccurate in detail is evident in Fig. 6b. Here the centromeres of homologous chromosomes are clearly misaligned. A further complexity facing the pairing mechanism is the pairwise alignment of long and apparently intertwined cores in the narrow confines of a nucleus (Fig. 7). The evidence for interlocks at metaphase I is rare and one of the possible explanations holds that homologous chromosomes are prealigned (Chapter 9). Observations on the actual pairing of cores suggest that there is a mechanism for breakage and rejoining of interlocked cores (von Wettstein *et al.*, 1984).

IV. RECOMBINATION

The physical exchange between non-sister chromatids at meiosis can be observed directly where there are morphological differences between the homologous chromosomes. In Fig. 8, one of the #8 chromosomes of *C. conspersa* carries a supernumerary segment (B). When the #8 homologues become paired at meiotic prophase, there is a synaptonemal complex between the standard segments of the #8 chromosomes but the B segment remains as an unpaired core (Fig. 8a, b). At the diplotene stage of meiotic prophase, the B segment is evident

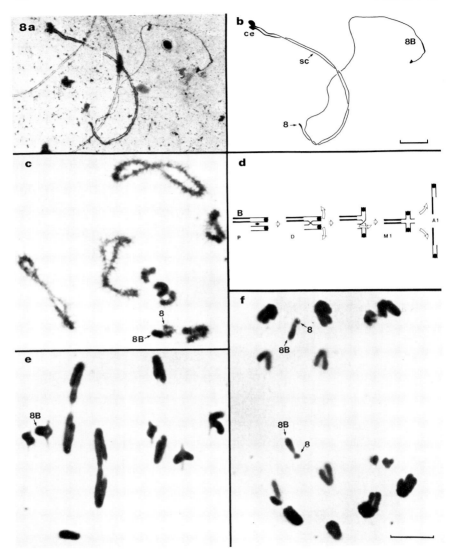

Fig. 8. Recombination. Demonstration of reciprocal exchange, a crossover, in a heteromorphic bivalent. (a, b) In a male grasshopper (*Chloealtis conspersa*) with one normal chromosome #8 and one chromosome #8 with an extra segment B, there is a synaptonemal complex (SC) between the normal portions and unpaired axial core of the 8B segment. Bar = 2 μm. (c) The heteromorphic 8-8B bivalent at the diplotene stage of meiotic prophase. (d) Diagram of a crossover event. At P the differential #8 chromosomes are paired as in Fig. 8a. As a result of breakage and repair, non-sister chromatids have exchanged parts at D. Each centromere now carries a short and a long chromatid. While the sister chromatids remain tightly associated, the centromeres orient toward the poles and the

1. Introduction to Meiosis

as a large heterochromatic knob on one of the chromosomes of the smallest bivalent (Fig. 8c). After an exchange event between non-sister chromatids in the standard region, each of the centromeres carries one long and one short arm (Fig. 8d). This is evident at metaphase I (Fig. 8e), where each #8 centromere has a standard arm on the right and a standard arm and a B segment on the left. The evidence is complete at anaphase I (Fig. 8f), where each #8 centromere has a long and a short arm. In summary, at the start of meiosis, one chromosome #8 has two short arms and the other has two long arms. As a result of breakage and rejoining of non-sister chromatids, each centromere has a long and a short arm at anaphase I. The mechanism and regulation of genetic exchange are discussed in Chapters 4, 5, and 6.

As a result of chromosome reduction, segregation, and genetic recombination, the genotype of the parent is not passed on as such to the offspring in sexually reproducing organisms. To what extent genetic differences between parent and offspring do or do not have selective advantages is a function of the interaction between individual and environment (Chapter 2). In the case of parthenogenesis, the organism uses the sexual reproduction machinery, but with such modifications that identical or near-identical offspring are produced (Chapter 3). A less thorough suppression of differences is found in permanent translocation heterozygotes as in *Rhoeo spatheceae* and some *Oenothera* species, where parental sets of chromosomes stay together from generation to generation. There is no meiotic recombination in males of a number of dipteran species and in females of some lepidopteran species. More subtle adjustments involve a regulation of the amount and position of crossing-over per bivalent at meiosis (Chapter 7).

A comparison between metaphase I in *C. abdominalis* (Fig. 1e) and *C. conspersa* (Fig. 8e) shows that the distribution of chiasmata differs between the two species. In *C. abdominalis* the two large metacentric bivalents each have three chiasmata along the length of the bivalent. In *C. conspersa*, on the other hand, there are two chiasmata in each of the three metacentric bivalents and they are strictly localized at the ends of the bivalents. The effect is that in *C. conspersa* the genes of a metacentric chromosome are inherited together, not subject to

---

chromosomes rotate to produce the metaphase I asymmetric cross configuration (see Fig. 8e). The crossover point or chiasma, together with the sister chromatic adhesion, holds the bivalent intact. At anaphase I, chromatids are released and the centromeres move to the poles. The original long chromosome #8 and the short chromosome #8 have clearly exchanged parts so that each chromosome now has a long and a short chromatid (see also Fig. 8f). (e) Metaphase I in *C. conspersa* showing the asymmetric bivalent #8 with the 8B segment to one side. The figure also illustrates the genetic regulation of chiasma positions. Unlike in *C. abdominalis* (Fig. 1), where chiasmata occur anywhere along the bivalent, the chiasmata in *C. chloealtis* are highly localized at the ends of the chromosomes. (f) At anaphase I, the exchange of non-sister chromatids is evident since centromeres #8 now carry a short and a long arm. Figure 8c, e, and f have the same magnifications. Bar = 10 μm.

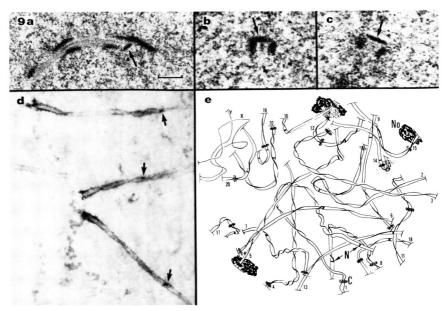

Fig. 9. Recombination nodules. (a, b, c) RNs in the synaptonemal complexes of a rat spermatocyte (arrows at the RNs). Bar = 0.5 μm. (d) RNs of the grasshopper *Chloealtis conspersa* are localized in the ends of the SCs, which correlates with the localization of chiasmata in the ends of the bivalents (Fig. 8c, e). (e) Distribution of 19 RNs (N) among the 20 SCs of a rat spermatocyte nucleus. The nonrandom distribution correlates RNs with reciprocal recombination events. It also is evidence for a regulatory mechanism. No = nucleolus, C = centromere, and SCs are numbered in decreasing order of length.

recombination as they are in *C. abdominalis*. The selective advantage is not obvious. The two species are morphologically similar and they both occur across North America with *C. conspersa* favoring a more humid habitat.

In organisms that have genetic exchange at meiosis, a minimum of one crossover per bivalent appears to be necessary to assure proper metaphase I orientation and normal anaphase I segregation. Short bivalents #4 to #9 in Fig. 1f all have one crossover. Considering that a crossover can form anywhere along the bivalent, the intriguing question arises how the one crossover prevents others from forming. This phenomenon of genetic or chromosome interference is also evident in long bivalents where chiasmata rarely form close together—statistically less frequently than if they were randomly distributed along the length of the bivalent (Chapter 7). The process of reciprocal recombination appears to be correlated with a specific structure, the recombination nodule (RN; Fig. 9). In general, RNs are distributed among the bivalents as reciprocal recombinational events are. The reconstructed rat spermatocyte nucleus in Fig. 9d has 19 RNs

1. Introduction to Meiosis

evenly distributed among the 20 bivalents. The fact that there are fewer RNs than expected is attributed to the short life span of a RN. Where the position of the crossover is predictable, the RN is found in the corresponding position. In the onion *Allium fistulosum,* the crossovers occur exclusively next to the centromere.

G. Jones (Chapter 7) has shown that in *A. fistulosum* the RN is present next to the centromere. In the grasshopper *C. conspersa* the chiasmata are localized in the ends of the chromosomes, which coincides with the positions of the RNs (Fig. 9c). The RN presumably is an enzyme complex active in breakage, annealing, and repair of DNA at a crossover site. The factors that regulate the numbers and positioning of RNs would also regulate interference, but these factors have not been defined as yet.

REFERENCES

von Wettstein, D., Rasmussen, S. W., and Holm, P. B. (1984). The synaptonemal complex in genetic segregation. *Annu. Rev. Genet.* **18,** 331–413.
Weith, A., and Traut, W. (1980). Synaptonemal complexes with associated chromatin in a moth, *Ephestia kuehniella* Z. *Chromosoma* **78,** 275–291.

I
Evolution

2

Genetic Transmission and the Evolution of Reproduction: The Significance of Parent–Offspring Relatedness to the "Cost of Meiosis"

MARCY K. UYENOYAMA

Department of Zoology
Duke University
Durham, North Carolina 27706

I. INTRODUCTION

At the heart of any theory of heredity lies the resemblance between parent and offspring. This fundamental phenomenon occurs in all forms of life and in fact makes Darwinian evolution possible. Just as the meiotic mechanism directs evolution through its effects on the pattern of inheritance, the process of genetic transmission itself evolves by natural selection. In this chapter I discuss the operation of natural selection on genetically based variation in reproductive mode. Parent–offspring relatedness determines in large part the relative selective advantages of biparental reproduction and uniparental reproduction by selfing or parthenogenesis.

Biparental reproduction with random mating bears a twofold "cost of meiosis" (Williams and Mitton, 1973) relative to uniparental reproduction. Heuristic explanations of this enormous selective disadvantage fall into two major categories: ecological and evolutionary (Williams, 1980). Ecological explanations suggest that anisogamous organisms which invest more resources in the gamete or sex that limits offspring production reproduce at higher rates (Maynard Smith, 1971a,b; 1976; 1977; 1978, Chapter 1). For example, if males are sufficiently abundant to ensure insemination of all females then offspring production in the subsequent generation is limited by the number of females in the present generation. Evolutionary explanations attribute the advantage of uniparental reproduction to higher parent–offspring relatedness (Williams, 1971, 1975, pp. 8–9; Williams and Mitton, 1973; Maynard Smith, 1974). For example, parent–offspring relatedness under selfing or parthenogenesis is twice as high as that under biparental reproduction with random mating. Both ecological and evolutionary explanations indicate that uniparental reproduction enjoys a considerable selective advantage, which, in view of the obvious success of biparental reproduction, gives rise to the "paradox of sexuality" (Williams, 1975, p. 7): Why don't all anisogamous sexual species evolve toward uniparental reproduction?

Possible resolutions to this paradox have been proposed on several levels of selection: species, group, and individual. Significant factors concerning the limited question of the relative advantages of uniparental and biparental reproduction (apart from the broader issue of sex, which includes the evolution of recombination and meiosis itself) may be classified as multigenerational or immediate in their effects. Advantages associated with biparental reproduction that are realized only over the span of several generations include the ability to evolve in response to deteriorating or fluctuating conditions caused by a variety of factors, including accumulation of deleterious genes, increasingly effective competitors or predators, abiotic changes resulting in diminishing habitable ranges, or persistent pathogen cycles (Maynard Smith, 1978, Chapters 2–4; Williams, 1975, Chapters 12 and 13; Glesener and Tilman, 1978; Shields, 1982, Chapter 6; Hamilton, 1982). Immediate advantages may accrue from increased offspring variability itself as a result of various forms of sib competition (Williams, 1975, Chapter 2; Maynard Smith, 1978, Chapter 6). Inbreeding biparental reproduction may reduce the cost of meiosis by increasing parent–offspring relatedness (Solbrig, 1979; Williams, 1980; Bernstein *et al.*, 1981; Shields, 1982, Chapter 6). Excellent critical reviews of this extraordinarily complex problem are available (Maynard Smith, 1978; Williams, 1975). I restrict the present discussion to a mathematical description of the genetic basis of the cost of meiosis within populations, with particular attention to the significance of parent–offspring relatedness; I do not consider the various mechanisms by which the cost may be paid.

II. THE COVARIANCE BETWEEN FITNESS AND GENOTYPE

While the numbers of offspring produced by means of the various reproductive modes represent readily observable phenotypes in an appropriate experimental

2. Genetic Relatedness and the "Cost of Meiosis"

system, elucidation of the relationship between phenotype and fitness may prove problematic. Fisher (1941) showed that the propensity to self is strongly favored, in populations in which male function is not impaired by selfing, even if the genotypes are identical with respect to visability and other qualities commonly recognized as components of fitness. He concluded that in such systems attempts to interpret evolutionary trends through the construction of adaptive topographies in fitness would be doomed to failure because the relevant measure of fitness is open to question. This problem of interpretation also arises in sex ratio selection and other forms of frequency-dependent selection in which the relationship between observable traits and fitness is complex. [Heuristically useful fitness functions have been proposed for such cases (Charlesworth and Charlesworth, 1978; Uyenoyama and Bengtsson, 1979), but those constructions do not appear in the basic recursions describing evolutionary changes in genotype frequencies.]

Li (1967) and Price (1970) independently introduced a means of representing evolutionary change which has proved to have great heuristic value in the analysis of frequency-dependent selection. This covariance method, as it has come to be known, involves the calculation of the covariance between a tautological measure of fitness and the additive genotypic value associated with an observable character, and offers a concise description of the selective significance of the trait. Upon its introduction, the method provided a simple means of obtaining Fisher's (1930) fundamental theorem of natural selection in modified form. The fundamental theorem states that the change between generations in mean fitness in an evolving population is equal to the additive genetic variance in fitness. The "theorem" itself is in fact an approximation which holds strictly only in the absence of dominance or if the genotypes assume Hardy–Weinberg proportions even after selection (Ewens, 1979, p. 52). Nevertheless, two corollaries of the fundamental theorem, that the additive genetic variance is zero at equilibrium and that the change in mean fitness is always nonnegative, are strictly true under nonepistatic viability selection (Kingman, 1961; Ewens, 1969) and have retained their roles as cornerstones of population genetics. The covariance method modifies the fundamental theorem by replacing the change in mean fitness by the change in mean additive genotypic value as the central symbol of adaptation.

In general, natural selection at a single multiallelic locus may be represented by recursions in genotype frequencies:

$$Tu_{ij}' = g_{ij} \qquad (1)$$

where u_{ij} is the frequency of genotype A_iA_j, g_{ij} is a function of the fitness parameters and genotype frequencies, and T is the normalizer that ensures that the genotype frequencies in the next generation sum to unity. A tautological measure of genotypic fitness is the increase in the number of individuals of that genotype relative to the initial number:

$$Tu_{ij}'/u_{ij} \qquad (2)$$

Using this quantity as the character measurement, one obtains the additive genotypic values associated with the genotypes by minimizing the dominance variance (Fisher, 1930; see Kempthorne, 1957, Chapter 16). Dominance is defined as the deviation between the trait value in question (fitness, in this case) and additive genotypic value, and the dominance variance (V_D) is the mean squared deviation between these measures:

$$V_D = \sum_i \sum_{j \geq i} u_{ij}\{[Tu_{ij}'/u_{ij} - T] - (\alpha_i + \alpha_j)\}^2 \quad (3)$$

where the first bracket within the square represents the trait value centered relative to the mean and the quantity $(\alpha_i + \alpha_j)$ represents the additive genotypic value of genotype A_iA_j. Differentiation of Eq. (3) with respect to the α_i produces the normal equations whose simultaneous solution produces the additive genotypic values that minimize V_D. Because the additive and dominance components of phenotype are uncorrelated, the covariance between additive genotypic value and fitness is equal to the additive genetic variance:

$$Cov(AF) = Cov[A(A + D)] = Var(A) \quad (4)$$

where A represents the additive genotypic value, F the central fitness measure, and D the dominance value. The various measures associated with genotype A_iA_j are summarized in Table I.

The central quantity of the covariance method is the covariance between fitness and additive genotypic value. [Li and Price originally considered the covariance between fitness and frequency of one of the alleles at a biallelic locus, but the extension to multiple alleles through the additive genotypic value is straightforward (Uyenoyama et al., 1981).] This covariance is in fact equal to both the change in mean additive genotypic value and the additive genetic variance:

$$\begin{aligned} Cov(AF) &= \sum_i \sum_{j \geq i} u_{ij}[Tu_{ij}'/u_{ij} - T](\alpha_i + \alpha_j) \\ &= 2T \sum_i \Delta p_i \alpha_i \\ &= Var(A) \end{aligned} \quad (5)$$

TABLE I

Trait Values Associated with Genotype A_iA_j

F	A	D
$Tu_{ij}'/u_{ij} - T$	$\alpha_i + \alpha_j$	$[Tu_{ij}'/u_{ij} - T] - \alpha_i - \alpha_j$

where p_i represents the frequency of allele A_i and Δp_i the change in gene frequency between generations. The equality between the covariance and the change in mean additive genotypic value represents a simple rearrangement of the definition of the covariance, and the equality between the covariance and the additive genotypic value is immediate from Eq. (4). Equation (5) yields Li's (1967) and Price's (1970, 1972) modified version of the fundamental theorem of natural selection: the change in mean additive genotypic value is equal to the additive genetic variance.

The covariance method finds its greatest use as a heuristic device which permits the investigation of the relationship between fitness and a simpler, observable component of fitness. For example, in Fisher's (1941) model the propensity to self, the basic character under selection, evolves according to the way in which it determines fitness. To each genotype, one associates two characters, fitness and selfing propensity, and each trait value is decomposed into its additive and dominance components. The covariance between fitness and additive genotypic value with respect to the propensity to self (rather than with respect to fitness) is again equal to the change in mean additive genotypic value:

$$\text{Cov}(A_\beta F) = \sum_i \sum_{j \geq i} u_{ij}[Tu_{ij}'/u_{ij} - T](\beta_i + \beta_j) \qquad (6)$$

$$= 2T \sum_i \Delta p_i \beta_i$$

where $\beta_i + \beta_j$ represents the additive genotypic value with respect to selfing propensity of genotype $A_i A_j$. Because selfing propensity represents but one component of fitness, the value of the covariance does not necessarily reduce to the additive genetic variance in fitness, but depends on the interaction between selfing and other components of fitness, which may include inbreeding depression or reduction in mating success through male gametes. In subsequent sections, I apply the covariance method to a number of well-known mathematical models, whose dynamic behavior was studied in greater depth in the original papers, in order to describe the nature of the selective forces operating under selfing and parthenogenesis within a common framework.

III. ON THE EVOLUTION OF SELFING

A. Genetic Variation in Selfing Propensity

Morphology strongly influences mating systems in self-compatible plants. For example, Moore and Lewis (1965) demonstrated that a natural population of

Clarkia xanthia that reproduced primarily by selfing differed from its outcrossing parent in the relative timing of stigma maturation and anther dehiscence, two correlated traits that can effectively promote or preclude selfing. A number of studies have documented the existence of genetic variation in the production of open, chasmogamous flowers and closed, cleistogamous flowers which self in the bud (see review by Lord, 1981). Bates and Henson (1955) estimated a broad sense heritability of about 36% for the percentage of seeds produced that were derived from chasmogamous flowers in greenhouse populations of *Lespedeza cuneata*. Clay (1982a) obtained broad sense heritability estimates of 8.7 and 3.1% for the numbers of cleistogamous and chasmogamous flowers per plant, respectively, in *Danthonia spicata*, which, like other grasses, can produce at most one seed per flower. Breese (1959) imposed artificial selection on the relative position of stigma and anthers in *Nicotiana rustica* and produced significant differences in selfing rates measured in the field under pollination by bees. Variation in a number of morphological traits that influence selfing propensity has been shown to include a genetic component (see reviews by Jain, 1976; Lord, 1981; Holsinger *et al.*, 1984).

B. Costs Associated with Selfing and Outcrossing

Interspecific comparisons of reproduction by selfing and outcrossing indicate that outcrossing incurs substantial energetic costs (Daly, 1978; Solbrig, 1979; Lloyd, 1980). Self-fertilizing species characteristically bear flowers of reduced size compared to their outcrossing ancestors (Stebbins, 1970). Solbrig (1979) observed that the thousandfold difference in pollen/ovule ratios between flowers of self-incompatible outcrossers and cleistogamous flowers represents a heavy energetic cost of outcrossing in the production of protein-rich pollen grains. Other energetic costs borne by outcrossing plants include the production of special morphological features that promote pollen movement and of attractants and rewards for pollinators (Solbrig, 1979; Lloyd, 1980).

Of primary importance to this discussion are the pleiotropic costs that confront changes in the breeding system within evolving populations. Interspecific comparisons may reflect responses or adaptations to inbreeding rather than factors that promote or oppose its evolution. For example, the mild or nonexistent inbreeding depression observed in plants that normally engage in selfing (Grant, 1975, p. 124; Clay and Antonovics, 1985), in contrast with the severe inbreeding depression caused by forcing selfing or incestuous matings of animals that normally outbreed (see reviews by Grant, 1975, pp. 124–126; Shields, 1982, pp. 60–65), represents a consequence of, rather than a preadaptation to, selfing. Lande and Schemske (1985) have described the expected negative relationship between the magnitude of inbreeding depression caused by deleterious mutations and the current level of inbreeding. A meaningful assessment of the trade-offs and options open to variants requires within-population comparisons of the reproductive success of individuals practicing different levels of inbreeding and outcrossing. In *Danthonia spicata* variation in the production of cleistogamous and chasmogamous flowers includes a genetic component (Clay, 1982a) and furthermore offspring derived from the two types of seeds differ with respect to a number of characters including seed weight, germination rate, and seedling survival (Clay, 1983): reproductive mode appears to be correlated with components of reproductive success. In addition, reproductive success through pollen may also be negatively correlated with selfing propensity, particularly in species in which mode of reproduction is enforced by flower morphology (Wells, 1979). For example, pollen from cleistogamous flowers, which never achieve anthesis, fails to enter the pool from which outcrossing ovules are fertilized. These studies demonstrate that components of the costs of reproduction, including seed and pollen fertility and offspring viability, can vary with selfing propensity; further, the magnitudes of those costs may themselves evolve (Lande and Schemske, 1985) and come to depend on the history of inbreeding in the population.

C. Models of the Evolution of Selfing

If the ability to self entails no reduction in male gamete (pollen) production, then genes that promote selfing are clearly favored at all frequencies (Fisher, 1941; Moran, 1962, pp. 58–59; Antonovics, 1968). This selective advantage can overcome considerable inbreeding depression (up to a twofold difference in viability) and other deleterious traits associated with selfing (Crosby, 1949; Jain and Workman, 1967; Maynard Smith, 1977; Charlesworth, 1980; Lloyd, 1979, 1980). Reductions in male gamete production as a consequence of selfing can neutralize in part or in whole the segregational advantage of selfing (Nagylaki, 1976; Wells, 1979; Charlesworth, 1980; Feldman and Christiansen, 1984; Holsinger *et al.*, 1984). Trade-offs between the genetic segregational advantage of selfing on the one hand and inbreeding depression and lowered pollen success on the other determine the adaptive significance of selfing.

To each genotype A_iA_j at the locus that determines selfing propensity associate four characters: the numbers of uniparental (π_{ij}) and biparental (σ_{ij}) offspring derived from ovules, relative pollen success (τ_{ij}), and fitness. Table II expresses each trait value as a deviation from the mean in the population and displays the additive genotypic value obtained for each character by minimizing the dominance variance (see Eq. (3)). Offspring numbers are censused at reproductive age and reflect viability and fertility differences associated with mode of reproduction. For example, Charlesworth (1980) set σ_{ij} equal to his $1 - \alpha_i$, the outcrossing rate of the ith genotype; π_{ij} to his $\alpha_i(1-\delta)$, the selfing rate discounted by

TABLE II

Four Characters Associated with Genotype A_iA_j

Character	Deviation	Additive genotypic value
Biparental offspring	$\sigma_{ij} - \sigma$	$\alpha_i + \alpha_j$
Uniparental offspring	$\pi_{ij} - \pi$	$\beta_i + \beta_j$
Pollen success	$\tau_{ij} - \tau$	$\gamma_i + \gamma_j$
Fitness	$Tu_{ij}'/u_{ij} - T$	

inbreeding depression; and τ_{ij} to his $1 - \alpha_i\lambda$, the outcrossing rate discounted by the reduction in pollen success. Holsinger et al. (1984) generalized Charlesworth's assignment of pollen success. Fitness, the fourth character, assumes the tautological representation in Eq. (2), which compares the number of individuals of each genotype in the next generation to the number in the present generation. Expressions for fitness are taken directly from the recursions

$$Tu_{ii}' = u_{ii}\pi_{ii} + \sum_{j \neq i} u_{ij}\pi_{ij}/4$$
$$+ m_i[u_{ii}\sigma_{ii} + \sum_{j \neq i} u_{ij}\sigma_{ij}/2]$$

$$Tu_{ij}' = u_{ij}\pi_{ij}/2$$
$$+ m_i[u_{jj}\sigma_{jj} + \sum_{k \neq j} u_{jk}\sigma_{jk}/2] \quad (7)$$
$$+ m_j[u_{ii}\sigma_{ii} + \sum_{k \neq i} u_{ik}\sigma_{ik}/2]$$

$$m_i = \frac{u_{ii}\tau_{ii} + \sum_{j \neq i} u_{ij}\tau_{ij}/2}{\sum_k \sum_{j \neq k} u_{jk}\tau_{jk}}$$

where m_i represents the frequency of allele A_i in the pollen pool from which outcrossed ovules are fertilized.

The covariance between fitness and additive genotypic value associated with biparental offspring production reflects the nature of the interactions among the components of fitness:

$$\text{Cov}(AF) = \sum_i \sum_{j \geq i} u_{ij}[Tu_{ij}'/u_{ij} - T](\alpha_i + \alpha_j)$$

$$= 2T \sum_i \Delta p_i \alpha_i \qquad (8)$$

$$= \text{Var}(A_\alpha)\{[1 + \phi_P \sigma/\tau]/2 + \phi_U\}$$

where p_i represents the frequency of allele A_i; $\text{Var}(A_\alpha)$ the additive genetic variance in biparental offspring production among the parents; σ the average number of ovules produced; τ the average pollen success; and ϕ_P and ϕ_U the linear regression coefficients of pollen success and uniparental reproduction on biparental reproduction. The terms within the bracket in Eq. (8) represent covariances between the maternal trait and the additive genotypic values of uniparental offspring and of biparental offspring derived from ovules and through pollen. Those covariances represent parent–offspring relatedness (Orlove and Wood, 1978; Michod and Hamilton, 1980; Uyenoyama et al., 1981).

The linear regression coefficient ϕ_U is

$$\phi_U = \frac{\text{Cov}(A_\alpha A_\beta)}{\text{Var}(A_\alpha)} \qquad (9)$$

where A_α and A_β represent the additive genotypic values of genotype $A_i A_j$ with respect to the characters of biparental and uniparental offspring production. Direct estimation of ϕ_U requires an analysis of covariance between the two characters among relatives (Falconer, 1981, Chapter 19). The similarity between relatives with respect to the product of the two character measurements (numbers of biparental and uniparental offspring censused at reproductive age) associated with each individual provides a means of estimating the additive genetic correlation, and ϕ_U, the regression, can then be obtained if the heritabilities of the characters are known. No estimates of this kind are available. Clay's (1982b, Chapter 4) variance–covariance matrix yields a phenotypic correlation between the numbers of cleistogamous and chasmogamous flowers produced by the same individual of 0.2; this value may be greatly inflated by environmental effects that promote positive correlations between the numbers of the two kinds of progeny produced by the same plant. The regression coefficient ϕ_P expresses the relationship between the additive genotypic values of individuals with respect to biparental offspring production and pollen success.

Inbreeding depression and reduced pollen success in selfers tend to create negative ϕ_U and positive ϕ_P: the uniparental component of fitness is negatively correlated with the other two components. For example, under the parameter assignments adopted by Holsinger et al. (1984) for their model of proportional rates of pollen discounting,

$$\phi_U = -(1-s) \quad \text{and} \quad \phi_P = \delta \qquad (10)$$

where $(1-s)$ represents the viability of uniparental offspring relative to biparental offspring, and δ reflects the reduction in pollen success of selfers. Charlesworth's (1980) parameters also produce Eq. (10) with s replaced by his δ and δ replaced by his λ. Under Eq. (10), Eq. (8) becomes proportional to

$$\text{Var}(A_\alpha)\{(1 - s) - [1 + \delta\sigma/\tau]/2\} \tag{11}$$

fitness and outcrossing are positively correlated if the intensity of inbreeding depression is sufficiently high relative to a function involving the rate of pollen discounting. In the biallelic model of Holsinger et al. (1984) Eq. (8) reduces to

$$2T\Delta p = (\alpha_1 - \alpha_2)(4pq - v)[2s - (1 - \delta)/(1 - \delta\bar{a})]/2 \tag{12}$$

where p represents the frequency of A_1, q the frequency of A_2, v the frequency of the heterozygote, and $\bar{a}$ the average rate of selfing. The frequency of A_1 increases monotonically if $(\alpha_1 - \alpha_2)$, the average effect of gene substitution (Fisher, 1930, 1941) on the character of outcrossing, and the second bracket both remain positive, or if both remain negative. In the absence of overdominance and underdominance in selfing rate, the average effect of substitution never changes sign and the gene promoting selfing monotonically increases in frequency to fixation if

$$(1 - \delta)/(1 - \delta a_L) > 2s \tag{13}$$

where a_L is the lowest selfing rate among genotypes. Charlesworth (1980) obtained Eq. (13) using the "phenotypic fitnesses" method. If the inbreeding depression parameter lies in an intermediate range, such that

$$(1 - \delta)/(1 - \delta a_H) > 2s > (1 - \delta)/(1 - \delta a_L) \tag{14}$$

where a_H is the highest selfing rate among genotypes, then the gene promoting selfing increases only if it is sufficiently common (Holsinger et al., 1984). The frequency-dependent nature of the advantage of selfing derives from competition within the pollen pool.

In the presence of overdominance or underdominance, changes in gene frequencies depend on both the effect of the genes on the rate of outcrossing $(\alpha_1 - \alpha_2$, the average effect of substitution) and whether outcrossing contributes positively to fitness. The average effect of substitution is itself influenced by the degree of inbreeding depression and the selfing rates through its dependence on the frequencies of the various genotypes (see Kempthorne, 1957, Chapter 16). Equations (8) and (11) indicate that a necessary condition for the existence of a polymorphic equilibrium is that either the additive genetic variance in outcrossing rate or the second bracket be zero. In the model of Holsinger et al. (1984) an equilibrium corresponding to zero additive genetic variance exists only if the two homozygotes have identical selfing rates; otherwise, any polymorphic equilibrium that may exist (possibly none exist) corresponds to equality between the average rate of selfing and a function of the level of inbreeding depression and the pollen discounting rate (see Eq. (12)). The evolutionary dynamics depend on

2. Genetic Relatedness and the "Cost of Meiosis" 31

both the correlation between outcrossing and fitness (inequalities (13) and (14)) and the average effect of substitution on the rate of outcrossing.

IV. ON THE EVOLUTION OF PARTHENOGENESIS

A. Genetic Variation in Parthenogenetic Ability

Thelytoky, the development of females from unfertilized eggs, generally requires profound disruption of the normal meiotic process in diploids. Upon its introduction in an imperfect, unrefined state, it may represent a much less efficient system of reproduction than either pure sexuality or pure parthenogenesis (Williams, 1975, pp. 106–108; White, 1978, pp. 317–318). I specifically exclude from this discussion cases in which reproductive isolation accompanies the appearance of the parthenogen (e.g., *Cnemidophorus uniparens;* Cuellar, 1976), such that the entire species may derive from a single individual. The question of the persistence of such parthenogenetic clones involves ecological factors including competition and direct "reproductive wastage" (White and Contreras, 1979) caused by inviable hybridization with the bisexual ancestor. I also exclude cases in which the incipient parthenogenetic mechanism is "destabilized" (Lynch, 1984) by introgression with the sexual parent species. This discussion addresses the population genetics of the gradual evolution of thelytoky from tychoparthenogenesis within populations. While the hybrid origin of several parthenogenetic species in the genus *Cnemidophorus* is certain, the parthenogenetic capacity almost surely preceded the appearance of the alloploid individuals that gave rise to the parthenogenetic species (see reviews by Cuellar, 1974; White, 1978, Chapter 9). I consider the origin of parthenogenesis during that earlier stage, within species that primarily reproduce biparentally.

While White (1970, 1973, p. 743, 1978, p. 318) suggests that in certain cases cytogenetic mechanisms that resulted in clonal parthenogenesis may have arisen as macromutations, perhaps the more frequent evolutionary pattern involves the appearance of ill-formed tychoparthenogenetic mechanisms, followed by gradual improvement, and finally the transition to ameiotic parthenogenesis (Suomalainen, 1950; Stalker, 1954; Suomalainen *et al.,* 1976; Templeton, 1982). Parthenogenesis appears as a rule to require coordinated changes at a number of genetic loci in plants as well as in animals (Marshall and Brown, 1981), rather than arising as a syndrome of correlated expressions of a single developmental process as may be the case for certain mechanisms promoting selfing (Breese, 1959; Moore and Lewis, 1965). Artificial selection and breeding experiments have revealed polygenic variation for parthenogenetic rate (Stalker, 1954, 1956; Muntzig, 1958; Carson, 1967; Templeton *et al.,* 1976a,b). Nevertheless, instances involving the abrupt appearance of viable, self-perpetuating parthenogenetic lines have been reported (Stalker, 1954; see review by Cuellar, 1974).

B. Costs Associated with Parthenogenesis

Cytogenetic and developmental abnormalities occur with high frequency in parthenogenetic populations that are relatively recently derived from bisexual ancestors (see reviews by Uyenoyama, 1984; Lynch, 1984). For example, White (1973, pp. 713–714) suggests that the high mortality rates in the egg and embryo stages in *Lonchoptera dubia* and *Drosophila mangabeirai* may be attributed in part to fusion between the acentric and dicentric products of crossing-over within paracentric inversions. In *D. mangabeirai* all individuals are heterozygous for the same paracentric inversions, and offspring are regularly derived by fusion of the central nuclei which would hold the acentric and dicentric chromosomes (Murdy and Carson, 1959).

In addition to lowered viability of uniparental offspring, reduced fertility as a result of the production of triploid offspring upon insemination may oppose the rise of parthenogenesis. Experimentally induced hybridization between *Warramaba virgo*, a clonally reproducing parthenogen, and a close bisexual relative results in the formation of sterile triploids (White and Contreras, 1979). In contrast, uniparental reproduction usually occurs only in the absence of insemination in *Drosophila* (Stalker, 1954) and in *Solenobia* (see White, 1973, p. 721). The importance of fertilization of unreduced parthenogenetic eggs as a factor opposing the evolution of thelytoky appears to depend on the cytogenetic mechanism involved and varies among groups.

An empirical assessment of the pleiotropic costs summarized by the genetic regression coefficient between the number of biparental and uniparental offspring produced requires the study of populations in which individuals are capable of both modes of reproduction. Stalker (1954, Table 11), working with *Drosophila parthenogenetica*, and Carson *et al.* (1969), working with *Drosophila mercatorum*, reported the apparent simultaneous production of the two kinds of offspring. (Because impaternate offspring were recognized by recessive sex-linked visible markers, the exceptional offspring may in fact have represented nondisjunction events.) Comparison of brood sizes among two groups of individuals, one reproducing entirely uniparentally and the other entirely biparentally, provides a rough estimate of the relative costs of the two kinds of offspring. Obligate parthenogenetic species, in which reallocation of resources formerly invested in primary and secondary sexual characters into uniparental reproduction may have evolved, often produce broods of sizes comparable to or greater than those of their bisexual relatives (see examples in Uyenoyama, 1984). However, comparisons between more closely related biparental and uni-

2. Genetic Relatedness and the "Cost of Meiosis"   33

parental forms indicate that twofold or greater reductions in brood size have often accompanied the transition to parthenogenesis (Uyenoyama, 1984; Lynch, 1984).

C. Models of the Evolution of Parthenogenesis

Uniparental reproduction, if unopposed by factors such as inbreeding depression or cytogenetic abnormalities, maintains a selective advantage over biparental reproduction at all gene frequencies (Jaenike and Selander, 1979; Charlesworth, 1980; Marshall and Brown, 1981). Inbreeding depression must reduce the viability of uniparental offspring by a factor of $\frac{1}{2}$ in hermaphroditic or dioecious populations with equal sex ratios in order to neutralize the segregational advantage of parthenogenesis (Charlesworth, 1980; Lloyd, 1980; Bulmer, 1982; Harper, 1982; Uyenoyama, 1984). In dioecious populations with sex ratios differing from unity among biparentally derived broods, the level of inbreeding depression that balances the advantage of parthenogenesis is equal to the inverse of the frequency of females in bisexual broods (Triesman and Dawkins, 1976; Charlesworth, 1980; Lloyd, 1980; Bell, 1982, p. 69; Uyenoyama, 1984).

The dependence of the advantage of uniparental reproduction on the sex ratio in broods derived by biparental reproduction called into question the importance of parent–offspring relatedness as a determinant of the cost of meiosis because, with respect to autosomal loci, parents are equally closely related to sons and daughters (Triesman and Dawkins, 1976; Charlesworth, 1980; Lloyd, 1980). Charlesworth (1980) and Lloyd (1980) analyzed models indicating that the mating system in a biparentally reproducing population has no effect on the relative advantage of a reproductively isolated parthenogenetic population and concluded that relatedness does not contribute to the cost of meiosis. Their assumption of reproductive isolation transforms the genetic problem into an ecological one: interspecific competition replaces genotypic selection as the primary evolutionary force. Williams (1980) emphasized that comparisons of parent–offspring relatedness among reproductive modes is relevant to the question of the evolution of parthenogenesis by individual selection only when such comparisons involve individuals within the same genetically variable population. An analysis of the change in gene frequency at a locus controlling parthenogenesis in a single, genetically variable population in which sibmating occurs under biparental repreduction indicates that inbreeding has little effect on the cost of meiosis in diploids but promotes the evolution of thelytoky in haplodiploids (Uyenoyama, 1985). I assess the importance of parent–offspring relatedness to the cost of meiosis through the application of the covariance method to these models of the evolution of parthenogenesis in dioecious populations.

Each genotype A_iA_j at the locus controlling parthenogenetic capacity influences three characters: the numbers of uniparental (π_{ij}) and biparental (σ_{ij})

offspring and fitness (see Table II; pollen success is not relevant here). Minimization of the dominance variance associated with the first two phenotypes produces expressions for the additive genotypic value of A_iA_j with respect to biparental $(\alpha_i+\alpha_j)$ and uniparental $(\beta_i+\beta_j)$ offspring production. Offspring numbers are censused at reproductive age and reflect maternal fertility, the ability of fertilized and unfertilized eggs to initiate and complete development, and the relative viabilities of the two types of offspring. The proportions of males and females among the biparental offspring produced by each genotype are r and $(1-r)$, respectively. Among the uniparental offspring of heterozygous mothers, a proportion k are identical in genotype to their mother, with $(1-k)/2$ the proportion of each of the two homozygotes that can be formed from the mother's alleles (see Asher, 1970; Templeton and Rothman, 1973). This formulation subsumes ameiotic parthenogenesis $(k=1)$, pronuclear duplication $(k=0)$, and other cytogenetic mechanisms falling between these extremes. If biparentally reproducing individuals mate at random, the recursions in genotype frequencies among females and males become

$$T_f f_{ii}' = f_{ii}\pi_{ii} + (1-k)\sum_{j\neq i} f_{ij}\pi_{ij}/2$$
$$+ (1-r)y_i[f_{ii}\pi_{ii} + \sum_{j\neq i} f_{ij}\pi_{ij}/2]$$

$$T_f f_{ij}' = k f_{ij}\pi_{ij}$$
$$+ (1-r)y_i[f_{jj}\pi_{jj} + \sum_{k\neq j} f_{jk}\pi_{jk}/2] \qquad (15)$$
$$+ (1-r)y_j[f_{ii}\pi_{ii} + \sum_{k\neq i} f_{ik}\pi_{ik}/2]$$

$$T_m y_i' = ry_i \sum_j \sum_{k\geq j} f_{jk}\pi_{jk}/2$$
$$+ r[f_{ii}\pi_{ii} + \sum_{j\neq i} f_{ij}\pi_{ij}/2]/2$$

where f_{ij} is the frequency among females of genotype A_iA_j, y_i the frequency among males of A_i, and T_f and T_m the normalizers that ensure that the frequencies of the genotypes among females and among males sum to unity.

Under dioecy, differences in reproductive value between the sexes (Fisher, 1930) affect the overall covariance between genotype and fitness. While all females are assumed to reproduce at rates determined only by their genotypes, males contribute to future generations only to the extent that they succeed in inseminating biparentally reproducing females. The expected number of matings

2. Genetic Relatedness and the "Cost of Meiosis"

by a male is equal to the inverse of the sex ratio in the population at large, but only a fraction of those matings will result in offspring that bear paternal genes. The proportion of females that reproduce biparentally evolves under the selective process under study; perhaps the best estimate of the current value of this proportion is the proportion of females that are themselves biparental. In total, the covariance among males between genotype and fitness is weighted, relative to the covariance among females, by the number of biparentally derived females divided by the number of males (see Lloyd, 1980; Uyenoyama, 1984, 1985).

Under this assignment of male reproductive value, the covariance method indicates that the adaptive significance of parthenogenesis depends on the trade-off between numbers of offspring of the two types and on the frequency of females in bisexual broods:

$$\text{Cov}(A_\alpha F) = \text{Cov}_F(A_\alpha F) + \text{Cov}_M(A_\alpha F)(1 - r)\sigma/T_m$$
$$= 2 \sum_i [T_f \Delta x_i + (1 - r)\sigma \Delta y_i] \alpha_i \quad (16)$$
$$= \text{Var}(A_\alpha)[(1 - r) + \phi]$$

where x_i represents the frequency among females of A_i, $\text{Var}(A_\alpha)$ the additive genetic variance among females, $(1-r)\sigma$ the average number of biparental daughters produced, and ϕ the regression over females of additive genotypic value with respect to uniparental reproduction on additive genotypic value with respect to biparental reproduction (see Eq. (9)). The regression coefficient ϕ is likely to be negative because allocation of a greater fraction of resources to the one mode of reproduction probably entails lower investment in the other mode. For example, if $\pi_{ij} = (C - \sigma_{ij})s$, where C represents the maximum brood size and s the viability of uniparental offspring relative to biparental offspring, then

$$\phi = -s \quad (17)$$

(see Bulmer, 1982). Substitution of Eq. (17) into Eq. (16) indicates that natural selection favors biparental reproduction only if the viability of uniparental offspring is lower than the frequency of females in bisexual broods.

Parent–offspring relatedness enters into this result through covariances between maternal offspring production and offspring additive genotypic value for each offspring type. Relatedness between parent and uniparental offspring is equal to unity. Biparental offspring of both sexes are related to the parent by $\frac{1}{2}$; the mother–son relatedness, like the male covariance, is weighted. The covariance between the additive genotypic value of biparental offspring and the maternal phenotype with respect to biparental offspring production is proportional to

$$[(1 - r) + r\sigma(1 - r)/T_m]/2 \quad (18)$$

The weight assigned to male contributions relative to female contributions in the model under consideration is equal to $(1 - r)/r$, and the expression in Eq. (18)

reduces to $(1 - r)$. Expression (16) represents a comparison between offspring numbers, weighted by relatedness.

The hypothesis that the cost of meiosis derives from parent–offspring relatedness implies that the cost of meiosis should vary with the mating system under biparental reproduction (Barash, 1976; Solbrig, 1979; Williams, 1980; Bernstein et al., 1981). Inbreeding promotes investment in females by increasing the reproductive value of females relative to males (Hamilton, 1967; Maynard Smith, 1978, p. 161), and indirectly reduces the cost of meiosis by reducing the disparity in female production between individuals practicing biparental and uniparental reproduction (Charlesworth, 1980, Shields, 1982, p. 126; Bell, 1982, p. 70). Shields (1982, Chapter 6) argued that resolution of the paradox of sexuality lies in inbreeding biparental reproduction, which incurs little or no cost of meiosis, a lower mutational load than asexuality, and a lower recombinational load than outbreeding. Application of the covariance method to a model describing the evolution of parthenogenesis in a population undergoing sibmating permits one to distinguish between the direct effects of inbreeding on relatedness and reproductive value.

Consider a population in which maternal genotype controls reproductive mode as described earlier, with the additional stipulation that a proportion t of all females, regardless of mode of reproduction, are paired with brothers. The full description of genetic change requires the construction of recursions in frequencies of mating pairs, in which reciprocal matings are distinguished. Genotypic frequencies, obtained by summing over the appropriate mating types, are incorporated in the usual way in the covariance method. As in Eq. (16), the overall covariance between fitness and genotype is formed from the male covariance, weighted by the expected fraction of matings that result in genetic contribution by males, and the female covariance:

$$\text{Var}_M(A)\{(1 - r)b_{M \to S}^{(B)} + b_{M \to B}^{(B)} [(1 - t) + t\sigma(1 - r)/T] (1 - r) \quad (19)$$
$$+ \phi[1 + tb_{M \to B}^{(U)} \sigma(1 - r)/T]\}$$

where $b_{M \to S}^{(B)}$ and $b_{M \to B}^{(B)}$ are the mother-to-daughter and mother-to-son relatedness measures with respect to biparental offspring production, $b_{M \to B}^{(U)}$ the mother-to-son relatedness with respect to uniparental offspring production, and ϕ the regression defined as in Eqs. (16) and (9) (Uyenoyama, 1985). The relatedness between mother and son with respect to biparental offspring production is

$$b_{M \to S}^{(B)} = \frac{\text{Cov}(S_A M_G)}{\text{Cov}(M_A M_G)} \quad (20)$$

where $\text{Cov}(S_A M_G)$ represents the covariance between the additive genotypic value of biparental daughters and their mothers' phenotype with respect to bi-

2. Genetic Relatedness and the "Cost of Meiosis"

parental reproduction, $b_{M \to B}^{(B)}$ a similar quantity involving biparental sons, and $b_{M \to B}^{(U)}$ a quantity involving the covariance between the additive genotypic value of sons and their mothers' phenotype with respect to uniparental reproduction. In the absence of sibmating ($t = 0$), the parent–offspring relatedness measures reduce to $\frac{1}{2}$ and Eq. (19) reduces to Eq. (16).

Equation (19) indicates that while the correlation between fitness and biparental reproduction is a complex, frequency-dependent function, relatedness between parents and offspring of both types retains its central role. As before, ϕ, the regression measure which expresses the trade-off in offspring number between uniparental and biparental reproduction, is expected to be negative. Factors that increase the magnitude of the term in brackets multiplied by ϕ promote parthenogenesis by reducing the covariance between fitness and biparental reproduction, and factors that increase the other terms promote biparental reproduction.

Relatedness between parent and biparental offspring of both sexes increases with the rate of inbreeding in diploids, as expected. In haplodiploids mother-to-daughter relatedness increases with inbreeding, but mother-to-son relatedness is independent of the similarity between mates because males make no genetic contribution to their sons. Inbreeding increases the reproductive value of females relative to males. Production of a daughter by either reproductive mode represents the eventual production of a brood of grandoffspring; further, if the daughter sibmates the relatedness between the controlling parent and its grandoffspring will increase to the extent that its sons contribute to the grandoffspring. Production of an additional son does not increase grandoffspring production by sibmating daughters because a single son is assumed to be capable of inseminating all females that sibmate. Production of an additional son does increase grandoffspring production through randomly mating females, but the number of such females declines as inbreeding increases.

These two direct effects of inbreeding act antagonistically in determining the adaptiveness of biparental reproduction: higher relatedness between parent and biparental offspring promotes biparental reproduction but lower male reproductive value promotes parthenogenesis. In diploids, the two effects cancel each other out in the conditions for initial increase and the cost of meiosis remains near its value in randomly mating populations. In haplodiploids, the relatedness effect is weaker because it occurs in only one sex, and the reproductive value effect predominates: inbreeding promotes thelytoky in haplodiploids.

Hamilton's (1967) survey of closely inbreeding haplodiploid arthropods indicated a high incidence of thelytoky within the genus or family of the species listed. The evolution of thelytoky was regarded as the completion of the extreme female-bias in brood sex ratio that Hamilton predicted would evolve under local mate competition. The analysis described here suggests that thelytoky may represent an evolutionary response to inbreeding itself.

V. SUMMARY

Parent–offspring relatedness emerges as a central quantity that determines the selective advantage of biparental reproduction relative to selfing or parthenogenesis. The association between biparental reproduction and fitness depends on the trade-off between numbers of offspring of the two types weighted by the parent–offspring relatedness. Uniparental reproduction maintains a segregational advantage that derives from the higher fidelity of replication of the parental genotype. Under inbreeding, this advantage interacts with the reproductive value of females and males. Because thelytoky alters both parent–offspring relatedness and the sex ratio within broods, inbreeding does not always reduce the cost of meiosis. Inbreeding has little effect on the conditions permitting the maintenance of biparental reproduction in diploids and promotes parthenogenesis in haplodiploids.

ACKNOWLEDGMENTS

Keith Clay graciously made available several unpublished manuscripts and gave generously of his time. I thank Janis Antonovics and Peter B. Moens for suggestions on the manuscript. Research supported by PHS grant HD 17925.

REFERENCES

Antonovics, J. (1968). Evolution in closely adjacent plant populations. V. Evolution of self-fertility. *Heredity* **23,** 219–238.
Asher, J. H. (1970). Parthenogenesis and genetic variability. II. One-locus models for various diploid populations. *Genetics* **66,** 369–391.
Barash, D. P. (1976). What does sex really cost? *Am. Nat.* **110,** 894–897.
Bates, R. P., and Henson, P. R. (1955). Studies of inheritance in *Lespedeza cuneata* Don. *Agron. J.* **47,** 503–507.
Bell, G. (1982). "The Masterpiece of Nature." Univ. of California Press, Berkeley.
Bernstein, H., Byers, G. S., and Michod, R. E. (1981). Evolution of sexual reproduction: Importance of DNA repair, complementation, and variation. *Am. Nat.* **117,** 537–549.
Breese, E. L. (1959). Selection for differing degrees of out-breeding in *Nicotiana rustica. Ann. Bot. (London)* **23,** [N.S.] 331–344.
Bulmer, M. G. (1982). Cyclical parthenogenesis and the cost of sex. *J. Theor. Biol.* **94,** 197–207.
Carson, H. L. (1967). Selection for parthenogenesis in *Drosophila mercatorum. Genetics* **55,** 157–171.
Carson, H. L., Wei, I. Y., and Niederkorn, J. A., Jr. (1969). Isogenicity in parthenogenetic strains of *Drosophila mercatorum. Genetics* **63,** 619–628.
Charlesworth, B. (1980). The cost of sex in relation to mating system. *J. Theor. Biol.* **84,** 655–671.
Charlesworth, D., and Charlesworth, B. (1978). A model for the evolution of dioecy and gynodioecy. *Am. Nat.* **112,** 975–997.

Clay, K. (1982a). Environmental and genetic determinants of cleistogamy in a natural population of the grass *Danthonia spicata*. *Evolution (Lawrence, Kans.)* **36,** 734–741.
Clay, K. (1982b). Ecological and genetic consequences of cleistogamy in the grass *Danthonia spicata*. Ph.D. Dissertation, Duke University, Durham, North Carolina.
Clay, K. (1983). The differential establishment of seedlings from chasmogamous and cleistogamous flowers in natural populations of the grass *Danthonia spicata* (L.) Beauv. *Oecologia* **57,** 183–188.
Clay, K., and Antonovics, J. (1985). Demographic genetics of the grass *Danthonia spicata:* Success of progeny from chasmogamous and cleistogamous flowers. *Evolution (Lawrence, Kans.)* **39,** 205–210.
Crosby, J. L. (1949). Selection of an unfavourable gene-complex. *Evolution (Lawrence, Kans.)* **3,** 212–230.
Cuellar, O. (1974). On the origin of parthenogenesis in vertebrates: The cytogenetic factors. *Am. Nat.* **108,** 625–648.
Cuellar, O. (1976). Intraclonal histocompatibility in a parthenogenetic lizard: Evidence of genetic homogeneity. *Science* **193,** 150–153.
Daly, M. (1978). The cost of mating. *Am. Nat.* **112,** 771–774.
Ewens, W. J. (1969). Mean fitness increases when fitnesses are additive. *Nature (London)* **221,** 1076.
Ewens, W. J. (1979). "Mathematical Population Genetics." Springer-Verlag, Berlin and New York.
Falconer, D. S. (1981). "Introduction to Quantitative Genetics," 2nd ed. Ronald, New York.
Feldman, M. W., and Christiansen, F. B. (1984). Population genetic theory of the cost of inbreeding. *Am. Nat.* **123,** 642–653.
Fisher, R. A. (1930). "The Genetical Theory of Natural Selection." Oxford Univ. Press (Clarendon), London and New York.
Fisher, R. A. (1941). Average excess and average effect of a gene substitution. *Ann. Eugen. (London)* **11,** 53–63.
Glesener, R. R., and Tilman, D. (1978). Sexuality and the components of environmental uncertainty: Clues from geographic parthenogenesis in terrestrial animals. *Am. Nat.* **112,** 659–673.
Grant, V. (1975). "Genetics of Flowering Plants." Columbia Univ. Press, New York.
Hamilton, W. D. (1967). Extraordinary sex ratios. *Science* **156,** 477–488.
Hamilton, W. E. (1982). Pathogens as causes of genetic diversity in their host populations. *In* "Population Biology of Infectious Diseases," (R. M. Anderson and R. M. May, eds.), Dahlem Konf. 1982, pp. 269–296. Springer-Verlag, Berlin and New York.
Harper, A. B. (1982). The selective significance of partial apomixis. *Heredity* **48,** 107–116.
Holsinger, K. E., Feldman, M. W., and Christiansen, F. B. (1984). The evolution of self-fertilization in plants: A population genetic model. *Am. Nat.* **124,** 446–453.
Jaenike, J., and Selander, R. K. (1979). Evolution and ecology of parthenogenesis in earthworms. *Am. Zool.* **19,** 729–737.
Jain, S. K. (1976). The evolution of inbreeding in plants. *Annu. Rev. Ecol. Syst.* **7,** 469–495.
Jain, S. K., and Workman, P. L. (1967). Generalized F-statistics and the theory of inbreeding and selection. *Nature (London)* **214,** 674–678.
Kempthorne, O. (1957). "An Introduction to Genetic Statistics." Wiley, New York.
Kingman, J. F. C. (1961). A mathematical problem in population genetics. *Proc. Cambridge Philos. Soc.* **57,** 574–582.
Lande, R., and Schemske, D. W. (1985). The evolution of self-fertilization and inbreeding depression in plants. *Evolution (Lawrence, Kans.)* **39,** 41–52.
Li, C. C. (1967). Fundamental theorem of natural selection. *Nature (London)* **214,** 505–506.
Lloyd, D. G. (1979). Some reproductive factors affecting the selection of self-fertilization in plants. *Am. Nat.* **113,** 67–79.

Lloyd, D. G. (1980). Benefits and handicaps of sexual reproduction. *Evol. Biol.* **13,** 69–111.
Lord, E. M. (1981). Cleistogamy: A tool for the study of floral morphogenesis, function and evolution. *Bot. Rev.* **47,** 421–449.
Lynch, M. (1984). Destabilizing hybridization, general-purpose genotypes and geographical parthenogenesis. *Q. Rev. Biol.* **59,** 257–290.
Marshall, D. R., and Brown, A. H. D. (1981). The evolution of apomixis. *Heredity* **47,** 1–15.
Maynard Smith, J. (1971a). The origin and maintenance of sex. *In* "Group Selection" (G. C. Williams, ed.), pp. 163–175. Aldine-Atherton, Chicago, Illinois.
Maynard Smith, J. (1971b). What use is sex? *J. Theor. Biol.* **30,** 319–335.
Maynard Smith, J. (1974). Recombination and the rate of evolution. *Genetics* **78,** 299–304.
Maynard Smith, J. (1976). A short-term advantage for sex and recombination through sib-competition. *J. Theor. Biol.* **63,** 245–258.
Maynard Smith, J. (1977). The sex habit in plants and animals. *In* "Measuring Selection in Natural Populations" (F. B. Christiansen and T. M. Fenchel, eds.), Springer-Verlag, Berlin and New York.
Maynard Smith, J. (1978). "The Evolution of Sex." Cambridge Univ. Press, London and New York.
Michod, R. E., and Hamilton, W. D. (1980). Coefficients of relatedness in sociobiology. *Nature (London)* **288,** 694–697.
Moore, D. M., and Lewis, H. (1965). The evolution of self-pollination in *Clarkia xanthia*. *Evolution (Lawrence, Kans.)* **19,** 104–114.
Moran, P. A. P. (1962). "The Statistical Processes of Evolutionary Theory." Oxford Univ. Press (Clarendon), London and New York.
Muntzig, A. (1958). The balance between sexual and apomictic reproduction in some hybrids of *Potentilla*. *Hereditas* **44,** 145–160.
Murdy, W. H., and Carson, H. L. (1959). Parthenogenesis in *Drosophila mangabeirai* Malog. *Am. Nat.* **93,** 355–363.
Nagylaki, T. (1976). A model for the evolution of self-fertilization and vegetative reproduction. *J. Theor. Biol.* **58,** 55–58.
Orlove, M. J., and Wood, C. L. (1978). Coefficients of relationship and coefficients of relatedness in kin selection: A covariance form for the rho formula. *J. Theor. Biol.* **73,** 679–686.
Price, G. R. (1970). Selection and covariance. *Nature (London)* **227,** 520–521.
Price, G. R. (1972). Fisher's "fundamental theorem" made clear. *Ann. Hum. Genet.* **36,** 126–140.
Shields, W. M. (1982). "Philopatry, Inbreeding, and the Evolution of Sex." State Univ. of New York Press, Albany.
Solbrig, O. T. (1979). A cost-benefit analysis of recombination in plants. *In* "Topics in Plant Population Biology" (O. T. Solbrig, S. Jain, G. B. Johnson, and P. H. Raven, eds.), pp. 114–130. Columbia Univ. Press, New York.
Stalker, H. D. (1954). Parthenogenesis in *Drosophila*. *Genetics* **39,** 4–34.
Stalker, H. D. (1956). On the evolution of parthenogenesis in *Lonchoptera* (Diptera). *Evolution (Lawrence, Kans.)* **10,** 345–359.
Stebbins, G. L. (1970). Adaptive radiation in angiosperms. I. Pollination mechanisms. *Annu. Rev. Ecol. Syst.* **1,** 307–326.
Suomalainen, E. (1950). Parthenogenesis in animals. *Adv. Genet.* **3,** 193–253.
Suomalainen, E., Saura, A., and Lokki, J. (1976). Evolution of parthenogenetic insects. *Evol. Biol.* **9,** 209–257.
Templeton, A. R. (1982). The prophecies of parthenogenesis. *In* "Evolution and Genetics of Life Histories" (H. Dingle and J. P. Hegmann, eds.), Springer-Verlag, Berlin and New York.
Templeton, A. R., and Rothman, E. D. (1973). The population genetics of parthenogenetic strains of *Drosophila mercatorum*. *Theor. Appl. Genet.* **43,** 204–212.

Templeton, A. R., Sing, C. F., and Brokaw, B. (1976a). The unit of selection in *Drosophila mercatorum*. I. The interaction of selection and meiosis in parthenogenetic strains. *Genetics* **82,** 349–376.

Templeton, A. R., Carson, H. L., and Sing, C. L. (1976b). The population genetics of parthenogenetic strains of *Drosophila mercatorum*. II. The capacity for parthenogenetic strains of *Drosophila mercatorum*. II. The capacity for parthenogenesis in a natural, bisexual population. *Genetics* **82,** 527–542.

Triesman, M., and Dawkins, R. (1976). The "cost of meiosis": Is there any? *J. Theor. Biol.* **63,** 479–484.

Uyenoyama, M. K. (1984). On the evolution of parthenogenesis: A genetic representation of the "cost of meiosis." *Evolution (Lawrence, Kans.)* **38,** 87–102.

Uyenoyama, M. K. (1985). On the evolution of parthenogenesis. II. Inbreeding and the cost of meiosis. *Evolution* **39,** 1194–1206.

Uyenoyama, M. K., and Bengtsson, B. O. (1979). Towards a genetic theory for the evolution of the sex ratio. *Genetics* **93,** 721–736.

Uyenoyama, M. K., Feldman, M. W., and Mueller, L. D. (1981). Population genetic theory of kin selection: Multiple alleles at one locus. *Proc. Natl. Acad. Sci. U.S.A.* **78,** 5036–5040.

Wells, H. (1979). Self-fertilization: Advantageous or deleterious? *Evolution (Lawrence, Kans.)* **33,** 252–255.

White, M. J. D. (1970). Heterozygosity and polymorphism in parthenogenetic animals. *In* "Essays in Evolution and Genetics in Honor of Theodosius Dobzhansky" (M. K. Hecht and W. C. Steere, eds.), pp. 237–262. Appleton, New York.

White, M. J. D. (1973). "Animal Cytology and Evolution," 3rd ed. Cambridge Univ. Press, London and New York.

White, M. J. D. (1978). "Modes of Speciation." Freeman, San Francisco, California.

White, M. J. D., and Contreras, N. (1979). Cytogenetics of the parthenogenetic grasshopper *Warramaba* (formerly *Moraba*) *virgo* and its bisexual relatives. V. Interaction of *W. virgo* and a bisexual species in geographical contact. *Evolution (Lawrence, Kans.)* **33,** 85–94.

Williams, G. C., ed. (1971). "Group Selection." Aldine-Atherton, Chicago, Illinois.

Williams, G. C. (1975). "Sex and Evolution." Princeton Univ. Press, Princeton, New Jersey.

Williams, G. C. (1980). Kin selection and the paradox of sexuality. *In* "Sociobiology: Beyond Nature/Nurture? Reports, Definitions and Debate" (G. W. Barlow and J. Silverberg, eds.), pp. 371–384. Westview Press, Boulder, Colorado.

Williams, G. C., and Mitton, J. B. (1973). Why reproduce sexually? *J. Theor. Biol.* **39,** 545–554.

3

The Evolution of Parthenogenesis: A Historical Perspective[1]

ORLANDO CUELLAR

Department of Biology
University of Utah
Salt Lake City, Utah 84112

> Les problèmes que pose la Parthénogenèse naturelle s'intriquent d'ailleurs si intimement avec la plupart des grandes questions biologiques actuelles: sexualité, fécondation, variabilité, origine des espèces, etc., qu'une mise au point de la Parthénogenèse naturelle pourra rendre service, non seulement aux naturalistes qui s'occupent spécialement de cette question, mais encore à tous ceux qui s'intéressent aux problèmes de la Biologie générale. (Vandel, 1931)

I. INTRODUCTION

The study of parthenogenesis is very old, with major papers appearing on the subject as early as 1774 (see Bell, 1982, for a brief history). Several extensive reviews were written during the next century, particularly between 1840 and 1870. By 1890 artificial parthenogenesis was an active field of research, and studies of the cytology of egg development had been started. The decade between 1900 and 1910 marked the beginning of the modern period, and also a time of greatly accelerated research into all essential aspects of origin; cytology, genetics, polyploidy, hybridization, and geographical distributions. Many general reviews in both plants and animals have appeared since then, the most recent being those by Grant (1971), Richards (1973), White (1973, 1978), Williams (1975), Maynard Smith (1978), and Bell (1982). Particularly well documented and useful for the specialists are the older ones by Winkler (1920), Seiler (1923,

---

[1]This chapter is dedicated to Åke Gustafsson.

1960), Vandel (1931), Gustaffson (1936, 1946, 1947a,b, 1948), Matthey (1945), Rostand (1950), Stebbins (1950), Suomaleinen (1950), White (1954), and Narbell-Hofstetter (1964).

This paper is a review of the cytological origins of parthenogenesis. To a large extent, it is the story of parthenogenesis as told by the workers themselves, particularly the old and forgotten ones, and those involved in controversial or novel findings. To accomplish this, I have quoted liberally many of their key or concluding remarks. The origin of parthenogenesis has been a controversial topic from the very beginning, but this is seldom appreciated from reading the recent literature, most of which gives but a fleeting glance of the older works. Indeed, some of the most outstanding contributions to this field today are now largely forgotten. In certain endeavors, such as the study of meiosis (see Narbel-Hofstetter, 1964, for a review) and the formation of polyploidy (see Gustafsson, 1936, 1946, 1947a,b), we may never surpass these classical works. The study of meiosis requires a primitive tool, the microtome, and that of polyploidy an ancient practice, the crossing of species, and yet in this modern day of theoretical and molecular biology we seem to know less about the origins of parthenogenesis than we did more than half a century ago.

II. MEIOSIS AND PARTHENOGENESIS

True parthenogenesis is a very special type of reproductive process by which populations of organisms exist exclusively as females in nature. The term parthenogenesis is derived from two words: "parthenos" (Greek) meaning virgin and "genesis" (Latin) meaning birth, so it means virgin birth, but the common interpretation means the development of eggs without fertilization. Other commonly used names for parthenogenesis are thelytoky (female birth) and, in plants, apomixis, apogamy, and agamospermy. Where appropriate, all are used interchangeably throughout this paper, but mostly I use parthenogenesis. How eggs can develop successfully into adults without actually being fertilized by sperm is related to the early stages of egg development, specifically to the complex maturational divisions of meiosis that prepare the gametes of sexual species for fertilization. Since all parthenogenetic species have evolved directly from bisexual progenitors then the origin of parthenogenesis must be explained in relation to the need for meiosis in sexual reproduction.

In all bisexual species, meiosis reduces the number of chromosomes in the gametes by one-half so that in fertilization the resulting embryo formed by the fusion of sperm and egg will have the same number as the parents. But why have sexuality in the first place? Why not reproduce by some asexual means, as many organisms do, and therefore avoid the need for fusion, which in turn eliminates

3. Evolution of Parthenogenesis

the need for meiosis, thereby eliminating the need for males? A major premise of evolutionary biology since the time of Darwin is that sexuality evolved to increase genetic variability so that organisms can adapt rapidly to the ever-changing environment (Fisher, 1930; Muller, 1932; Darlington, 1939; Dobzhansky, 1941; Stebbins, 1950; Mayr, 1969; Williams, 1975; Maynard Smith, 1978; Bell, 1982).

According to Mayr (1969), "The production of different genotypes can be vastly increased by the simple device of permitting exchange of material among sexual individuals. This is the biological significance of sex. . . . The amount of genetic variability released by such crossing over is enormous. Through recombination, a population can generate ample genotypic variability for many generations without any new genetic input (by mutation or gene flow) whatsoever." Since recombination is not possible without meiosis, it follows that meiosis led directly to the evolution of sex. Therefore, meiosis is a very ancient evolutionary mechanism designed to amplify genetic variation, but requiring as a consequence the elimination of one-half of the chromosomes in the gametes of both parents. The need for such a reduction is obviously related to the fact that without it the number of chromosomes would double in every generation, eventually reaching a lethal volume.

But if meiosis is the key to bisexuality, it is the ruin of parthenogenesis, for in the absence of fertilization, reduction would now deplete one-half of the chromosomes in each succeeding generation, eventually eliminating the entire set as well as the budding species. Clearly, it is this rigorous reductional force that parthenogenesis must first overcome to generate a successful species. Hence the fundamental basis of parthenogenesis hinges on the ability of eggs to circumvent meiosis so that every generation achieves a constant somatic number of chromosomes. In this regard, Maynard Smith (1971) correctly points out that "selection acts not by eliminating parthenogenetic varieties when they arise, but by favoring genetic and developmental mechanisms which cannot readily mutate to give a parthenogenetic variety." Apparently acknowledging the rigor of such constraints, White *et al.* (1977) recently referred to a possible restorational process in a parthenogenetic grasshopper as "this cytological tour-de-force."

According to Suomaleinen (1950), "It is clear that the most essential feature of parthenogenesis, absence of fertilization, causes important alterations of basic nature in cytological as well as in genetic respects. As this is in many forms accompanied by lack of chromosome conjugation and reduction, the differences as compared to forms having normal fertilization become greater. The gametes of such animals are not haploid as they are in bisexual animals but contain the diploid chromosome set as do the somatic cells. On the other hand, in those parthenogenetic forms whose gametes have undergone chromosome reduction and are thus haploid, the diploid number is not, in contrast to the bisexual forms restored in fertilization, but in some wholly abnormal manner."

The precise manner by which a mechanism of meiotic restitution arises in a parthenogenetic species remains a mystery. For instance, with regard to insects, Suomaleinen (1962) noted that "the origin of parthenogenesis is, so far, obscure in its details." Similarly, Maslin (1971) stated with regard to lizards that "the actual transition from bisexuality to unisexuality remains unexplained particularly with regard to the mechanisms wherein meiosis is retained and restitution of the somatic complement is achieved." This is undoubtedly the most complex of all the factors involved in the origin of parthenogenesis, and is bedeviled with numerous problems. First of all, the meiotic process itself must be dramatically altered. Second, the alterations must lead to cytogenetic mechanisms that actually restore the normal number of chromosomes. Third, the mechanisms, once evolved, must become genetically fixed in order to recur consistently in future generations. Fourth, they must be varied enough to include the many different forms of parthenogenesis. Fifth, they must accommodate chromosomal complements greater than the diploid set, for many parthenogenetic species are polyploid (3N, 4N, 5N, 6N. . .). Sixth, some mechanisms are presumably triggered by hybridization, for many species are hybrids. Seventh, the eggs destined to develop parthenogenetic strains must not be fertilized, and, finally, such eggs must divide autonomously without the cleavage-activating influence of sperm.

But how does a mechanism arise in the first place? What is the original cellular trigger? Despite the presumed inflexibility inherent in meiosis, there is nevertheless a substantial body of genetic and developmental evidence suggesting that meiotic abnormalities may be a common feature of sexual reproduction, and that such failures in the normal sequence of meiotic events have provided the cellular basis for the restorational mechanisms of parthenogenesis. Concerning such modifications of meiosis, White (1948) stated: "In certain groups of animals however, the usual type of meiosis has been replaced by some entirely new mechanism. . . . most of these aberrant variants of the normal process represent secondary simplifications, one or more of the usual stages or processes being omitted. . . . so long as there is no evidence to the contrary we may assume that all anomalous types of meiosis . . . were genetically caused. . . . It is mutations of this type which must have given rise to most of the naturally occurring meiotic mechanisms which are unusual in type." In this context, parthenogenesis has evolved by taking advantage of the reproductive mistakes of sexuality.

The evidence for this viewpoint is based on the many observations of the spontaneous origin of self-perpetuating parthenogenetic lines produced by virgin females. Particularly striking examples of spontaneous origin have been documented in worms (Hertwig, 1920), locusts (Nabours, 1925), moths (Seiler, 1942), grasshoppers (Hamilton, 1953), and phasmatids (Bergerard, 1954). Another line of evidence stems from the tendency in numerous bisexual species, both vertebrate and invertebrate, to undergo rudimentary parthenogenetic devel-

3. Evolution of Parthenogenesis

opment—tychoparthenogenesis (see Vandel, 1936; Suomaleinen, 1962; White, 1973; Cuellar, 1974, for reviews). Also, such a rudimentary tendency can be greatly increased by artificial selection (Stalker, 1954; Olsen, 1965; Carson, 1967; Olsen *et al.*, 1968). According to Stalker (1954), in many species of *Drosophila* "the very low rates of parthenogenesis observed are of some potential value to the species in supplying the first step toward thelytoky." Likewise, Carson (1961) believes that such rudimentary parthenogenesis "may serve as an example of an evolutionary stage through which the obligatory parthenogenetic species . . . may have passed during the evolution of its parthenogenesis." Suomaleinen (1962) coined this transitional state "*in situ nascendi,*" for the fact that the sequence from bisexuality to thelytoky has been directly observed and experimentally demonstrated in many species. Consequently, many authors have concluded that tychoparthenogenesis is the original source for thelytoky (see Vandel, 1936; Whiting, 1945; Suomaleinen, 1950; White, 1954).

III. THE HYBRIDIZATION HYPOTHESIS IN PLANTS

Unraveling the complex nature of the cellular basis of parthenogenesis would be vastly simplified if it were not for the fact that many parthenogenetic species are polyploid and that polyploidy in such forms is frequently associated with hybridization. This is particularly striking in plants in which the majority of apomictic species are hybrid polyploids-allopolyploids (Gustafsson, 1947a; Grant, 1971; Stebbins, 1971; Richards, 1973). In this regard, Gustaffson (1948) states: "Agamospermous plants, as well as viviparous apomicts of numerous genera, almost exclusively consist of polyploids. Some retain a certain ability of sexual seed formation. In other instances the sexual facilities are entirely extinguished." Similarly Stebbins (1971) notes: "In some groups . . . apomixis has developed in diploid species or hybrids. Much more commonly, however, it is associated with polyploidy. As with sexual polyploids, the genetic constitution of apomictic polyploids forms a complete spectrum from strict 'autopolyploidy' to 'allopolyploid' origin resulting from crossing between widely different diploid parents. Furthermore, when groups containing apomictic polyploids are studied in their entirety, the entire spectrum is usually found within the same circle of relationships."

Even among sexual plants a very high percentage of the flowering species are considered polyploid (Stebbins, 1971). Such a dominant relationship between apomixis, polyploidy, and hybridization prompted Winge (1917) and Ernst (1918) to propose that parthenogenesis arises through hybridization, i.e., that hybridization causes parthenogenesis. According to Harrison (1927), "In his classical work Ernst (1918) urges that hybridity and parthenogenesis are related in the way of cause and effect. As his critical example he addresses the case of the Stonewort *Chara crinita,* which exists in two forms. One of these is dioecious, and as he calls it, haploid in both sexes, and the other, termed by him diploid, for present purposes may be regarded as existing solely in the female sex. Further the former reproduces itself sexually and the latter parthenogenetically (apogamously). Since the apogamous plant has double the chromosome complement of the second, and diploidy (in Ernst's sense) may arise from a crossing between two distinct haploid *chara* species, that worker asserts that the diploid plant has been produced by hybridization, and also that its apogamy is the direct result of such an origin."

Because this proposition by Ernst was theoretical, Harrison (1927) further states: "In spite of this, it cannot be denied that the explanation covers the observed facts, and besides, these are strengthened by parallel and pertinent observations made both in the field and on cytological preparations. For instance, Lundstrom (1909), Dingler (1908) and myself (Harrison, 1920) have all demonstrated that the majority of European polyploid roses are apomictical. Furthermore, Täckholm (1922), my colleague Dr. Blackburn and myself (Blackburn and Harrison, 1921) have shown that in their meiotic phase these roses manifest the cytological behavior of hybrids. Similar observations have been made by Rosenberg (1909, 1917) and Ostenfeld (1910, 1912) on *Taraxacum* and *Hieracium*. In no case, however, it is necessary to emphasize, has an apogamous (apomictical) hybrid plant been built up experimentally by crossing two sexual species lower in the polyploid series characterizing the genera in question. In spite of these objections, in my opinion, Ernst's views are correct. . . ."[2]

The major cytogenetic difficulty with the assumption of hybridity resulting in parthenogenesis is the lack of a sound explanation regarding the genetic origins of parthenogenesis, as well as the meiotic mechanisms that control it. This concern prompted Peacock (1925) to ask; "Is there any connection between the emergence, through hybridization, of new forms and the emergence, through hybridization of a new method of propagation of these novelties? Stated in cytological language, have the cytoplasmic and chromosomal constitution of the oocytes of a new hybrid form or the reactions between the two complexes entering into the composition of these oocytes endowed these reproductive cells with the faculty of segmenting and developing without the stimulus of fertilization?" Attempting to provide the essential cytogenetic connection between parthenogenesis and hybridization, Muntzing (1940) proposed what we may now call the "genetic imbalance" hypothesis. According to Muntzing, apomixis in plants "is due to a rather delicate genetic balance. This balance may be upset in

[2]The reader is referred to the original source for complete references of the works cited within quoted passages.

3. Evolution of Parthenogenesis

various ways by crosses with other types or merely by quantitative changes in chromosome number either in minus or plus direction.''

Much earlier, however, Holmgren (1919) had already proposed that the tendency for apomixis in plants resided in the gametes themselves, rather than upon extrinsic factors. Even though Rosenberg (1930) had also demonstrated a hybrid constitution in many apomictic plants, he concluded; "Man kommt vielleicht den wahren Verhältnissen näher, wenn man die Sache so ausdruckt, dass die Bastardierung nicht die Ursache der Parthenogenesis ist, sondern sie kann sehr wohl ein wishtiges Moment sein für die Entstehung parthenogenetischer Formen und vielleicht sogar eine Voraussetzung dafür, ohne dass jedoch ein eigentliches Kausalverhältnis zwischen den beiden Erscheinungen bestände.'' ("Perhaps one would come closer to the real circumstances if one were to express the case such that hybridization is not the cause of parthenogenesis, but rather an important factor in the genesis of parthenogenetic forms and perhaps even a prerequisite for it, without, however, proving an exact causal relationship between the two phenomena.") Similarly, Peacock (1925) remarked: "From the botanical evidence of to-day the idea that hybridization is an occasion, but not a cause of the appearance of parthenogenesis, is acceptable, the species crossing, as it were, releasing a latent or slight tendency to parthenogenesis in one or both parents as well as enabling it to develop more fully and to be exploited more freely.'' According to Gustaffson (1948), "Apomictic groups form the top of the cytological system, polyploidy proceeding upwards. . . . polyploidy in its turn, enhances an already existing reproduction or may, incidentally induce it.''

Although Ernst is usually regarded as the founder of the hybridization hypothesis, essentially the same idea was advanced and documented earlier by Ostenfeld (1912) working on the plant genus *Hieracium*. Prior to that, Ostenfeld and Raunkiaer (1903) had found that most of the species in *Hieracium* and *Taraxacum* were apogamous. According to Ostenfeld (1912), the members of *Hieracium* are of exceptional interest in several other features. They possess hundreds of different species and numerous races, many of recent origin and some probably appearing at the present time. There is evidence of extensive hybridization among them. The majority are strictly apogamous, but some are partly apogamous and a few are strictly sexual, so all stages in the development of apogamy are represented. In a paper entitled "Experiments on the Origin of the Genus *Hieracium* (Apogamy and Hybridization),'' Ostenfeld (1912) reported the results of several crosses conducted "to find out the causes of this polymorphy.''

In a cross between two partly apogamous species, *H. excellens* and *H. aurantiacum*, he found that "the hybrids are often wholly or nearly sterile. . . . However some of the hybrids are partially or wholly fertile and these are apogamous as are one or both parents; and consequently the later generations remain like their offspring of F_1. In other words, by means of crossing, new forms do actually arise; and they are at once perfectly constant, and behave like ordinary species." In another cross between *H. auricula* (a sexual species) and *H. aurantiacum*, 29 different hybrid individuals were produced. "Most of them were sterile and weak and have since died, but a few of them proved to be fertile and at present have reached the third generation. The individuals of the F_2 and F_3 generations remain like the individual of the F_1 generation from which they sprang. It is thus evident that hybridization plays a role as a factor in the origin of new forms."

I should clarify, however, that Ostenfeld did not regard apogamy the same as parthenogenesis, because according to him the embryo sac resulting either from the mother cell or from the nucellus "has the unreduced number of chromosomes. Therefore I speak of apogamy not of parthenogenesis." In any event, the end result between apogamy and parthenogenesis would be the same, namely, the production of progeny identical to the parent. Despite the convincing nature of his crosses, Ostenfeld was careful not to conclude that hybridization gives rise directly to apogamy: "My position may be summarized in a few words as follows: *in Hieracium new forms arise by means of hybridization and also by means of single variations; in both cases the prevailing apogamy supports their existence and constancy;* thus polymorphism in the genus is correlated with apogamy, but it is not allowable to draw any conclusion as to causality between them." The production of apogamy in hybrids between different species of *Hieracium* should not be surprising since apogamy is the normal method of reproduction in this genus, and the species used to produce the hybrids were partly or entirely apogamous. Therefore the tendency for apogamy in the hybrids was undoubtedly inherited from the parents, rather than being induced by hybridization.

Almost 50 years ago, Gustafsson (1936) wrote his first review of the origin of apomixis in plants entitled "Studies on the Mechanisms of Parthenogenesis." Here he presented a brief history of what he coined "Ernst's hypothesis" (p. 58), pointing out the great controversy created by that idea prior to 1930, but also revealing that Ernst may not actually have implied what his critics thought.

> At about the same time Rosenberg (1917), Ernst (1917,1918) and Winge (1917) advanced their theory, based on cytological facts, that apomixis is conditioned by or associated with heterozygoty and polyploidy. As Schnarf (1929) points out Ernst never intended to maintain that hybridisation was in itself sufficient to produce apomixis. Instead he asserted only that in all carefully examined cases apomixis was associated with so many cytological and genetical disturbances that intense heterozygoty must be postulated. It is then a matter of little importance whether one assumes crossings within or without the systematic species, for the boundaries of the systematic species are conventional and theoretically rather arbitrary.
>
> In fact the importance of Winkler's criticism (1920) of Ernst's hypothesis does not lie so much in this sphere but rather in an examination of nomenclative questions and in a classification of zoological and botanical data. Not more than two years after Winkler's criticism Täckolm succeeded in showing its defect, and besides that it was undoubtedly necessary to

3. Evolution of Parthenogenesis

postulate crosses between distantly related biotypes as the cause of the constant hybrid complement within the apomictic *Rosa* groups. Although Erlandson (1933) has in certain aspects re-examined Täckholm's results and conclusions, the main result still holds good.

To summarize it may thus be said that the genus *Taraxacum* furnishes another instance of parthenogenesis correlated with polyploidy and cytological hybridity. With regard to the actual method of origin of apomixis there is a great deal of evidence in support of Schnarf's conception (1929). Sexual diploids may conceal genes that condition apomixis. By crosses these genes are brought together. The sexual maturation divisions bring about sterility, and in the absence of competing sexual embryo-sacs and embryos the tendency to apomixis can find expression.

In the second of his three later classic reviews of "Apomixis in Higher Plants," Gustafsson (1947a) discussed in considerable detail the relationship between apomixis hybridization and polyploidy. Concerning the genus *Alchemilla* he states (p. 117):

> According to Buser and Strasburger (see Latter 1905) the sexual species *A. pentaphyllea, glacilis* and *gelida* hybridize with one another and with apomictic form-circles, yielding hybrids which, contrary to the parents, have bad pollen and multiply apomictically. . . .
>
> If these results are accurate, which must be confirmed by re-examinations, agamospermy would be able to arise direct as an F_1 phenomenon in *Alchemilla*. Ernst (l.c., p. 252) considers that these observations support his hybridization theory, while Strasburger presumes that the hybrids combine special factors for apomixis, latent in the sexual parents. The last method of explanation is perhaps the correct one: The more or less sterile F_1 hybrids and back-cross products cannot form seed by the sexual method, and therefore only those biotypes which possess some tendency to apomictic seed-formation can go on living and spread. Secondarily, a series of modifiers may intensify the apomictic reproduction. The same can possibly take place through an increase in the chromosome number. Selectionally, of course, those forms are favoured which set the most abundant seed.
>
> It must however be stressed that in the cases stated the hybrid nature has been postulated on the basis of morphological characters alone. There are no experimental investigations on which to fall back, and Strasburger's analysis is far from definite. The postulated hybrids may represent, instead of F_1 individuals, segregates and back-crosses, in which case the mode of origin becomes far more complicated.

More recently Stebbins (1950) stated: "the usual close association between apomixis, on the one hand, and polyploidy, interspecific hybridization, and polymorphy, on the other, must be explained on an indirect rather than a direct basis. . . . The first hypothesis developed to explain apomixis was that of Ernst (1918), who believed that it is caused by hybridization between species. Nevertheless, there is no evidence at all that hybridization by itself can induce apomixis. Hybrids between different sexual species . . . have in no instance shown any clear indication of apomictic reproduction, even though their parental species may be closely related to known apomictic forms. . . . A number of hybridizations have now been performed between sexual types and related facultative or obligate apomicts in the same genus. Segregation in later generations from these crosses has shown in every case that apomixis of the species is genetically controlled."

One of the few recent reviews on the origin of parthenogenesis in plants, although from the experimental standpoint, is that of Richards (1973) concerning the genus *Taraxacum*. All members of this genus consist of perennial herbaceous species, the majority of which are obligate apomicts. Almost 2000 of these agamospecies have now been described. By comparison only 50 species are considered sexual. The sexuals are always diploid (2N = 16), and the apogamous are polyploid, approximately 60% being triploid, 30% tetraploid, 8% pentaploid, and less than 1% hexaploid. While the polyploids are largely apomictic, a few can also reproduce sexually. As in *Hieracium* most of the polyploids are considered to be hybrids. Experimental crosses among the various species of *Taraxacum* have produced approximately 40 different interspecific hybrids. Such hybrids formed between sexual species are usually very fertile, and are themselves sexual. Those between sexuals and apomicts are also fertile, but are always apomictic. However, hybridization between the sexuals and the apomicts in nature is very rare, as these two reproductive modes seldom occur together, though in certain localized areas hybridization may be common.

Although Richards (1973) favors a hybrid origin for all of the agamospecies of *Taraxacum*, he did not specifically address the causal factors responsible for inducing apomixis in this genus. Nevertheless, one can vaguely interpret that apomixis came first and was later preceded by hybridization. "The species partaking in this original advance would have been diploid sexuals, represented today by the primitive species still found in the area . . . a single primitive apomictic tetraploid is known from the mountains of Afghanistan and it is possible that this is close to the original apomicts . . . once facultative apomixis had thus been established in a primitive tetraploid with an irregular meiosis and parthenogenesis, selection pressure would doubtless favor the establishment of asynaptic behavior in the female meiosis. Such asynaptic plants would have wholly restitutional egg cells with resultant obligate apomixis, enabling the plant to set perfect seed in all conditions and stabilize a suitable (perhaps hybrid and heterotic) genotype." Regarding the mechanism of restitution in the apomictic *Taraxacum*, there is evidence suggesting it is under genetic control (Richards, 1970), is present on only one chromosome, and is dominant (Richards, 1973). In this connection, Gustaffson (1946, p. 63) remarked:

> In passing two phenomena of sporadic nature may be touched upon here. When different species within a genus or even different biotypes of one and the same species are crossed with each other, products with an increased chromosome number are often formed, e.g. triploids from crosses between diploid species, and hexaploids from crosses between tetraploids. In these cases unreduced egg-cells generally function, and the unreduced chromosome number in the egg-cell depends either on premeiotic disturbances, by which the EMC has already obtained the double somatic number, or on meiotic disturbances in the first or second division, by which the uninucleate gametophyte gets the unreduced number. The egg-cell must, however, be fertilized in the usual manner. Cases in which unreduced female gametes have functioned are reported in such numbers in the literature that an enumeration is superfluous. In hybrids

3. Evolution of Parthenogenesis

with a disturbed meiosis unreduced egg-cells and pollen-grains are, naturally enough, considerably more readily formed, but even in them fertilization is necessary. Hybrid nature does not in itself initiate any apomictic reproduction.

In reviewing the origin of parthenogenesis in plants, Maynard Smith (1978) concluded: "In plants apomixis is commonly associated with polyploidy, and with evidence of hybrid origin, between species or at least between ecotypes (Stebbins, 1971). There is no reason to think that hybridization by itself causes apomixis in plants any more than it does in animals. The association with hybridization probably exists because a hybrid genotype may adapt a plant to a new habitat, so that if apomixis happens to arise in such a plant it has a good chance of surviving. The association with polyploidy may exist because if a polyploid genotype does arise in an apomict it will not cause infertility as it would in a sexual species."

To date, the most comprehensive and insightful review on the relationship between hybridization and polyploidy in plants is the paper "On Ö. Winge and a Prayer: The Origins of Polyploidy," by Harlan and deWet (1975). The rather amusing title of this paper has a double meaning, which merits clarification. On the one hand, it implies "barely making it," taken from the World War II song "Coming in on a wing and prayer," and, on the other, it refers to the author Ö. Winge (1917), who, along with Ernst (1918) first proposed the hybridization hypothesis. According to Harlan and deWet, Winge proposed that in the case of hybrids produced from very different parents, the chromosomes would not pair in meiosis, but instead would split, each making a pair, thereby producing a hybrid with twice the number as the parents, and thereby leading to polyploidy. The influence of this idea on later generations of botanists was profound, giving rise to what Harlan and deWet coined the "hybridization followed by chromosome doubling" cliché, which "has become more than a cliché; it is a veritable incantation whose magic is calculated to be strong enough to split chromosomes by constant repetition."

The evidence presented by Harlan and deWet (1975), from the literature and their own findings, is based on some 85 genera and several hundred species. They conclude that polyploidy in plants arises in two different ways, spontaneously and through hybridization. The most common source of origin is spontaneous, resulting first in the production of autotriploids formed by the pollination of unreduced eggs (N + 2N). These triploids then backcross with the parent, forming a tetraploid (N + 3N). "Once a series of primary tetraploids of related species is established, however, intricate polyploid complexes become possible and autoploids can be alloploidized by crossing with sibling species." Hybridization can also lead to the formation of unreduced gametes due to drastic disturbances of meiosis, which cause lack of pairing. However, upon selfing they produce different types of polyploids (4N, 5N, 6N) or aneuploids (lacking certain chromosomes). A high percentage of these hybrids are sterile, or otherwise produce a poor seed set. Crosses between widely different species are usually lethal. Most of the evidence indicates that unreduced gametes arise from failures in either the first or second meiotic divisions, and that such failures are largely under genetic control. "Meiosis is a complex process and is probably never perfectly executed 100% of the time in any sexual species."

They suggest the following sequence leading to polyploidy in plants: (1) Most polyploids result from unreduced gametes. (2) The most frequent event is the formation of autotriploids. (3) Such triploids yield 4N plants on backcrossing and 6N plants on selfing. (4) Spontaneous polyploids appear frequently but are not always successful. (5) Most polyploids are strictly autoploids but lack competitive ability. (6) Wide crosses are rare in nature. (7) Wide crosses resulting in triploidy (2N + N) may be more common. (8) The most likely pathway to natural polyploidy is via crosses between closely related populations within a species. (9) The resulting more or less autopolyploids may then generate true alloploids by interspecific hybridization. Since, according to Harlan and deWet, "A wide cross disrupts meiosis sufficiently that the production of unreduced gametes becomes the rule," and could lead in some cases to allopolyploids, then the original proposition by Winge (1917) was not really all that wrong. After all, unreduced gametes resulting from asynapsis are just as unreduced as those from a genetically controlled abortive division. But then why do hybrids seem to have so many problems? This question relates to another more fundamental one; How will a normal meiosis with two divisions handle all of these randomly distributed unpaired chromosomes? Quite conceivably, the first division might split them evenly, as they were double in the first place. But how will the second handle the remaining set of singles? The evidence presented by Harlan and deWet (1975) suggests that it will scatter them at random, hence the reason for the different ploidies, aneuploidy, infertility, and inviability. This is the real problem with flying on Ö. Winge. It has little or no controls.

Gustaffson (1936) was apparently aware of Winge's ideas concerning hybridization and polyploidy for he states (p. 63):

> *Apomixis cannot be caused by hybridisation and increase in the chomosome number alone.* A few years ago (1930) I called attention to the fact that the North American *Rubi eubati*, although many of them are closely related to the European blackberries and like the latter hybridize intensely and possess high chromosome numbers, have nevertheless remained sexual. Thus apomixis must be a "Begleiterscheinung" and not a "Folgeerscheinung" of the hybridisation. Apomixis is a phenomenon, the origin of which is aided by the two processes mentioned above but is not determined by them. The only way in which the origin of apomixis can be explained is that it is caused by specific genes. Hybridisation and increase in chromosome number proves that the genes are often of a complementary nature and that apomixis is therefore a phenomenon determined by several genes. Nor is the possibility excluded that apomixis is conditioned by purely recessive factors. Certain facts relating to the genus *Poa* are considered by Muntzing (1933) to argue in favour of this view. In 1930 I myself considered that in *Rubi eubati*, judging from Lidforss's results, pseudogamy is conditioned by a series of recessive factors.

3. Evolution of Parthenogenesis

The most recent study of interspecific crossings in plants is that of Hiesey and Nobs (1982) on the partly apomictic bluegrass genus *Poa*. Their main objective was to produce new self-perpetuating apomictic strains of potential agronomic value through hybridization between facultatively apomictic species. The genus *Poa* contains ~200 taxa, which are widely distributed throughout the world. The crossings were conducted among four species (*P. pratensis, P. nevadensis, P. scabrellae, P. caespitosa*) differing substantially in morphology and geographic distribution and being reasonably apomictic in seed production. In summarizing their exhaustive work, these authors remarked:

> We must acknowledge that the overall results from the present studies have fallen short of initial expectations. None of the synthesized hybrid derivatives have been clearly superior in growth performance to both parental lines when tested in all the different environments. In retrospect this result is perhaps not surprising in view of the multitude of exacting requirements needed to produce a truly outstanding agronomic product. One cannot but be impressed with the effectiveness of the natural evolutionary processes, which have produced the various end-products of the species-complex now growing in the wild and so well fitted to the native environments.
>
> We emphasize that despite our extensive effort in crossing *Poa* species, only a small fraction of the recombinations possible in this large agamic complex have actually been attempted. From this point of view, the fairly promising new derivatives that were obtained suggest the possibility of greater success by using different parental materials.
>
> To obtain a synthetic product that attains successful equilibrium between the disruptive effects of wide hybridization and the stabilizing effects of chromosomal balancing of the constituent genomes is a result which, from a statistical point of view, seldom occurs. We conclude, however, that the experimental evidence is clear that new synthetic combinations can be obtained through considerable effort, and that, theoretically at least, it should be possible to produce self-replicating derivatives with almost any recombination of characteristics of species that can be hybridized.

IV. MUTATIONS AS THE SOURCE OF PARTHENOGENESIS IN ANIMALS

In contrast to the hybrid origin hypothesis, most of the evidence of polyploidy in parthenogenetic animals (excluding vertebrates) supports the view that parthenogenesis arises first and is later preceded by polyploidy. In some cases the polyploidy results from hybridization between the unisexual and a bisexual species (allopolyploidy), but in others it develops intrinsically in the unisexuals (autopolyploidy). In any event, the most widely accepted opinion is that spontaneous diploidization (tychothelytoky) is the first step toward constant thelytoky (Suomaleinen 1950). For instance, regarding the origin of parthenogenetic weevils Suomaleinen (1940a) remarked; "It seems to me most probable that the parthenogenetic curculionids originally arose in diploid forms. We may further assume that a diploid parthenogenetic female in exceptional cases may pair with a male of either the species from which it has risen or some other related species." Later, Suomaleinen (1961) added; "Fertilization of an egg of a diploid parthenogenetic weevil by haploid sperm will give rise to a triploid race and fertilization of a triploid egg will produce a tetraploid."

Accordingly, Stalker (1956) proposed that a bisexual ancestor first produced a parthenogenetic diploid, which later gave rise to the autotriploid fly *Ochthiphila polystigma*. Essentially the same mode of origin was proposed by Basrur and Rothfels (1959) for the triploid black fly *Cnephia mutata:* "Evidence is presented that parthenogenesis in the triploids is meiotic and automictic, and that hybridization is probably not involved in the evolution of anisopolyploid forms in black flies." Likewise, regarding the tetraploid moth *Solenobia triquetrella,* Seiler (1923) concluded that parthenogenesis arose first in a diploid form, which later became tetraploid by automixis (see Suomaleinen, 1940a). In this connection Seiler states: "der Gedanke einer Arbastardierung als Urasache der Parthenogenese ausgeschlossen ist." ("The notion of a species hybridization as a cause of parthenogenesis is out of the question.") This same method of origin was found by Artom (1912) in the brine shrimp *Artemia* and by Vandel (1931) in the triploid isopod *Trichoniscus elisabethae*. Regarding the work of Artom (1912), Peacock (1925) remarked: "this worker found that the Cagliari race of the brine shrimp *Artemia salina* was bisexual and had a diploid chromosome number of 42 while the Capodistra race was parthenogenetic with a tetraploid chromosome number of 84 . . . as is generally admitted for animals, species now parthenogenetic have evolved from a former bisexual condition and such appears to be the case in *Artemia salina*. Hence hybridization and its sequelae of cytological disturbances are ruled out." Concerning *Trichoniscus* Vandel states: "On voit donc que jusqu'ici l'idée de l'origine hybride des formes parthénogénétiques constitue, au moins dans le règne animal une simple hypothèse de travail."("One can therefore see that up until now the idea of the hybrid origin of the parthenogenetic forms constitutes, at least in the animal kingdom, merely a working hypothesis.")

Recently, Lamb and Willey (1975) discovered the first known parthenogenetic populations of saltatorial Orthoptera, these being the cave crickets *Hadenoecus cumberlandicus* and *Euhadenoecus insolitus*. In their review of the cave crickets of North America, Hubbell and Norton (1978) advanced the following hypothesis to explain the existence of these two parthenogenetic species. "It assumes that a very small proportion of their virgin females produce viable young which are always female and diploid. Such occasional parthenogenesis has been observed in most groups of insects, including numerous species of Saltatoria; its occurrence in the *Hadenoecini* would provide a basis for the formation of parthenogenetic populations from sexual ones."

Although Suomaleinen (1940a) had originally proposed that fertilization by haploid sperm was the source of polyploidy in the parthenogenetic weevils, he

3. Evolution of Parthenogenesis

later suggested (Suomaleinen, 1947) that it could also result from the loss or addition of one or several metaphase plates during meiosis. This new idea was advanced in response to the frequent observation during meiosis of unusually aberrant orientations of the metaphase spindles in the egg nucleus. Recently, Takenouchi *et al.* (1981) tested this idea experimentally using cold treatment and obtained haploid, diploid, tetraploid, hexaploid, and octaploid embryos, thereby supporting the "multiple spindle hypothesis." More recently, Takenouchi *et al.* (1983) electrophoretically analyzed the enzyme phenotypes of two natural polyploid populations of *Scepticus insularis* from Japan, one triploid and the other pentaploid. Despite the very different locations of these two polyploid races of *S. insularis,* one from Honshu Island and the other from Hokkaido, their enzyme phenotypes were identical. Therefore Takenouchi *et al.* (1983) concluded: "It is highly improbable that a chance fertilization of a diploid parthenogenetic female by a haploid sperm would have resulted into a triploid type and this triploid having been fertilized by one male to give a tetraploid and therefore still one male to give a pentaploid would have preserved an identical genotype at over twenty enzyme loci. . . ." Hence, Takenouchi's experimental studies support the multiple spindle hypothesis of Suomaleinen, demonstrating that different levels of polyploidy may be achieved without fertilization. Similar results were obtained by Saura *et al.* (1977) in diploid and triploid populations of the parthenogenetic weevil *Polydrosus mollis,* leading them to refute "the contention that triploidy has originated by a chance fertilization of a diploid parthenogenetic female." Moreover, earlier attempts to cross the parthenogenetic forms with the same or closely realted species were unsuccessful (Takenouchi *et al.*, 1983), suggesting that hybridization was not the cause of parthenogenesis in these weevils.

In a related study, Suomaleinen (1961) measured morphological variability in different parthenogenetic polyploid populations of a weevil to determine if such populations have remained evolutionarily unchanged or have differentiated. The results revealed large differences, which he attributed to the genetic capacity of parthenogenetic species to undergo change. This view was later challenged by White (1970) and Smith (1971), who proposed that the differences were mainly due to polyphyletic hybrid origins (see Suomaleinen and Saura, 1973), rather than to changes within the parthenogenetic populations themselves. To further settle this point, Suomaleinen and Saura (1973) conducted a more extensive study comparing 10 enzyme systems in two bisexual species (*Otiorrhincus scaber* and *Strophosomus capitatutus*) and four related polyploid parthenogenetic species (3N and 4N *O. scaber,* 3N *O. singularis,* and 3N *S. melanogrammus*). In general, the bisexuals revealed a high degree of heterozygosity whereas the unisexuals were predominantly homozygous, but "homozygosity and heterozygosity are not mutually exclusive." In two loci ADK-1 and Amy all weevils were homozygous for the same allele in all populations of the two unisexual species. Also in virtually all cases of heterozygosity the enzyme phenotypes were identical between the triploid and tetraploid species.

Summarizing these findings, Suomaleinen and Saura state: "The results indicate that apomictic parthenogenetic populations can differentiate genetically. The genotypes within a population resemble each other more than genotypes belonging to different populations. It is evident that evolution still continues—even if slowed down—in parthenogenetic weevils. A comparison between the allele relationships in geographically isolated polyploid parthenogenetic populations and related diploid bisexual forms does not support the hypothetical hybrid origin of parthenogenesis and polyploidy in weevils. Parthenogenesis within a parthenogenetic weevil species is evidently monophyletic." In a later paper Suomaleinen *et al.* (1979) once again contest the contention by White (1970) and Smith (1971) that parthenogenesis in the weevils originated by multiple hybridizations: "The bisexual, that is, the ancestral population, has by far most of the genes, which occur in parthenogenetic populations. The alleles present in the parthenogenetic populations, but which have not been found in bisexuals, can be explained as results of gene mutations. If the parthenogenesis had been accompanied with hybridizations, the resulting populations should contain foreign alleles, which have not been found in the bisexual race. There is no evidence of such strange genes. This excludes, at least in part, the hybridization hypothesis."

In contrast to the monophyletic origin of the weevils, the various tetraploid parthenogenetic races of the flightless moth *Solenobia triquetrella* from Central Europe are believed to have arisen repeatedly from different bisexual lineages (Lokki *et al.*, 1975; Suomaleinen *et al.*, 1979). This is based on the very different enzyme patterns possessed by each of these populations, which could not have resulted from single mutations (Lokki *et al.*, 1975). According to Suomaleinen *et al.* (1979), bisexual populations of this moth still reside in the Alps and surrounding areas and give rise to new diploid unisexuals from which the tetraploid parthenogenetic lineages originate. Moreover, two other tetraploid parthenogenetic species, *S. lichenella* and *S. fenicella,* are monophyletic and have diverged through mutations (Suomaleinen *et al.*, 1981). Also, crosses between *S. lichenella* and several related bisexual congeners do not produce hybrids (Suomaleinen *et al.*, 1981), so the current evidence does not favor a hybrid origin for the parthenogenetic *Solenobia*.

V. HYBRIDIZATION AS THE SOURCE OF PARTHENOGENESIS IN INSECTS

Against this formidable body of evidence demonstrating a mutational origin for parthenogenesis in invertebrates there nevertheless remain two seemingly

3. Evolution of Parthenogenesis

convincing cases in insects showing that hybridization is indeed the cause of parthenogenesis. The first was presented by Harrison (1920) in a study concerning the inheritance of melanism in bisexual moths. Crosses between the species *Tephrosia crepuscularia* and *Tephrosia bistortata* yielded intermediate and fully fertile hybrids of both sexes. Among the many F_1 and F_2 broods produced, parthenogenesis was detected in four females, but only one produced viable offspring, two males and one female. Unfortunately, Harrison did not describe the experimental conditions for detecting parthenogenesis, nor the fate of the three virgin individuals.

Oddly enough, Harrison's results were later clarified by another worker (Peacock, 1925) in a little theoretical paper entitled "Animal Parthenogenesis in Relation to Chromosomes and Species." Peacock explained that the number of F_1 hybrids that had actually laid parthenogenetically was very small, that the number of eggs laid by these hybrids was fewer than that laid by the parents, that hatching was reduced, larvae were difficult to rear, and mortality was high, and that there was evidence of anomalous chromosome behavior. The latter prompted Peacock to query "Have we here a case in which the reactions between the parental chromatins in the parthenogenetic egg have been so violent as to cause parthenogenetic development or, at least, to stimulate anew any inherent latent capacity for parthenogenesis?" Yet Peacock (1925) seemed cautious in accepting Harrison's meager findings: "we must await further experimental results in Lepidoptera and other animals before asserting that hybridization in these moths evoked parthenogenesis *de novo*."

In an appendix to his paper, Peacock added that he had just obtained corroborative material from Harrison from a cross between a melanic male of *Tephrosia bistortata* and a white *T. crepuscularia* female. "One of these light colored females laid 200 to 300 unfertilized eggs though only 6 larvae emerged. These larvae newly hatched, were handed to me . . . but only two reached the pupal stage; the others for some unaccountable reason died." According to Peacock the two surviving pupae wintered but produced two dead melanic moths, a male and a female. Finally, Peacock states: "Dr. Harrison also informs me that despite repeated attempts since the specimens mentioned in the above contribution were initiated six years ago, he has never again obtained parthenogenetically laid eggs until May 1924."

Two years later, Harrison (1927) reported an additional example of parthenogenesis in the F_1 hybrids, but again was vague in describing his results, and apparently ignored Peacock's findings regarding the six larvae he had sent him. "Remembering that similar occurrences in the earliest *Tephrosia* work (Harrison, 1920, p. 77) had ended in some of the eggs developing parthenogenetically, I enclosed twenty-four females singly in chip boxes, and with exactly the same result. Of the egg masses so obtained, most collapsed without development. In two cases, however, the ova assumed the opaque appearance and faint lutescence characterizing the fertile and developing eggs. One of these batches was handed to my friend, Professor A. D. Peacock, who is making a special study of parthenogenesis, whilst the second was retained by me."

As mentioned above, Peacock (1925) had already published the results of this batch, which were not very convincing (two dead moths hatched out of 200–300 eggs) and complicated by the appearance of a male, as in the first experiment of 1920. Concerning the batch retained by him, he remarked: "After about a dozen days my batch assumed a dirty colour . . . after which about 50% hatched." However, he did not state how many larvae were in the batch nor the number of eggs from which they were derived. In another statement one can deduce that the batch consisted of six larvae based on the 50% figure. "The mortality became great so that in the end only three moths were bred viz. two females, one typical and one melanic, and one melanic male."

There are two problems with Harrison's results: One concerns the high percentage of males, derived supposedly from unfertilized eggs, suggesting that the isolated F_1 hybrids were not really virgin. The second concerns the low levels of parthenogenetic development, suggesting tychoparthenogenesis rather than hybridization. Harrison was keenly aware of both. Regarding the first he states; "Thus it becomes very difficult indeed to understand how, on any sex chromosome basis and under the circumstances postulated, both sexes should be represented in the hybrid parthenogenetic cultures." Regarding the second; "This necessarily implies that it has been proved beyond all doubt that neither *Tephrosia bistortata* nor *T. crepuscularia* are facultatively parthenogenetic. . . . I have employed *T. bistortata* for numerous experiments in which great numbers of its females have been caged alone without depositing a normal batch of ova, much less one developing parthenogenetically. . . . It thus appears certain that by crossing two distinct species of *Lepidoptera* known from exact observations, to be parthenogenetic, there have been developed F_1 insects which were capable of reproducing themselves parthenogenetically. In other words, there has been supplied experimentally, although from the animal side, proof of the correctness of Ernst's position."

Despite Peacock's (1925) earlier skepticism of Harrison's findings, they coauthored a review paper in *Nature* (Peacock and Harrison, 1926) entitled "Hybridity Parthenogenesis and Segregation," in which they defend the previous contentions of parthenogenesis in the *Tephrosia* hybrids: "From these facts we concluded that there was some relation between parthenogenesis and hybridity which, in our opinion, was one of cause and effect." In addition they used this paper to allege that the parthenogenesis described by Nabours (1919) in the grouse locust *Apotettix* had resulted from hybridization, a proposition that Nabours himself had not advanced. Like the studies by Harrison on *Tephrosia*, Nabours was also concerned with the inheritance of color patterns in *Apotettix*, but apparently was not at first familiar enough with the specific status of his

3. Evolution of Parthenogenesis 61

locusts. Consequently specimens were examined by two different authorities, whose opinions Nabours (1919) included in a footnote (p. 131). One authority (Mr. Rehn) identified the specimens as *Apotettix eurycephalus* and the other (Dr. Hancock) as more closely related to *A. convexus* than to *A. eurycephalus.* Hancock, however, added: "Inasmuch as you have used material from both Texas and Mexico in your experiments, it is possible you have hybridized the two." This is what prompted Peacock and Harrison above to suspect that parthenogenesis in the locusts was caused by hybridization.

Upon reading this and other related papers by Nabours, Peacock and Harrison (1926) state: "Recently however we received papers from Prof. Nabours . . . which recount his painstaking genetical experiments with grouse locusts . . . and afford welcome support to our work. This aspect, however, does not seem to have occurred to Prof. Nabours. . . . Investigations into Prof. Nabours' pedigree tables of the females used in parthenogenesis show that the ancestors can be traced to both stocks, the Texan and the Mexican." Addressing the statement by Hancock regarding the specific status of *Apotettix,* they conclude: "Recognizing the implication in this footnote, we submit, therefore, that Prof. Nabours, in supplying from a different group of insects this additional example of segregation under parthenogenesis, has also furnished, although he has not appreciated the fact, a case of parthenogenesis consequent upon hybridity."

Returning now to Nabours' actual work, by 1925 the taxonomic status of his locusts was apparently no longer in question, as he concludes: "The members of the species *Apotettix eurycephalus,* are bisexual, the fertilized eggs producing males and females in equal numbers, and parthenogenetic, the unfertilized eggs, with rare exceptions hatching females." In the earlier paper (Nabours, 1919) he states that "there have been discovered in nature eleven factors for color pattern" in *Apotettix.* Further, he notes: "In nature individuals of the pattern AA of a mottled ground color, in contrast with, and much less conspicuous than, the rest, exceed in numbers all the others combined. . . . The form AA seems to correspond to the so called 'normal' or 'wild type' though all the others have also been found exclusively in nature, none (in *Apotettix*) having so far originated in the laboratory."

The patterns designated by Nabours (1919) are AA, GG, KK, MM, OO, RR, TT, WW, XX, YY, and ZZ. Except for the mottled pattern of AA, all the others are "sharply defined and distinct, each from any other . . . any two make a readily recognizable hybrid pattern with the elements of each parent seemingly equally represented." The discovery of parthenogenesis resulted from crosses between KK females of *Apotettix* and males of another genus, *Paratettix.* Invariably such crosses resulted in all-female offspring exactly like the KK mother. Then it was discovered that virgin KK females isolated from males of any kind also produced female progeny exactly like the mothers and like themselves. In no instance, however, was copulation observed between members of the two genera, although crosses between the various patterns of *Apotettix* produced the expected heterozygous ("hybrid") combinations with progeny of equal sexes. For instance, an F_2 female KK mated to an RR male produced 49 male and 39 female KR intermediate offspring.

Summarizing his initial findings of parthenogenesis in *Apotettix,* Nabours (1919) remarked: "Subsequently from KK females, individually and in groups, but not exposed to males of any kind at any time, there have been given 2,726 female and 4 male offspring, all KKs, some of them having arrived at the fifth parthenogenetic generation. Including the KK females . . . exposed to males without effect there have been produced parthenogenetically from KK females a total of 3,289 females and 5 males, all of the KK pattern. Other females than KK . . . have produced 1,181 females and 2 males making a total of 4,470 females and 7 males of various patterns produced parthenogenetically (Aug. 1, 1918)."

Although the exact ancestry of the original *Apotettix* used by Nabours, vis-à-vis Hancock's concerns that he might be using hybrid material, may never be resolved, the proposition of hybridity by Peacock and Harrison, enticing as it might be, is nevertheless open to question. For one, there is virtually no connection between the rudimentary levels of parthenogenesis described in *Tephrosia* and the prolific ones in *Apotettix* leading to several generations of self-perpetuating all-female lines. Despite Harrison's (1927) assurance that facultative parthenogenesis has never been observed in the bisexual species, the vagueness of his reports does not rule out the possibility that tychoparthenogenesis was overlooked.

But even accepting the assertion by Peacock and Harrison (1926) that "the ancestors can be traced to both stocks," the fact remains that one of the two authorities (Mr. Rehn) identified the species as pure *A. eurycephalus*. Peacock and Harrison (1926) were not only unfamiliar with this genus, but writing from abroad never saw the American material. Also, it is difficult to imagine that Nabours, on the one hand, carelessly overlooked the hybridization of two different species, and on the other, meticulously left a record of their ancestry. Except for Hancock's footnote, there is no other mention of localities in Nabours' original paper (1919). Even assuming that some of the *Apotettix* in the colony were indeed hybrid, what is the likelihood that two different species could have been mixed, by crossing and backcrossing, so thoroughly that the resulting swarm would appear as one, and yet inherit the same orderly polymorphism in color pattern characteristic of the bisexual species? As Nabours stated: "These patterns are as sharply defined and distinct, each from any other, as are those of *Paratettix*." In summarizing the works of Harrison and Peacock, one gets the impression that there is something inherently tantalizing—almost mesmerizing—in an origin by hybridization, which compels its proponents to uphold such a method regardless of the conflicting evidence.

VI. THE PRIME CASE OF A HYBRID ORIGIN IN INSECTS

The best example of hybridization giving rise to parthenogenesis was presented recently by White et al. (1977) in the Australian parthenogenetic grasshopper *Warramaba virgo*. Cytologically this is one of the best studied insect species today. Earlier studies by White et al. (1963), White (1966), White and Webb (1968), White (1974), and Webb and White (1975) have shown that *W. virgo* has a peculiar diploid chromosomal complement with $2N = 15$. It is a fixed heterozygote for various chromosomal rearrangements and has a small metacentric chromosome (M_2) without a homologue. These studies had also suggested that the heterozygosity had developed as a result of mutations accumulated since the origin of parthenogenetic reproduction. However, White and Webb (1968) remarked: "But although this formulation may be acceptable in a general sense, it is not easy to decide just how much of the cytogenetic heterozygosity of virgo has arisen since the acquisition of thelytokous reproduction and how much has been handed down from bisexual times. . . ."

Regarding the origin of *W. virgo* White and Webb (1968) originally believed that the transition from bisexuality to thelytoky "came into existence by a single mutation affecting the behavior of the chromosomes in the premeiotic oocytes." They did not suggest a hybrid origin. There are at least four bisexual species of *Warramaba* closely related to *virgo*. Except for *W. picta*, the remaining are as yet undescribed and are designated by the symbols P125, P169, and P196. *Warramaba picta* was previously known as P151 (White et al., 1977). The three undescribed species are characterized by highly variable color patterns, whereas *W. picta* is uniform like *W. virgo* (White and Webb, 1968). "On the basis of its striking colour resemblance to virgo, P151 must be regarded as close to the ancestry of that species. P125 and P169 are quite differently coloured and, both on this ground and because of their divergent karyotypes must be assumed to be much more distantly related to the phyletic lineage leading to virgo." However, the range of *W. virgo* at this time was known only from southeastern Australia, being separated from the bisexuals, including *W. picta*, by over 1500 kilometers. According to White and Webb (1968) this wide gap resulted from the inability of the parthenogen to adapt itself to environmental changes during the Pleistocene. "It is possible that parthenogenetic biotypes became extinct over a wide area during these climatic changes," leaving the present population remotely isolated from the bisexuals.

In 1974 White discovered another population of *W. virgo*, this one adjacent to the bisexuals in southwestern Australia, specifically next to P196, which also was found to share part of its range with P169, suggesting now that either one of these two bisexuals, or their hybrids, may have been the progenitors of *W. virgo*. In addition, previous hybridization experiments had demonstrated that all of these bisexuals can be hybridized easily in the laboratory as well as in nature (White and Cheney, 1966; White *et al.*, 1969). Because of the peculiar chromosomal complement of *W. virgo*, White *et al.* (1973) and Webb and White (1975) had suggested its derivation from some ancestral X_1X_2Y sex-determining system, which P196 was found to possess, along with a small M_2 chromosome similar in size to that of *W. virgo*, further suggesting that *W. virgo* and P196 were closely related and derived from a common ancestral species possessing an X_1X_2Y system. These suggestions prompted Hewitt (1975) to first propose a hybrid origin for *W. virgo*. Surprisingly, in making the case for hybridization, Hewitt remarked: "A common origin for parthenogenesis is hybridization. Quite convincing examples exist in the simuliid *Cnephia mutata* (Basrur and Rothfels, 1959)," when in fact Basrur and Rothfels specifically denied such an origin for this species (see Section IV). Either Hewitt did not interpret their manuscript correctly or overlooked its meaning to support his own hypothesis. In any event, he further states: "Consequently one might imagine that sporadic hybridization could have occurred between the bisexual relatives of *M. virgo*, and that on one such occasion the hybrid underwent premeiotic doubling to produce a form stabilized by thelytoky. The hypothesis that follows shows that the peculiar karyotype and late replication patterns can be simply explained if *M. virgo* is assumed to have originated in such a manner following hybridization between two forms similar to the extant West Australian relatives, P196 and P169."

Despite White and Webb's (1968) previous belief that *W. virgo* had evolved by mutation and that P169 was too dissimilar to be considered a potential parent, all of this new evidence on distribution and karyology, and the ingenious suggestion by Hewitt (1975), prompted a reinterpretation of the origin of *virgo*. According to White *et al.* (1977), "It must be pointed out that this hybrid origin could not have been suspected at the beginning of our studies on *W. virgo*, when it was only known from eastern Australia and the species P169 and P196 were yet known. . . . If, as we now believe, *W. virgo* had a hybrid origin, the heterozygosities would be due in part . . . to differences between the haploid karyotypes and the ancestral P196 and P169 populations." One way to test the hypothesis of hybrid origin seemed to be the construction of hybrids between P169 and P196 that would resemble synthetic *virgo* individuals.

In the first series of crosses, a total of 38 pairs, using P169 males and P196 females, were mated, producing 1919 eggs, all of which died prior to hatching or during the first instar stage. The reciprocal cross was more successful. Four P169 females mated to P196 males produced 200 eggs, 32 of which hatched, produc-

3. Evolution of Parthenogenesis

ing 30 females and 2 males. The males died within three days, as did many females, but 18 survived to maturity. These hybrid females were intermediate morphologically between the parental species but did not exhibit the usual green coloration of *W. virgo*. However, cytological analysis of the "synthetic" *virgo* revealed the "expected karyotype of 15 chromosomes, 8 from P169 and 7 from P196." Regarding fertility, they were all characterized by rudimentary ovaries and were largely sterile. Nevertheless, they managed to lay a few eggs (17 total), of which three developed to the embryo stage. These embryos apparently did not develop further and were used to determine their karyotypes, which were the same as the mother's. Unfortunately White *et al.* (1977) were not specific regarding the fate and anatomical condition of the 18, first generation, "synthetic *virgo*." It is known that they were "largely sterile," but they did not elaborate on the extent of the fertility, nor where the 27 eggs originated, whether from one or several of the females. However, they state: "when this experiment is repeated on a larger scale, it should be possible to obtain some second generation diploid hybrids which had arisen by parthenogenesis. The result that was obtained obviously speaks in favor of the hypothesis that *W. virgo* arose . . . from a cross between ancestors of P196 and P169 having essentially the karyotypes and genetic constitution of the modern representatives of those species."

In general, these results in *Warramaba* very closely resemble those of Harrison (1920) in *Tephrosia*. In both cases, the mortality of the F_1 progeny was exceptionally high. Many of the dead embryos examined from the first cross were haplo–diploid mosaics, reminiscent of the anomalous chromosome behavior mentioned by Peacock (1925). Those that survived from the second cross were essentially sterile, most of which died, and the few eggs laid by these F_1 progressed only to the embryo stage. Finally, White *et al.* (1977) did not mention the potential extent of parthenogenesis in the bisexuals, nor any previous experiments to demonstrate it or rule it out. In a previous paper, White and Webb (1968) state: "The reproduction of *virgo* is exclusively parthenogenetic, no males having ever been encountered either in nature or in laboratory stocks. In contrast to this, the sex ratio in the related bisexual species appears to be normal and there is no evidence that any of them reproduce parthenogenetically in nature." Clearly, one of the most important tests remaining to be done in *Warramaba*, and other parthenogenetic–sexual complexes for that matter, is a precise examination of potential rudimentary levels of parthenogenetic development in the related bisexuals. Except for the above hint by White and Webb, the true extent of tychothelytoky in *Warramaba* remains unknown.

But even if future workers, as White *et al.* (1977) have optimistically suggested, were ultimately successful in producing a healthy second generation of parthenogenetic hybrids, such "synthetic" *W. virgo* must still acquire a genetically controlled meiotic mechanism that will circumvent the asynaptic problems posed by single or very different homologues. In other words, how will meiosis divide the single M_2 of these newly created *W. virgo?* Will it sometimes cast it to the polar body and other times retain it in the egg? And what about the other chromosomes with greatly differing homologues? Should they sometimes synapse with each other, at other times with nonrelated ones, and still at others with none? This is why a stabilizing meiotic mechanism is essential, to ensure chromosomal constancy in every generation, and this is also why a mechanism of "genetic imbalance" as proposed by Muntzing (1940) above is not a viable alternative.

White *et al.* (1977) were fully aware of this major cytogenetic impediment facing their fledgling parthenogenetic hybrids, and they discussed it in detail. "The mechanism of *W. virgo* depends entirely on the premeiotic doubling of the chromosome number in the oocyte. . . . It was originally suggested that this cytological tour-de-force in the *W. virgo* oocyte could only have arisen as a single mutation which was itself the factor initiating parthenogenetic reproduction (White *et al.*, 1963). It has been difficult to reconcile this mechanism with a hybrid origin for *W. virgo*, since it is not obvious why it should appear spontaneously in a species hybrid when it is certainly not present in the parent species. . . . Uzzell and Goldblatt (1967), in the case of the parthenogenetic *Ambystomas,* have speculated that 'difficulties in synapsis . . . , or genetic imbalance due to unmatched sets of chromosomes (in some way) resulted in premeiotic chromosomal division not accompanied by cytokinesis.' However, . . . there is not the slightest reason to suppose that meiotic abnormalities would lead to a radical change in chromosome behaviour at an *earlier* stage. Thus, even if we are right in assuming the spontaneous manifestation of the premeiotic doubling mechanism in the P169 × P196 hybrids, the cellular causes for this remain mysterious."

VII. THE SPONTANEOUS VERSUS HYBRIDIZATION CONTROVERSY IN VERTEBRATES

> In spite of much evidence, the "hybrid theory" has not received universal acceptance. Astaurov (1940) wrote that "hybridization can hardly be regarded as the main universal reason for parthenogenesis, but at the same time, we have certain reasons to believe the secondary transition from sexual reproduction to parthenogenesis was in a number of cases accompanied by hybridization." Without touching upon the mechanisms of such a connection, which are not yet clear, it should be noted that, according to Astaurov, hybridization does not result directly in parthenogenesis, but only favors it through heterosis. In any case the connection between hybridization and parthenogenesis seems to be rather complicated. (Darevsky, 1966)

In 1974 I wrote an article proposing that hybridization may not be the direct cause of parthenogenesis in vertebrates, as is the current belief by most authorities of this group. My suggestion was based on four premises: (1) the

3. Evolution of Parthenogenesis

numerous observations of rudimentary parthenogenesis in invertebrates, (2) the spontaneous production in these forms of self-perpetuating parthenogenetic lines derived from virgin eggs, (3) the proposition by the majority of these workers that parthenogenesis evolves first and is later preceded by hybridization and polyploidy, and (4) the failure in both vertebrates and invertebrates to produce hybrid unisexuals despite repeated crossings of putative parents. In that article I also suggested that although the evidence of hybridity was overwhelming in some species, the original source of unisexuality may have been spontaneous, and that a meiotic mechanism of asynapsis ("genetic imbalance"), as was then upheld by some of those authorities, was not a viable option, especially in the case of triploids.

Since that time, little has changed. Those who believed in a hybrid origin have strengthened their positions, and additional evidence and cases of hybrid origin are being reported (Cole, 1975; Uzzell and Darevsky, 1975; Bickham *et al.*, 1976; Parker and Selander, 1976; Leslie and Vrijenhoek, 1977; Downs, 1978; Wright, 1978; Brown and Wright, 1979; Parker, 1979; Vrijenhoek, 1979; Turner *et al.*, 1980; Echelle and Mosier, 1981; Sessions, 1982; Cole *et al.*, 1983; Wright *et al.*, 1983; Dessauer and Cole, 1984; Good and Wright, 1984; Moore, 1984). According to Cole (1978), "Many who investigate parthenogenetic lizards agree that the data on external morphology, ecological and geographical distribution, karyotypes, and biochemistry indicate that evolution of parthenogenesis in many instances involved hybridization between bisexual species." Similarly Wright (1978) commented: "Needless to say, the issue is still open as to whether hybridization can give rise to parthenogenesis, but if the parthenogenetic species, as they live and breathe in nature, have the morphology of a hybrid, the karyotypes of a hybrid, and/or the allozymes of a hybrid, there would be reason to suspect that they also might have hybrid influenced ecologies. To use ecological observations for these hybrid-appearing organisms to support a model of their origin that excludes the possibility of hybrid involvement is absurd at best."

In his recent review of parthenogenesis, Bell (1982) states: "The origin of parthenogenetic vertebrates has been the subject of some controversy. The prevailing view has been that they arise as a consequence of interspecific hybridization: hybridization yields allodiploid hybrids, which reproduce parthenogenetically, and which by back-crossing to the ancestral amphimicts (sexuals) may in turn give rise to parthenogenetic allotriploids. This view has been urged for fish . . . amphibians . . . reptiles." The other point of view ("the lonely opposition") "maintains that spontaneous genetic change within an amphimictic diploid may produce a parthenogenetic diploid, which by crossing with the same or another amphimict in turn gives rise to a parthenogenetic autotriploid or allotriploid. To decide between these conflicting opinions we must first recognize their common ground. In the first place the controversy is restricted to vertebrates; it is not alleged that hybridization is usually or commonly involved in the origin of parthenogenetic invertebrates.'' And later he stresses again that the "objection loses much of its force when we recall that nobody has ever claimed that interspecific crosses always or even often give rise to parthenogenetic hybrids." I will not belabor the point, except to recall that the controversy germinated in plants in 1912 (Ostenfeld, 1912) was carefully tendered in the same group in 1917 and 1918 (Winge, 1917; Ernst, 1918), was very much alive in moths in 1926 (Peacock and Harrison, 1926), accelerated its pace in beetles in the seventies (White, 1970; Suomaleinen and Saura, 1973), and is thriving today in grasshoppers (Hewitt, 1975; White et al., 1977).

Despite the recent reports of new parthenogenetic hybrid lizards (Cole et al., 1983; Moritz, 1984), the basic evidence relating to origin in the vertebrates has not really changed one way or the other. Unisexual hybrids have been synthesized in the hybridogenetic fish *Poeciliopsis* (Schultz, 1973), which despite claims to the contrary (Moore, 1984) is not parthenogenetic. Hybridogenesis is an odd form of unisexuality unrelated to obligatory parthenogenesis. According to Bell (1982), "In fact with the exception of the highly aberrant *P. monacha-lucida* there is no example among vertebrates of an interspecific cross yielding parthenogenetically reproducing hybrids." The recent suggestion by Moore (1984) that "there are great similarities between the ecologies of parthenogenetic fishes and lizards" overlooks the essence of thelytoky. All the purely parthenogenetic species of plants and animals throughout the world (Vandel, 1931; Gustafsson, 1947b; Stebbins, 1950; Suomaleinen, 1950; Grant, 1971; Richards, 1973) occur in habitats essentially devoid of bisexuals. The hybridogenetic as well as gynogenetic fish exist as parasites on the males of their bisexual hosts. If they indeed possess the power of rapid reproduction, they can never express it, as they would swamp their hosts into extinction, nor can they ever found new colonies by themselves, for they are veritable prisoners, inextricably tied to the sperm of their male hosts. As Vrijenhoek (1984) correctly described: "Unlike truly parthenogenetic species gynogenetic and hybridogenetic *Poeciliopsis* can never escape from their sexual hosts to invade new habitats, nor can they competitively exclude their hosts, for in so doing they lose their sperm source and ensure their own extinction."

According to Monaco et al. (1984), "*Poecilia formosa* is not a true thelytokous parthenogen . . . instead it reproduces by gynogenesis, a mechanism that is sperm dependent. Males of related species provide sperm which serve only to activate the egg. . . . The obligatory dependence on sperm that is a hallmark of gynogenetic reproduction compels *P. formosa* to behave in nature as a sexual parasite . . . the survival and success in nature of these all-female forms require their coincident distribution with at least one of the bisexual host species." Regardless of the fervid and eloquent arguments by Moore (1984), the advantages of parthenogenesis have no connection whatsoever with those of parasitism. According to this worker, "The most important population ecological question concerning unisexual fishes is: How do the unisexual species coexist with the bisexual species upon which they depend for sperm? . . . no one has bothered to find out whether or not unisexual fishes compete with their host species! The population ecology of unisexual fishes will not progress until the competition question is answered, and the critical experiments obviously remain to be done. . . . Such experiments are logistically difficult and probably would be impossible with *Poeciliopsis*. . . ." By insisting that the advantages of parthenogenesis apply to hybridogenetic and gynogenetic parasites, Moore (1984) placed himself in the difficult position of having to conclude that "the search for resource partitioning approaches a kind of limiting absurdity: i.e. by looking ever more intensely one is bound to find some ecological difference between any two genotypes, and if one argues that such a difference explains coexistence, then the conclusion that two species coexist by niche partitioning is inevitable."

To date, all attempts in vertebrates to produce unisexuality by hybridization have failed, even the most ambitious ones in the gynogenetic fish *Poecilia formosa* (Hubbs, 1955; Turner, 1982). Invariably, as one might accurately predict from the annals of hybridization, the hybrids are bisexual, intermediate, or sterile (Hubbs, 1955; Darevsky, 1966; Christiansen and Ladman, 1968; Cuellar and McKinney, 1976; Turner, 1982; Moore, 1984). Oddly enough, one of the foremost *aficionados* of the hybridization hypothesis in vertebrates has apparently refuted the second step in the vertebrate hybridization scheme, the backcrossing of unisexuals to bisexuals for the production of polyploids. According to Moore (1984), "an experiment conducted by Charles Cole suggests hybridization would be unlikely. Cole (personal communication) thus believes that natural hybridization between unisexual and bisexual whiptails is rare even when there is sympatry." This is strange in view of the various hybrids between sexual and unisexual *Cnemidophorus* that have been discovered in nature or formed in the laboratory (Cuellar and McKinney, 1976; Dessauer and Cole, 1984).

One of the most novel findings in unisexual vertebrates in the last decade is that of ameiotic restitution in the gynogenetic fish *Poecilia formosa* (Rasch *et al.*, 1982; Monaco *et al.*, 1984). Ameiotic restitution is also known as apomixis (Suomaleinen, 1950). It is the usual mechanism in plants and is relatively common in invertebrates, but this is the first report of its occurrence in a vertebrate. In the usual apomixis, there is no real meiosis. The chromosomes do not synapse to form bivalents and there is only one division, and therefore only one polar body. Also the chromosomes in meiotic prophase do not reveal the usual appearances of zygonema, pachynema, diplonema, diakinesis, and metaphase I, as these are determined by the double structure of the bivalents, each consisting of two synapsed homologues. The telltale appearances are particularly striking after pachynema, when the crossover loops become apparent and condensation produces the classic bivalent configurations. Unfortunately the analysis of meiosis in *P. formosa* by Rasch *et al.* (1982) and Monaco *et al.* (1984) is based mostly on cytophotometry, autoradiography, and electron microscopy. Evidence derived from the former two is indirect and the magnification of the latter is too great to see the meiotic apparatus, i.e., whole chromosomes, spindles, metaphase plates, etc. They did observe the earliest stages of oogonial development (zygonema and pachynema) in serial sections of ovaries, but not the critical ones, nor the actual divisions and formation of polar bodies.

The number of chromosomes observed in these early stages was the same as the somatic number, 46 in diploids and 69 in triploids, suggesting premeiotic doubling, if these were truly bivalents. The bisexuals revealed the expected haploid set of bivalents ($N = 23$). According to Monaco *et al.* (1984), "If one considers only the number of chromosome strands in pachytene oocyte nuclei from the unisexuals of *Poecilia,* these observations at first glance could be construed as evidence for a doubling . . . prior to the formation of pseudobivalents. . . . When the diameters and staining intensities of these chromosomes are considered, however, it is apparent that the 46 chromosome strands seen in pachytene oocytes from *P. formosa* do not represent 46 bivalents. Rather, there are 46 univalents, each of which contains a pair of sister chromatids." Apomixis may well be the mechanism of restitution in *P. formosa,* but it will not be confidently ascertained until all of the cytological details of meiosis are observed directly in a reasonable number of mature oocytes from a reasonable number of individuals of this species.

I have mentioned previously (Cuellar, 1976) that female meiosis is extremely difficult to analyze in animals. For one, eggs are rare relative to sperm. They are also difficult to section, especially the yolky ones of vertebrates. The divisions themselves are complex and they occur suddenly during a long interval of development. Staining the chromosomes is tedious, and they are difficult to count, especially from side view. Finally, polar bodies are small and delicate and they deteriorate rapidly. I agree with Monaco *et al.* (1984) that nonconventional methods would be practical, if only they were reliable. Molecular methods such as cytophotometry may well verify the DNA levels assumed from stained sections, but there is no substitute for the serial sections obtained through the microtome, which allow direct observation of the entire meiotic process.

A complication resulting from the discovery of apomixis in these unisexual fish is the occurrence now of two vastly different methods of restitution in two very similar gynogenetic forms, *Poeciliopsis* and *Poecilia*. According to Cimino (1971), the mechanism in *Poeciliopsis* is of the premeiotic type. To worsen matters, members of *Poeciliopsis* reproduce by two very different methods, gynogenesis and hybridogenesis. Presumably the mechanisms of restitution in these two forms are also different. A triploid gynogenetic goldfish (Cherfas, 1966) purportedly develops an abortive tripolar spindle, thereby resulting in only one division. On the other hand, diploid gynogenetic populations of a similar

3. Evolution of Parthenogenesis

goldfish supposedly undergo two divisions (Leider, 1959). Adding more confusion to this assortment of mechanisms are interpretational differences for the same mechanism. For instance, for Cimino, the 46 chromosomes observed during pachynema were bivalents, whereas for Monaco *et al.* (1984) these very same strands were univalents. Thus, one worker elected a premeiotic mechanism and the other apomixis.

With the possible exception of the mechanisms reported in the parthenogenetic lizard *Cnemidophorus uniparens* (Cuellar, 1971) and in the gynogenetic salamander *Ambystoma tremblayi* (Cuellar, 1976), there are still no studies in vertebrates providing a cytological analysis of the entire meiotic process. Therefore the true meiotic mechanisms operating in these various species remain uncertain (Leider, 1955, 1959; Darevsky and Kulikova, 1961; Cherfas, 1966; Macgregor and Uzzell, 1964; Cimino, 1971, 1972; Rasch *et al.*, 1982; Monaco *et al.*, 1984). But even the above seemingly complete studies in the lizard and the salamander are deficient in many respects. For instance, the actual act of premeiotic duplication has never been observed. It has merely been assumed from the presence, during prophase, of as many bivalent chromosomes as the somatic number, but as Monaco *et al.* (1984) caution, this of itself can be misleading. Also, the second division was never observed in either species, suggesting an abortive mechanism rather than endoduplication. In *C. uniparens* the close association of certain bivalents in diakinesis suggests the possibility of multivalent formation, and the few counts made of this stage did not reveal exact triploid numbers. Only one metaphase plate was observed from polar view and this too revealed clumping, reminiscent of multivalents. Hence, at least for *C. uniparens* the evidence remains skimpy at best. With such great difficulties presently facing an accurate analysis of meiotic restitution in parthenogenetic vertebrates, on the one hand, and with such a wealth of meiotic information in other parthenogenetic groups, on the other (Narbell-Hofstetter, 1964), it almost seems that our selection of an appropriate mechanism is determined more by convenience and availability than by the meiotic process itself. This apparent arbitrariness is particularly applicable to the mechanism of premeiotic doubling (endomitosis, endoduplication, endoreduplication—endoredundant!), which seems to be the raging fashion these meiotically depauperate days. As Bogart (1980) noted: "In spite of this limited information, the process of premeiotic endomitosis has been widely disseminated as the mechanism for maintaining somatic ploidy in virtually all parthenogenetic or gynogenetic amphibians and reptiles. . . . This same mechanism has even been speculated to occur in *Rana esculunta* for the production of diploid eggs or sperm. . . ."

The latest discoveries of parthenogenesis in vertebrates are two species of lizards, both in the family Gekkonidae, *Heteronotia binoei* from Australia (Moritz, 1983, 1984) and *Hemidactylus vietnamensis* from Southeast Asia (Darevsky *et al.*, 1984). They are similar in being triploid, but apparently have vastly different origins and ecological requirements. *Heteronotia binoei* supposedly evolved through hybridization, whereas *H. vietnamensis* did not, and the former inhabits predominantly continental deserts, whereas the latter inhabits tropical islands. The method of origin proposed by Moritz for *H. binoei* is virtually the same as that proposed for the majority of triploid vertebrates, except that in *H. binoei* the hybridizations supposedly occurred between closely related races rather than between species. In many areas, the diploid sexuals of *H. binoei* occur together with the unisexuals and are both very similar in appearance. Chromosomal analysis has revealed two different populations of sexuals (SM6 and A6) and three different unisexual clones.

The major chromosomal differences among the two bisexual "cytotypes" occur in chromosomes numbers 4 and 6. In SM6 the number 4 pair is predominantly telocentric (has very short arms) whereas in A6 it is predominantly acrocentric (no visible arms). However, both populations possess individuals or local races in which the number 4 consists of a combination of the above two types, resembling a hybrid pair. The number 6 is virtaully the same in both populations, consisting of similar-sized submetacentrics with terminal constrictions (NOR satellites). In one race, however, the number 6 is very different, lacking satellite and being acrocentric. There are actually four different chromosomal types in the A6 populations, but the others are relatively minor: number 6 in the Baroalba race and number 1 in the Bing Bong have acrocentric rather than a pair of submetacentrics and telocentrics, respectively. Therefore, these two races, like those of SM6, are also heterozygous for chromosome morphology (Table I).

The chromosomes of the triploid parthenogenetic clones are very similar to those of the bisexual species, but each clone differs from the other in having one or more of the parental hybrid pairs, suggesting multiple origins of the clones from the different sexual cytotypes. If these differences were truly acquired from the bisexuals, rather than having evolved independently in the unisexuals, then the origin of the unisexuals can be traced to some combination of the bisexual races. Moreover, the triplets in the number 6 in the clones could only have been derived by combining the major differences between sexuals, hence strongly suggesting hybridization. According to Moritz (1984), "Indeed, it seems highly probable that hybridization is the universal mode of origin of parthenogenetic vertebrates. . . ." To account for the three chromosomally distinct clones he proposed that parthenogenesis in *H. binoei* arose by two steps. "The first was the hybridization of the A6 and SM6 cytotypes to produce diploids capable of parthenogenesis, and the second was the subsequent backcrossing of these hybrids to bisexual males of each cytotype. Furthermore it was argued that the A6 cytotypes involved were the central and western karyomorphs that lack the secondary constriction on 6."

By assuming an initial hybridization and subsequent backcrossing, Moritz

3. Evolution of Parthenogenesis

TABLE I

The Heteromorphic Chromosomes of *Heteronotia binoei*[a]

LOCATION	BISEXUAL POPULATIONS OF SM6		
	CHROMOSOMES		
	1	4	6
Barrow Ck (W)	⋏⋏	∧⋏	⋋⋋
North Central	⋏⋏	⋏⋏	⋋⋋

	BISEXUAL POPULATIONS OF A6		
	1	4	6
Baroalba (N)	⋏⋏	∧∧	∧⋋
Bing Bong (N)	⋏∧	∧∧	⋋⋋
E + S + N	⋏⋏	∧∧	⋋⋋
Cordillo (C+W)	⋏⋏	∧⋏	∧∧

Clones	PARTHENOGENETIC POPULATIONS		
	1	4	6
A	⋏⋏⋏	∧∧∧	∧∧⋋
B	⋏⋏⋏	∧∧∧	∧⋋⋋
C	⋏⋏⋏	∧∧⋏	∧⋋⋋

[a] N, S, E, W, = North, South, East, and West Australia. The biarmed chromosomes of pair 1 are telocentric, those of pair 2 telocentric or submetacentric, and those of pair 6 submetacentric. Dots on 6 denote satellites.

(1983) was able to match each of the clonal karyotypes with a hybrid combination of the bisexuals. However, one can also assume an initial spontaneous diploidization and derive essentially the same triploid karyotypes without invoking the first step of the hybridization hypothesis. In Table I, I have formulated karyotypes for chromosomes 1, 4, and 6 from each of the bisexual and unisexual cytotypes as interpreted from actual spreads shown in Fig. 1 and 7 in Moritz (1984), as well as from verbal descriptions in the texts (1983, 1984). I present this table to simplify interpretation of the rather complex chromosomal reorganizations of *H. binoei* and to better visualize the possible modes of origin.

According to Moritz, all three major clones (ABC), as well as lesser derivatives determined from C banding, originated from crosses between A6 and MS populations.

On the other hand, a spontaneous mode of origin can produce all of the original diploids, as well as the C and B triploids, exclusively from the A6 population. Only the A triploid requires derivation from a cross between SM6 and A6. For example, clone C could have started as a spontaneous diploid of E + S + N (Table I), which then backcrossed to males of C + W bearing the telocentric 4. Exactly the same diploid backcrossed to a CW male with the acrocentric would produce the B clone. The A clone can be derived by diploidizing Cordillo and backcrossing to males of Barrow bearing the acrocentric 4. The latter synthesis is not perfect since clone A does not have a telocentric in the triplet. Nevertheless this telocentric may well have lost the short arms subsequently and now resembles an acrocentric. Also it is well known that arm length varies considerably within chromosomal groups and that traditional designations are often arbitrary (Levan *et al.*, 1964; Brown, 1972; John and Freeman, 1975; Swanson *et al.*, 1981) Considering the great diversity of chromosomal types in *H. binoei* and the possibility of long-term modifications through various mechanisms (Swanson *et al.*, 1981), some of the existing clones could even be derived by autopolyploidy. Clone A, for instance, may have evolved from a Baroalba spontaneous diploid fertilized by its own males bearing the acrocentric 6, which subsequently lost its satellites in the acrocentrics, just as the Cordillo population apparently has. In fact, allowing a reasonable amount of chromosomal reorganization all of the clones of *H. binoei* could just as easily be derived through autopolyploidization. The most certain proof of a hybrid origin would be, of course, to synthesize the presumed hybrid diploids by crossing, an aim that has failed in all organisms to date. Another certain proof would be to find the diploid hybrids in nature. According to Moritz (1983), "The postulated diploid parthenogen has not been found in central Australia despite intensive collecting." Likewise, for the spontaneous method, the ultimate proof would lie in producing instant parthenogenetic diploids from unfertilized eggs or to find such diploids in nature. Considering their probable identicity to the mother, they would be extremely difficult to detect, except by the production of all-female progeny or triploids upon fertilization. To my knowledge no one has yet attempted to determine the levels of fertility in virgin eggs of bisexual lizards.

In discussing the possibility of a spontaneous method of origin for the triploid parthenogenetic lizard *Leiolepis triploida,* Peters (1971) quotes from a letter by William P. Hall in which he suggests to Peters that the extra set in this species arose by autotriploidy: "There is no theoretical reason that requires parthenogenesis to have a hybrid origin, although this has been shown in all of the well documented cases. It may even be possible that the triploidy and parthenogenesis originated in a single individual. Triploidy may result in a sexually reproducing

3. Evolution of Parthenogenesis

species through either a double fertilization or failure of one of the meiotic divisions in the egg. The latter is required for parthenogenesis also, so it is not unreasonable to expect triploidy and parthenogenesis to be associated. However, I would think it more likely that there is or at least was a stage of diploid parthenogenesis first followed by a fertilization to produce the triploidy. I have karyotyped over 400 *Sceloporus grammicus* and one of the individuals turned out to be a triploid male showing that triploidy can occur without parthenogenesis. Of course, the triploids will be sterile unless they can avoid meiosis by reproducing parthenogenetically. So on the basis of what you say about the phenotypes of the triploids I would conclude that they did not have a hybrid origin.'' To my knowledge, this is the first reported case of spontaneous triploidization in a lizard. Similar autotriploids have been reported in fish and salamanders (Cuellar and Uyeno, 1972).

The latest discovery of parthenogenesis in natural populations of the lizard *Hemidactylus vietnamensis* by Darevsky *et al.* (1984), as well as its suspected occurrence in the related species *H. karenorum* (Darevsky *et al.*, 1984), now brings to five the number of species known to be parthenogenetic among geckos. The others are *H. garnoti* (Kluge and Eckardt, 1972), *Lepidodactylus lugubris* (Cuellar and Kluge, 1972), and *Heteronotia binoei* (Moritz, 1983). Among them only *L. lugubris* is diploid (2N = 44). *Hemidactylus garnoti* is 3N = 70, *H. vietnamensis* 3N = 60, and *H. binoei* 3N = 63. *Hemidactylus karenorum* has not been karyotyped (see Table II). Although Kluge and Eckardt (1969) proposed triploidy for *H. garnoti*, they did not mention the probable mode of origin, nor did they explain why the 3N = 70 departs from the expected 3N = 69 or 3N = 72 of a truly triploid set derived either from 2N = 46 or 2N − 48 bisexuals. Judging from the large area encompassed by one of their spreads, and by the conspicuous differences in size and shape displayed by the ''homologues'' of the two metacentric sets, it is possible one or two chromosomes may be either extra or missing. Also the third triplet contains two acrocentrics with secondary constrictions, suggesting that the odd one was acquired from another source through hybridization.

Hemidactylus vietnamensis is very similar chromosomally to *H. garnoti*. The first two triplets of both species consist of metacentrics and the third one possesses two chromosomes with secondary constrictions. If it were not for the great difference in size among the metacentrics in *H. garnoti*, the two species might appear identical. According to Darevsky *et al.* (1984), ''Therefore we have reason to assume that *H. karenorum* as well as *H. garnoti* and *H. vietnamensis* are morphologically similar parthenogenetic species. . . .'' Regarding the origin of *H. vietnamensis* Darevsky *et al.* do not believe their chromosomal or electrophoretic data favor hybridization. ''The third acrocentric element, which lacks the secondary constriction, might be derived from another parental species (2N = 40). However one cannot but take into consideration our present knowl-

TABLE II
Known or Suspected Parthenogenetic Lizards: Chromosome Number and Mode of Origin[a]

Species	Ploidy	Source	Parental species	Source
Family Teiidae				
Cnemidophorus velox	3N 69	Pennock (1965)	Sexlineatus? × inornatus × inornatus	Lowe et al. (1970)
C. uniparens	3N 69	Lowe and Wright (1966)	Gularis × inornatus × inornatus	Lowe et al. (1970)
C. exsanguis	3N 69	Pennock (1965)	Gularis × inornatus × sexlineatus?	Neaves (1969)
C. opatae	3N 69	Lowe et al. (1970)		Wright (1967)
C. flagellicauda	3N 69	Lowe and Wright (1966)	Sexlineatus? × sexlineatus? × sexlineatus	Lowe et al. (1970)
C. sonorae	3N 69	Lowe and Wright (1966)	Sexlineatus? × sexlineatus? × sexlineatus	Lowe et al. (1970)
C. tesselatus	3N 69	Pennock (1965)	Sexlineatus? × sexlineatus? × sexlineatus	Neaves (1969)
C. tesselatus	2N 46	Wright and Lowe (1967)	Tigris × septemvitatus × sexlineatus	Neaves (1969)
C. dixoni	2N 46	Scudday (1973)	Tigris × septemvitatus	Scudday (1973)
C. neomexicanus	2N 46	Pennock (1965)	Tigris × septemvitatus	Lowe and Wright (1966)
C. cozumela	2N 50	Fritts (1969)	Tigris × inornatus	Fritts (1969)
C. rodecki	2N 50	Fritts (1969)	Angusticeps × deppei	Fritts (1969)
C. laredoensis	2N 46	Bickham et al. (1976)	Angusticeps × deppei	McKinney et al. (1973)
C. lemniscatus	2N 48	Peccinini (1971)	Gularis × sexlineatus	Vanzolini (1970)
			(Spontaneous)	Peccinini (1971)
Gymnophthalmus underwoodi*	?		Lemniscatus × lemniscatus	Thomas (1965)
	?		?	Dessauer and Cole (1984)
Teius sp.*	?		Speciosus × pleei	Martori[a]

Taxon	Chromosomes	Parents	Reference
*Leposoma percarinatum**	?	?	Uzzell and Barry (1971)
*Kentropyx borckionus**	?	?	Hoogmoed (1973)
Family Lacertidae			
Larcerta dahli	2N 38	*Saxicola* × *derjugini*?	Darevsky and Kulikova (1961)	Darevsky (1966)
L. armeniaca	2N 38	*Saxicola* × *derjugini*?	Darevsky and Kulikova (1961)	Darevsky (1966)
L. rostombekovi	2N 38	*Saxicola* × *derjugini*?	Darevsky and Kulikova (1961)	Darevsky (1966)
L. unisexualis	2N 38	*Saxicola* × *derjugini*?	Darevsky and Kulikova (1961)	Darevsky (1966)
Family Gekkonidae			
Hemidactylus garnoti	3N 70	?	Kluge and Eckardt (1969)
H. vietnamensis	3N 60	(Spontaneous)	Darevsky et al. (1984)
*H. karenorum**	?	?	Darevsky et al. (1984)
Heteronotia binoei	3N 69	*Binoei* × *binoei* × *binoei*	Moritz (1983)	Moritz (1983, 1984)
Lepidodactylus lugubris	2N 44	(Spontaneous)	Cuellar and Kluge (1972)	Cuellar and Kluge (1972)
Family Xantusidae			
Leipdophyma flavimaculatum	2N 38	*Flavimaculatum*? × *tuxtlae*? × *pajapensis*?	Bezy (1972)	Bezy (1972)
Family Agamidae			
Leiolepis triploida	3N 54	(Spontaneous)	Hall (1970)	Peters (1971)
Family Chamaelonidae			
*Brookesia spectrum**	?			Hall (1970)

a Putative bisexuals are listed in alleged hybridization sequence. Question marks indicate bisexual species uncertain or information not known or given by author. Asterisk indicates parthenogenesis suspected from absence of males. In a letter dated May 28, 1986, Ricardo Martori from the Universidad Nacional de Rio Cuarto, Cordoba, Argentina, informed me that his group has apparently just discovered a parthenogenetic population of *Teius* in central Argentina.

edge bearing on the structure and the nature of secondary constrictions (King, 1980). . . . These data as well as those of genomic organization of the ribosomal RNA cistrons of parthenogenetic taxa which often lie in the constriction region (White *et al.*, 1982), suggest that the heteromorphism of the chromosomes of *H. vietnamensis* cannot be regarded as resulting only from hybridization. . . ." Concerning electrophoresis, Darevsky *et al.* state: "This analysis does not eliminate the possibility of a hybrid origin of *H. vietnamensis,* but only shows that the extant bisexual species did not participate in its formation."

In 1978 I received a letter from G. J. Ingram of the Queensland Museum, Brisbane, Australia, in which he informed me that the gecko *Gehyra variegata,* reported by Hall to be parthenogenetic, was actually *Lepidodactylus lugubris.* He states: "In your recent *Science* paper ('97:837) on page 842 you say *Gehyra variegata* is parthenogenetic quoting W. P. Hall's *Experientia* paper (26:1271). His conclusions are based on the taxon *G. v. ogasawarasimae* Okada 1930. Shigei (1971) [Reptiles of the Bonin Islands. *J. Fac. Sci. Tokyo Univ. (Zool.)* 12:145–166] placed *ogasawarasimae* in the synonymy of *Lepidodactylus lugubris."* Hall cited two reasons for suggesting parthenogenesis in *G. variegata:* Makino and Momma (1949) reported a count of 63 chromosomes in this species, and they based their material on embryonic ovaries because they found no males. Since 63 corresponds very closely to triploidy in geckos, and several species are 2N = 42, Hall assumed *G. variegata* was triploid and derived from a 2N diploid bisexual possessing 2N = 42. According to Makino and Momma (1949) "The embryonal ovaries comprise the material for the study. They were obtained in July 1936 at Chichishima of Ogasawara Islands, and fixed in Hermann's mixture. The number of chromosomes determined by examination of four nuclear plates of oogonia, is 63 without exception in every case. . . . The chromosomes are all rod-shape of various sizes, ranging from elongated rods to short dot-like ones, all being terminal in their spindle-fibre attachment. They show a gradual diminution in length so that it is impossible to sort them into such distinct groups as macro- and micro-chromosomes. . . . Unfortunately due to insufficient material, the present study failed to investigate the chromosomes of the male."

Based on this description, one can assume that the four oogonial plates were observed in a single ovary obtained from a single individual. They did not mention sample sizes. Therefore, the assumption of parthenogenesis by Hall was based more on the potential for triploidy than on the absence of males. Recently Kluge (1982) called attention to this confusion, clarifying for the first time Shigei's inclusion of *G. v. ogasawarasimae* into *L. lugubris.* He states: "The questionable interpretation, plus the unusually high chromosome number for geckos, led Hall to speculate that the race might be a parthenogenetic triploid derivative. Thus, *G. v. ogasawarasimae* has been widely cited as the third all-female gekkonid lizard. . . ." Despite the high number of 63 chromosomes

3. Evolution of Parthenogenesis

observed by Makino and Momma relative to other geckos, they nevertheless concluded the species was diploid. Matching the larger chromosomes at the periphery of the oogonial plates, these were "found to constitute homologous pairs of equal size." However, Makino and Momma were troubled by the inclusion of an unmated chromosome, which they suggested was evidence "of the heterogametic nature of the female cell in this form. . . ." If the individual was triploid, as Hall suspected, then the extra element would be one of a triplet, rather than of an XO sex-determining system. Future investigations of the Chichishima population may well reveal the first known triploid population of *L. lugubris*.

So far, the only recent analysis of the chromosomes of *L. lugubris* is that of Cuellar and Kluge (1972) from specimens of the island of Oahu, Hawaii. They reported a diploid number of 44, all consisting of acrocentrics and exhibiting a gradual diminution in length as reported by Makino and Momma (1949). The chromosomes were obtained from the intestinal epithelium of adults. Here, I present preliminary verification of the diploid number from populations of the island of Hawaii (Kapoho) (Fig. 1a) and the island of Moorea near Tahiti (Fig. 1b). Both populations revealed a diploid number of 44 as described from Oahu. As these latter specimens were part of a histocompatibility study requiring living specimens (Cuellar, 1984), I devised a nonlethal procedure to obtain chromosomes using regenerating tail tissue. For this, a small portion of the tail was pinched off, inducing the lizard to grow a new tip. After one or two weeks, the regenerating blastema was excised and incubated for approximately 4 hr in Eagle's or McCoy's medium with 0.03% colchicine, transferred to distilled water or 4% KCl for 30 min and fixed in 3:1 methanol:acetic acid. The cells were then dissociated and processed as described by Stock *et al.* (1972). Occasional counts of 42 chromosomes suggest the possibility of slight intrapopulational differences in Hawaii.

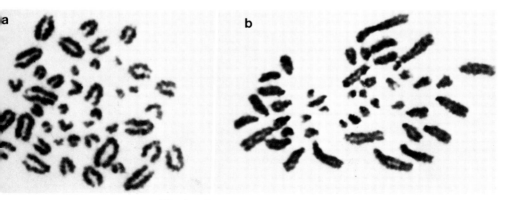

Fig. 1. See text for explanation.

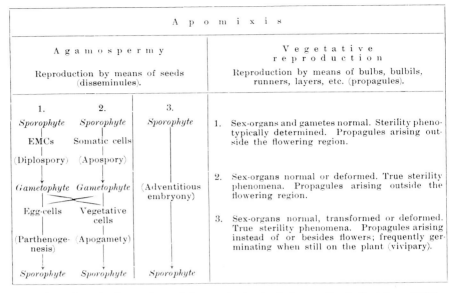

Fig. 2. The major forms of apomictic reproduction in plants according to Gustafsson (1936).

VIII. THE MECHANISMS OF MEIOTIC RESTITUTION

The cellular mechanisms by which parthenogenetic organisms maintain a constant number of chromosomes vary considerably depending on the different taxonomic groups, the different types of parthenogenesis, and the different levels of ploidy. In plants restitution is complicated by the alteration of generations, the sporophyte and gametophyte, and also because the unfertilized embryos may arise either from somatic cells or from unfertilized eggs (Gustaffson, 1946). The former originate through mitosis and the latter from a failure of meiosis, but the end result is the same, namely, the production of a plant identical to the parent and with the same number of chromosomes. However, the actual developmental sequence is quite complicated and many different terms have been used during the last century. Fortunately, Gustaffson (1946) provided an excellent description of these terms and their relation to the different forms of apomixis (Figs. 2 and 3). According to Gustaffson:

> The apomict terminology may seem unnecessarily complicated to an outsider. This is owing to the fact that an excessive number of terms have arisen in the course of years, that different investigators have followed different lines of approach in attacking the problem, and that the development of the gametophyte and sporophyte was not kept separate for a long succession of years and the terms became mixed up. . . .
>
> Two sub-groups of apomixis may be immediately distinguished. In the first group reproduction takes place by means of seed and is therefore called *agamospermy* (Tackholm, 1922). The

3. Evolution of Parthenogenesis

A. NORMAL SEXUAL LIFE CYCLE

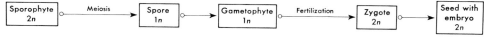

B. GAMETOPHYTIC APOMIXIS

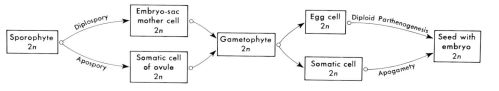

C. ADVENTITIOUS EMBRYONY

Fig. 3. Modern diagram of the various asexual reproductive modes in plants incorporating the classification and terminology of Gustafsson (1936). From Grant (1971). Grant's diagram simplifies the complex pathways and allows a comparison with the normal sexual cycle.

other group comprises species in which reproduction is not effected by means of seed but by vegetative formations such as bulbils, runners, etc., and may be suitably termed *vegetative reproduction*.

The sexual phanerogams exhibit an alternation of generations: sporophytic and gametophytic generations alternate, although the gametophyte is invariably strongly reduced. This alternation fails to appear in certain agamospermous species and the embryos arise direct in the ovules of the mother sporophyte in the form of nucellar or integumental outgrowths. The phenomenon is called *adventitious embryony*.

The gametophyte of the remaining agamospermous species can arise directly or indirectly from a macrospore mother-cell (EMC) or be derived from somatic cells in the nucellus or chalaza. The former process is known as *diplospory*, the latter as *apospory*. The gametophyte gets the unreduced chromosome number in both cases. In the gametophyte there develops in normal manner an egg-cell with its two synergids, further two polar nuclei and three antipodal cells. . . . If the egg-cell forms an embryo without fertilization (a new sporophyte) the phenomenon is spoken of as *parthenogenesis* (Siebold, 1856); if one of the other cells of the gametophyte does so, the process is called *apogamety* (Renner, 1916). In higher plants apogamety is an exceptional phenomenon, as a rule it is the egg-cell which develops into an embryo. In pteridophytes, on the other hand, apogamety is not at all rare.

In diplospory, the simplest mechanisms occur in species having a unicellular archespore and therefore only a single macrospore mother cell (EMC). Members of the Asteraceae (Compositae) are of this type. According to Gustaffson (1946), forms with multicellular archespores, such as some species of the Rosaceae, are very difficult to analyze cytogenetically. Therefore the following descriptions are based on those in which the embryo sac develops from a single macrospore

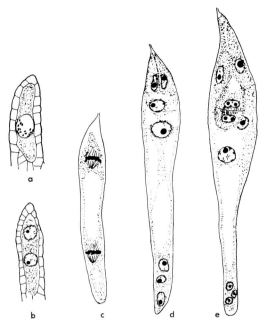

Fig. 4. The *Antennaria* type of embryo sac formation in *Erigeron ramosus*. From Gustafsson (1946) after Holmgren (1919). See text for explanation. (a,b) Formation of unreduced nuclei. (c) Dividing nuclei. (d) Ripe embryo sac. (e) Embryo formation.

mother cell. The diplosporous embryo sac develops in two rather different ways following the first meiotic division. These are known as the *Antennaria* and *Taraxacum* types for the genera in which they occur. In the first, the EMC forms two daughter nuclei that are not divided by a cell membrane. In the latter the EMC forms two distinct cells, one of which degenerates.

In both forms the first meiotic division is abortive and is therefore of the somatic type ("mitotisized meiosis"). The homologous chromosomes do not synapse to form bivalents at the first metaphase. During anaphase all of the chromosomes are enclosed by a common nuclear membrane forming the restitution nucleus. The second metaphase divides the chromosomes equally, forming two unreduced daughter nuclei. In the *Antennaria* type, the daughter nuclei are not divided and the embryo sac remains intact. Eventually the two divide equally giving rise to the final embryo sac containing the usual eight nuclei (Fig. 4). In the *Taraxacum* type, the second division forms two unreduced macrospore cells, one of which disintegrates and the other gives rise to the embryo sac (Fig. 5).

In apospory the embryo sac develops from a single cell of somatic origin. These cells may arise from the epidermal nucellus or the deeper chalaza, from latteral cells of the true archespore, or from purely somatic cells below the primary archespore (Gustaffson, 1946). Initially, however, a normal sexual em-

3. Evolution of Parthenogenesis

bryo sac begins to form from the macrospore mother cell (as in diplospory) except that the daughter nuclei are haploid. Although many variations occur, typically the EMC gives rise to four sexual macrospores called the macrospore tetrad (Fig. 6). At this stage, one of the somatic cells mentioned above begins to enlarge enormously through vacuolization, and encroches upon the four macrospores, eventually crushing them and causing their degeneration. The large invading somatic cell then forms a unicellular embryo sac which later forms the aposporous embryo (Fig. 6). Apospory occurs commonly in apomictic species of *Hieracium, Crepis,* and *Atraphaxis*, but some genera such as *Alchemilla, Potentilla,* and *Atraphaxis* exhibit both diplospory and apospory.

Unlike in plants, the cellular mechanisms of parthenogenetic animals have been discussed extensively during the last century, the most recent works being those of Suomaleinen (1950), White (1954), and Narbel-Hofstetter (1964). The best overall review, including genetical aspects, evolution, polyploidy, etc., is that of Suomaleinen (1950), but the review by Marguerite Narbel-Hofstetter surpasses all others in its specific treatment of meiosis and deserves the rank of a

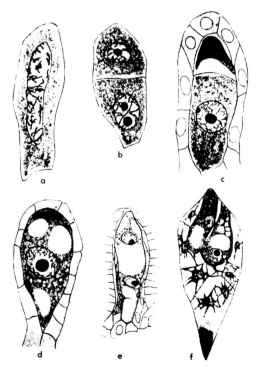

Fig. 5. The *Taraxacum* type of embryo sac formation in *Taraxacum albidum*. From Gustafsson (1946) after Osawa (1912). (a) Restitution nucleus. (b) Two macrospores. (c,d) One macrospore degenerating. (e,f) Embryo sac from single unreduced nucleus in d.

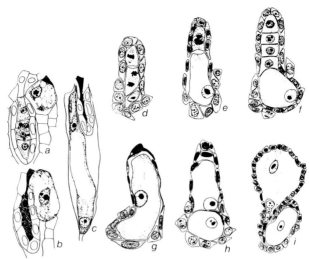

Fig. 6. Apospory in *Hieracium flagellare* (a–c). From Gustafsson (1946) after Rosenberg (1907). (a) Macrospore tetrad and initial somatic cell. (b) Enlarged somatic cell and degenerated tetrad. (c) Degenerated tetrad and somatic embryo sac. Apospory in *Crepis* (d–i). From Gustafsson (1946) after Stebbins and Jenkins (1939). (d) Dyads in second division and aposporous (somatic) initial below. (e) Degenerated macrospore mother cell with a uninuclear somatic embryo sac below. (f) Earlier phase of e. (g) Unicellular aposporous embryo sac. (h) Two aposporous embryo sacs. Degenerated tetrad above. (i) Mature sexual embryo sac above with antipodals and central nucleus. Uninuclear embryo sac below.

modern classic. Her review dates back to the early works of Weisman in 1886 and is elegantly documented with several hundred illustrations of the entire meiotic process.

Parthenogenesis in animals is commonly divided into two main types, arrhenotoky and thelytoky. Arrhenotoky is the production of haploid males from unfertilized eggs, also known as haploid or generative parthenogenesis. It is the normal form of reproduction in the Hymenoptera (bees, wasps, and ants) but it also occurs in rotifers, mites, thrips, white flies, some scale insects, and a beetle (see Whiting, 1945, for a review). Because of their larger size, the Hymenoptera have received more attention from cytologists, so members of this order, particularly the honey bee (*Apis mellifica*), have contributed more to the understanding of male haploidy. Haplodiploidy, as this process is also called, is a peculiar form of reproduction in which parthenogenesis is mixed with sexuality. The male drones are normally impaternate, developing from unfertilized eggs, while the female workers and queens develop through fertilization.

Despite the popular belief that male haploidy in the honey bee (the best studied species) has been thoroughly documented, most of the fundamental issues of development, such as meiosis, sex determination, and ploidy, remain in doubt, indeed confused. In the first place, the very premise of haplodiploidy itself may

3. Evolution of Parthenogenesis

be in error. There is good evidence demonstrating that the somatic and germinal tissues of the male are diploid rather than haploid as is currently assumed (see Whiting, 1945, for a review). Second, the somatic number of chromosomes is uncertain. Some workers have reported 16 (Petrunkewitsch, 1901; Whiting, 1945) and others 32 (Nachtscheim, 1913). Third, Nachtsheim regards 16 as haploid and 32 as diploid, whereas Whiting regards 8 as haploid and 16 as diploid. To complicate matters, Whiting in one part of his review refers to the somatic number as diploid and in another implies it is tetraploid. His descriptions of meiosis are equally confusing, and when the two are combined (ploidy and meiosis) the resulting summary boggles the mind:

> Diploid drone bees have tetraploid daughters, because the reduced eggs are diploid. Nevertheless, in the bee, eight remains the gametic number because reduction to eight takes place in the oocytes. The diploid drone, therefore, breeds as a haploid although he produces diploid sperms, and the tetraploid queen breeds as a diploid, although she produces diploid instead of haploid eggs. Eight dyads go to each pole in anaphase and oocyte I and there are eight "monads" in telophase of oocyte II. Whether these "monads" have already become dyads through regulative splitting, and just when the dyads divide to produce a diploid female pronucleus, is a problem for cytological technique and observation. Although somatic diploidy is here extended into the germ track, it does not produce genetic diploidy. If we had been studying genetics rather than cytology, we would not have been aware of this regulative splitting.

Fortunately, there is common ground of consensus. Everyone seems to agree that the meiosis of oogenesis is normal, while that of spermatogenesis is not. Regarding oogenesis the first division is reductional as usual, resulting in the formation of bivalents and crossing-over, although very few of these divisions have actually been observed (Petrunkewitsch, 1901; Doncaster, 1906; Whiting, 1945). The first division produces two polar nuclei with "dyads" and the second forms four with "monads." In the majority of insects these four nuclei are equal in size, equally spaced, and arranged in a line perpendicular to the egg surface (Petrunkewitsch, 1901; Doncaster, 1906; Thomsen, 1927; Comrie, 1938). The outer one is called "the first polar nucleus," the next is the "inner half of the first," the third is the "second polar nucleus," and the lowest toward the center is the "egg nucleus" or the "egg pronucleus." Little is certain after this stage in the honey bee, except that the first polar nucleus always disintegrates near the surface of the egg. Also the two central ones come together and fuse, forming the so-called richtungscopulationskern (RKK) or polar fusion nucleus or copulation nucleus (Petrunkewitsch, 1901).

According to Petrunkewitsch, in the eggs destined to become haploid drones, the copulation nucleus gives rise to a line of diploid cells that eventually form the testes. The testes would therefore be diploid as in normal males of a bisexual species, however, these observations have not been verified by anyone else. Initially the cells resulting from the copulation nucleus are distinguished by possessing a double nucleus, coined by Petrunkewitsch "Doppelkern." In the eggs destined to form females, a copulation nucleus is also formed, but the resulting diploid cells eventually disintegrate. The egg nucleus migrates deeper into the yolk where it apparently awaits fertilization. In the drone eggs, the egg nucleus supposedly doubles its chromosome number either by fusion of cleavage nuclei after meiosis or by endomitosis—"regulative splitting" (Whiting, 1945). According to Doncaster (1906), "In the parthenogenetic egg of the bee the somatic number of chromosomes is said to be restored by a process of doubling, which occurs before the first division of the egg nucleus, and which does not take place if the egg is fertilized." Anyway, the resulting diploid cells give rise to the blastoderm and eventually to the entire drone. According to Whiting (1945) some of the somatic tissues may even be polyploid. Obviously, if all of this proves to be true, male haploidy does not exist in the honey bee, even though the drones may arise from unfertilized eggs.

Regarding male meiosis in the drone, the first division is abortive. A spindle supposedly forms (Suomaleinen, 1950) but the chromosomes at first metaphase do not divide. According to Whiting (1945), "There are sixteen dyads in the nucleus of the abortive spermatocyte I division. The spermatocyte II nucleus is divided equally, eight dyads going to each pole in anaphase. The chromatids may be separated so that they already appear as eight pairs of monads." This description clearly suggests, although Whiting does not mention it, that each spermatid will eventually acquire 16 chromosomes, since each of the 8 dyads is a double element. If they actually acquire 16, then fertilization should double the number, thereby supporting Nachtsheim's observation of 32.

Thelytoky is divided into two main types depending on how the somatic number of chromosomes is restored: automictic and apomictic according to Suomaleinen (1950), or meiotic and ameiotic according to White (1954). In the meiotic type, the first division is very similar to that of bisexual reproduction: the homologous chomosomes pair, undergo crossing-over, condense, and exhibit the classic bivalent configurations such as diplonema and diakinesis (Fig. 7). The bivalents are always divided into two separate groups of dyads at anaphase I, but the groups may or may not remain separate depending on whether the first division is abortive or normal (Fig. 8). If the first division is abortive, the dyads return to a common position forming the second metaphase plate. The second division then divides them equally into two separate groups containing the original number of chromosomes. Hence, only two nuclei are formed. If the first division is normal, the end result is four nuclei, two from the first division and four from the second (Fig. 8). In the ameiotic type, meiosis is entirely suppressed, the chromosomes do not pair at metaphase 1, and there is only one division. For the most part, all meiotic characteristics are absent, except that in some forms the chromosomes become slightly condensed and prophase is slightly delayed. According to White (1954):

> We should expect the genetic consequences of ameiotic and meiotic thelytoky to be quite different. In ameiotic parthenogenesis genetic segregation will not occur. Recessive mutations

3. Evolution of Parthenogenesis

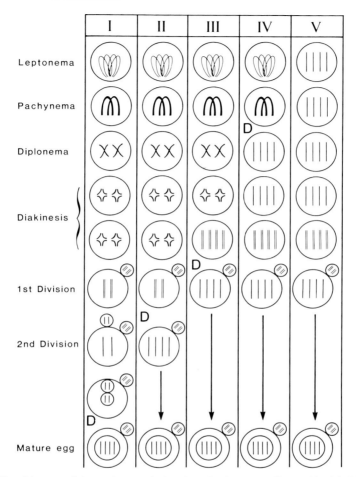

Fig. 7. Diagram of the various restitutional mechanisms according to Vandel (1931). All mechanisms are represented with four chromosomes. I. Normal meiosis with restoration by fusion of second polar nucleus with female pronucleus. Polar nucleus incorrectly shown as polar body. "Le nombre diploide ne réapparait que dans le noyau de segmentation. *Asterias, Artemia* (exceptionnel), *Lecanium.*" II. Meiosis with one division. Anaphase II abortive. "C'est à ce moment que réapparait le nombre diploide de chromosomes. *Solenobia pineti* quelques *Rhabditis.*" III. Meiosis with one division but dissociation of bivalents of diakinesis. "La division de maturation se produit avec le nombre diploide de chromosomes. *Marsilia, Houttuynia.*" IV. Same as III but dissociation prior to diakinesis. "*Alchemilla, Taraxacum, Aphis palmae.*" V. Ameiotic restitution. "Les phénomènes synaptiques ont entièrement disparu. Le cycle est diploide dans sa totalité. Presque tous les cas de parthénogenèse diploide répondent à ce type."

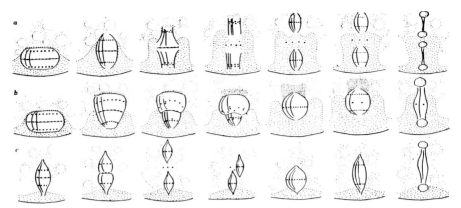

Fig. 8. Drawings of the meiosis of three parthenogenetic moths. From Narbel-Hofstetter (1964) based on the photographs of Seiler (1923) and Narbel (1946). (a) *Solenobia triquetrella* (4N), after Seiler. (b) *Solenobia lichenella* (previously *S. pineti*, see Fig. 7 for comparison with Vandel), after Narbel. (c) *Apterona helix*, after Narbel.

and structural rearrangements will tend to accumulate indefinitely in such organisms, only the ones which are immediately deleterious being eliminated by natural selection. Such forms must consequently be expected to become gradually more and more heterozygous, but all the offspring of a single female will resemble their mother exactly, except for newly arisen dominant mutations and differences due to the action of the environment. An ameiotic form evolving for a long period of time might be expected eventually to lose its diploid character in both the genetic and the cytological sense, its two chromosome sets having become almost completely unlike. Moreover, since no pairing of chromosomes takes place during the maturation of the eggs, there is no "mechanical" barrier to the establishment of any type of polyploidy in such forms and various forms of aneuploidy, due to irregular reduplication of some chromosome elements, must be expected to occur.

In meiotic parthenogenesis segregation may occur in the offspring of a single female, if she is heterozygous. Heterozygosity will, however, be very rare in such forms, irrespective of the stage at which the doubling of the chromosome number occurs, since in the absence of sexual reproduction it only arises by mutation and will be eliminated or greatly reduced at each successive meiosis. Thus all organisms with meiotic thelytoky will tend in practice to be homozygous for almost all loci; in this respect they possess a genetic system which is the antithesis of that found in species with ameiotic thelytoky. Except where they are also polyploid, ... forms with meiotic thelytoky are precluded from enjoying the advantages of heterosis, which may be an important factor in the adaptiveness of ameiotic mechanisms of thelytoky.

Figure 9 shows the genetic consequences of the various types of parthenogenesis discussed above by White. Restoration of the somatic number occurs in five major ways with parthenogenesis of the meiotic type, and will be described in order of their occurrence in the meiotic process as follows: the chromosomes are doubled premeiotically (Petrunkewitsch, 1901; Reisinger, 1940; Lepori, 1950; Omodeo, 1951, 1952; Muldal, 1952); the first division is abortive (Narbel, 1946; Narbel-Hofstetter 1950); the second division is abortive (Barigozzi, 1944);

3. Evolution of Parthenogenesis

polar nuclei fuse (Doncaster, 1906; Thomsen, 1927; Artom, 1931; Comrie, 1938; Smith, 1941; Seiler, 1960); and, finally, cleavage nuclei fuse (Seiler and Schaeffer, 1941).

Premeiotic doubling appears to be the dominant mechanism in parthenogenetic worms. It was first discovered in the turbellarian worm *Bothrioplana semperi* by Reisinger (1940), and was later described in the triploid planaria *Polycelis nigra* (Lepori, 1950) and several species of earthworms (Omodeo, 1951, 1952; Muldal, 1952). Lepori (1950) states that a chromosome doubling of endomitotic type occurs either in the final generation of the oogonia or just before the beginning of meiotic prophase (Fig. 10). According to Omodeo (1951) the last, oogonial mitosis fails, resulting in twice the number of chromosomes. A normal meiosis with production of two polar bodies then restores the somatic number. Similar mechanisms have now been reported in many other animals. Preliminary evidence suggests that premeiotic doubling is the standard mechanism in triploid vertebrates (Cuellar, 1976), but as mentioned previously, many of the critical stages have not been adequately documented.

The first abortive division can be divided into two types depending on the extent of chromosomal separation during anaphase I. In the less advanced type, a

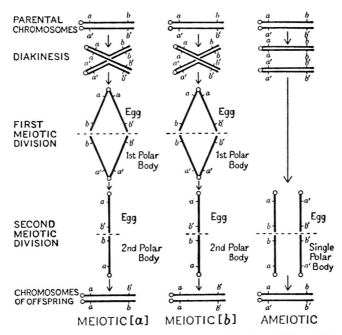

Fig. 9. The genetics of parthenogenesis. Diagram modified from White (1954). In type *a*, restoration is by fusion of the second polar nucleus with the egg nucleus (as in Fig. 16). In type *b*, restoration is by fusion of cleavage nuclei.

Fig. 10. Restoration by premeiotic doubling. Diagram based on one set of triploid chromosomes. (A) The chromosomes before meiosis. (B) Premeiotic doubling. (C) Replication. (D) Synapsis and bivalent formation at metaphase I. (E) First division. Lower group of dyads represents oocyte and upper group represents polar body. (F) Second division and restoration of somatic number.

spindle forms and begins to pull the bivalents apart. As soon as they separate, the spindle deteriorates and the two groups of dyads recombine into a second metaphase plate. The second division then divides the chromatids equally, forming two unreduced nuclei, of which the outer one degenerates and the inner forms the egg nucleus. This mechanism was first described by Narbel-Hofstetter (1950) in the tetraploid parthenogenetic moth *Solenobia lichenella* (Figs. 11 and 12). In the advanced type, the dyads are pulled much farther apart, but remain attached by a common spindle, which stretches greatly and eventually ruptures, forming two smaller ones. These two "daughter" spindles then migrate toward each other and coalesce. Their respective chromosomes reunite and form the second metaphase plate. As in the previous type, the second division forms two nuclei with the normal somatic number (Figs. 13 and 14). This mechanism has been described only in the diploid parthenogenetic moth *Apterona helix* (Narbel, 1946).

Restoration by means of an abortive second division has also been described in only one parthenogenetic species, the diploid brine shrimp *Artemia salina* (Barigozzi, 1944). During metaphase I, the diploid number of 42 chromosomes is

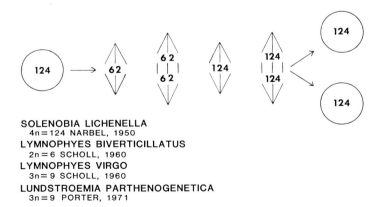

SOLENOBIA LICHENELLA
  4n=124 NARBEL, 1950
LYMNOPHYES BIVERTICILLATUS
  2n=6 SCHOLL, 1960
LYMNOPHYES VIRGO
  3n=9 SCHOLL, 1960
LUNDSTROEMIA PARTHENOGENETICA
  3n=9 PORTER, 1971

Fig. 11. Scheme representing an abortive first division. Restoration occurs with collapse of the first anaphase.

3. Evolution of Parthenogenesis

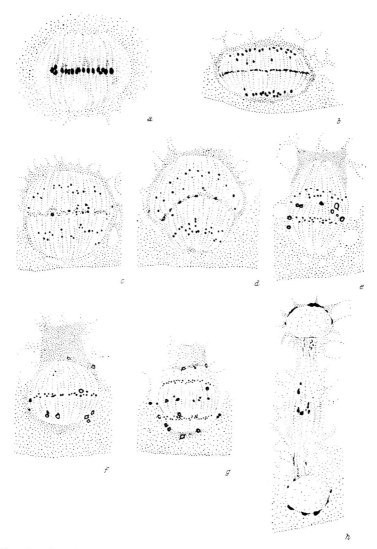

Fig. 12. Drawing of an actual first abortive division in the parthenogenetic moth *Solenobia lichenella*. (a) Metaphase I. (b) Anaphase. (c–e) Collapse of anaphase. (f) Metaphase II. (g) Anaphase II. (h) Telophase. After Narbel-Hofstetter (1950).

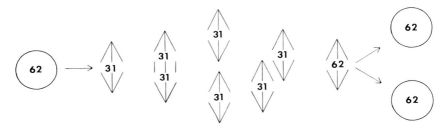

APTERONA HELIX 2N

NARBEL, 1946

Fig. 13. Restoration by fusion of spindles from an abortive first division. Both of the final nuclei form the embryo. No polar bodies are given off.

reduced to 21 bivalents, which are then split by a normal anaphase, forming two complete nuclei each containing 21 dyads. As usual, the outer nucleus disintegrates. The egg nucleus then initiates a second division, but anaphase collapses and the two separated haploid sets of chromatids reunite, restoring the diploid number of 42 in the final egg nucleus (Fig. 15). The exact details of this mechanism are uncertain and there are several conflicting interpretations. The cytology of *Artemia* has been analyzed by many investigators since 1894 and is complicated by the presence of diploid, tetraploid, and octaploid populations. The major disputes are between Artom (1931), Barigozzi (1944), and Stefani (1960), which are discussed by Narbel-Hofstetter (1964) and White (1978).

Fusion of polar nuclei is apparently the "normal" method of restitution among parthenogenetic invertebrates, occurring in many unrelated orders, such as the Lepidoptera, Hymenoptera, Homoptera, and Crustacea. This form of restitution involves a normal meiosis producing four haploid polar nuclei, which fuse in various combinations to restore the somatic number of chromosomes. The outer nucleus always degenerates below the surface of the egg. Some recent theoretical treatments of the genetic consequences of fusion (Asher, 1970; Templeton, 1982) have erroneously included this first polar nucleus in their fusion schemes. The most common form of fusion involves the lowest two, the second polar and the female pronucleus (Fig. 16). Since both stem from the same set of dyads, the resulting progeny are homozygous at all loci. The other form involves fusion of the two central ones, each derived from a different set of dyads (Fig. 17), so the progeny would be heterozygous for any mutations arising in the parent. In any event, it is not entirely certain that each of these modes of restitution is retricted to a particular phyletic line, or even species. For instance,

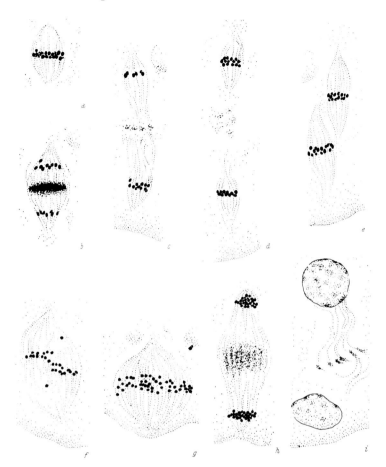

Fig. 14. Drawing of an actual fusion of spindle in the moth *Apterona helix*. (a) Metaphase I. (b) Anaphase I. (c,d) Formation of spindles. (e,f) Fusion of spindles. (g) Metaphase II. (h) Telophase. (i) Final nuclei with restored diploid number. See Fig. 12. After Narbel (1946).

the diploid forms of *Artemia* supposedly exhibit an abortive second division (Barigozzi, 1944), while the octoploids supposedly undergo fusion of polar nuclei (Artom, 1931). Narbel-Hofstetter (1964) discusses this topic in more detail, as well as the possible fusion of cleavage nuclei and other hypothetical modes of restitution.

According to White (1978), fusion of cleavage nuclei is believed to occur in the white fly *Trialeurodes vapororiorum* (Thomsen, 1927), in the stick insect *Bacillus rossii* (Pijnacker, 1969), and in the mite *Chyletus eruditus* (Peacock and

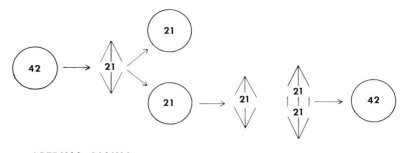

ARTEMIA SALINA

BARIGOZZI, 1944

Fig. 15. Scheme representing an abortive second division. Restoration occurs with collapse of anaphase II.

Weidmann, 1961). Fusion definitely occurs in the scale insect *Pulvinaria hydrangae* (Nur, 1963), in the coccid *Gueriniella serratulae* (Highes-Schrader and Tremblay, 1966), and in the phasmatid *Clitumnus extradentatus*. One of the most unusual restitutional mechanisms ever described, although of the ameiotic type, occurs in the triploid parthenogenetic worm *Lumbricellus lineatus* (Christensen, 1960). During metaphase, the unpaired chromosomes line up unevenly at

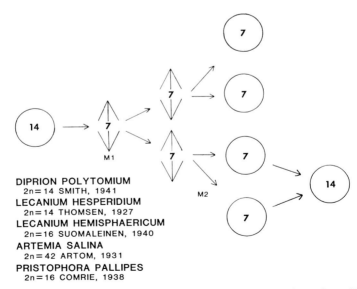

DIPRION POLYTOMIUM
2n=14 SMITH, 1941
LECANIUM HESPERIDIUM
2n=14 THOMSEN, 1927
LECANIUM HEMISPHAERICUM
2n=16 SUOMALEINEN, 1940
ARTEMIA SALINA
2n=42 ARTOM, 1931
PRISTOPHORA PALLIPES
2n=16 COMRIE, 1938

Fig. 16. Scheme representing restoration by fusion of the second polar nucleus with the egg pronucleus.

3. Evolution of Parthenogenesis

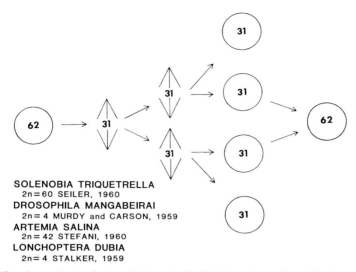

Fig. 17. Restoration by fusion of the inner half of the first polar nucleus with the second polar nucleus.

the plate, 22 on one side and 17 on the other (Fig. 18). These two groups then separate during anaphase, but as the spindle pulls apart it folds upon itself and simultaneously initiates another division, bringing the odd complements together as two separate triploid spindles, which then divide and produce four genetically identical nuclei. All four supposedly participate in the formation of the embryo.

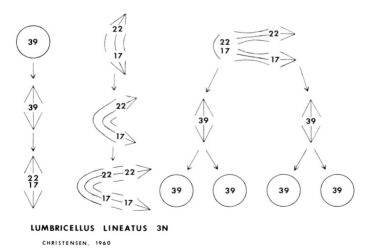

Fig. 18. Scheme of ameiotic restitution in a triploid parthenogenetic worm.

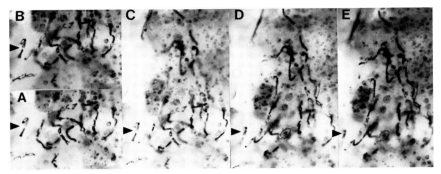

Fig. 19. A portion of a diplotene spread obtained from oocytes of the triploid salamander *Ambystoma tremblay* (author's unpublished data). The salamanders had been fixed in formaldehyde and preserved in alcohol for many years prior to obtaining the chromosomes. I first excised the germinal vesicles (nuclei) from the yolk by teasing with fine tungsten needles. The vesicles were then soaked in water for several days, after which they were stained and squashed in aceto orcein. Although the material was relatively "old," the details of diplonema show up remarkably well. With further refinement such a method could facilitate analysis of the prolonged diplotene stage of parthenogenetic vertebrates. Due to thickness of spread several photos were required to observe all of the chromosomes. The two photos on the left are focused on the bottom section of the spread. A represents the uppermost view, B slightly lower, C–E progressively lower but showing full spread. See chromosome with arrow for guidance. Spread may not be complete. Vesicular structures are yolk particles that could not be teased off from the G.V.

In a previous paper (Cuellar, 1971) I called attention to "the current but erroneous idea that polar bodies are capable of reentering the oocyte to fuse with the egg nucleus (Fritts, '69; Poole and Olsen, '57; Schultz, '67; Schultz and Kallman, '68; Uzzell, '70). Among invertebrates, however, where restitution by fusion of polar nuclei is known only, there is no evidence that once a polar body is extruded it may again reenter the oocyte." This notion probably stems from the first classification scheme of restitutional mechanisms by Vandel (1931, Fig. 7), in which he shows the second polar nucleus of an oocyte first resting on the outside, and later inside restoring the somatic number. The error has undoubtedly been confounded by two very similar but unrelated terms, polar bodies and polar nuclei. However, polar bodies are extruded from the oocyte, while polar nuclei either fuse with each other or disintegrate within the egg. This is not to say that polar nuclei may never be extruded (see Narbel-Hofstetter, 1964), but only that no one has ever documented an actual instance of reentry. The chromatin of extruded nuclear masses usually breaks down very rapidly after being expulsed from the oocyte. Unfortunately the notion of reentry remains entrenched in the modern literature (Monaco *et al.*, 1984), despite any cytological evidence to support it.

3. Evolution of Parthenogenesis

ACKNOWLEDGMENTS

I thank Mark Blackmore and Professor Richard Berchan for assistance with the translations, and Professor Robert K. Vickery for calling to my attention the paper by Hiesey and Nobs. I am also grateful to Professor Åke Gustafsson of the Department of Genetics, University of Lund, Sweden, for kindly allowing me to quote his works.

REFERENCES

Artom, C. (1912). Le basi citologische di una nuova sistematica del genere *Artemia*. *Arch. Zellforschu.* **1.**
Artom, C. (1931). L'origine e l'evoluzione delle partenogenesi attraverso i differenti biotipi di una specie collectiva (*Artemia salina*) con speciale riferimento al biotipo diploide partenogenetico di Sète. *Mem. Accad. Ital., Roma Cl. Sci.* **2,** 1–57.
Asher, J. H., Jr. (1970). Parthenogenesis and genetic variability. Ph.D. Dissertation, Dept. Zool., University of Michigan, Ann Arbor.
Barigozzi, C. (1944). I fenomeni cromosomici delle cellule germinale in *Artemia salina* Leach. *Chromosoma* **2,** 549–575.
Basrur, V. R., and Rothfels, K. H. (1959). Triploidy in natural populations of the black fly *Cnephia mutata* (Malloch). *Can. J. Zool.* **37,** 571–589.
Bell, G. (1982). "The Masterpiece of Nature." Univ. of California Press, Berkeley.
Bergerard, J. (1962). Parthenogenesis in the Phasmidae. *Endeavour* **21,** 137–143.
Bezy, R. L. (1972). Karyotypic variation and evolution of the lizards in the family *Xantusidae*. *Contrib. Sci. Nat. Hist. Mus. Los Angeles County* **227,** 1–29.
Bickham, J. W., McKinney, O., and Mathews, M. F. (1976). Karyotypes of the parthenogenetic whiptail lizard *Cnemidophorus laredoensis* and its presumed parental species. *Herpetologica* **32,** 395–399.
Bogart, J. P. (1980). Evolutionary implications of polyploidy in amphibians and reptiles. *In* "Polyploidy" (W. H. Lewis, ed.), pp. 341–378. Plenum, New York.
Brown, W. M., and Wright, J. W. (1979). Mitochondrial DNA analyses and the origin and relative age of parthenogenetic lizards (genus *Cnemidophorus*). *Science* **203,** 1247–1249.
Brown, W. V. (1972). "Textbook of Cytogenetics." Mosby, St. Louis, Missouri.
Carson, H. L. (1961). Rare parthenogenesis in *Drosophila robusta*. *Am. Nat.* **95,** 81–86.
Carson, H. L. (1967). Selection for parthenogenesis in *Drosophila mercatorium*. *Genetics* **55,** 151–171.
Cherfas, N. B. (1966). Analysis of meiosis in sexual and unisexual forms of the goldfish. *Tr. Vses. prudovogo rybn. Kh-va.* **14,** 119.
Christensen, B. (1960). A comparative cytological investigation of an amphimictic diploid and a parthenogenetic triploid form of *Lumbricellus lineatus*. *Chromosoma* **11,** 365–379.
Christiansen, J. L., and Ladman, A. J. (1968). The reproductive morphology of *Cnemidophorus neomexicanus* × *C. inornatus* hybrid males. *J. Morphol.* **125,** 367–378.
Cimino, M. C. (1971). Meiosis in triploid all-female fish, *Poeciliopsis*, Poeciliidae. *Science* **175,** 1484–1486.
Cimino, M. C. (1972). Egg production, polyploidization and evolution in a diploid all-female fish of the genus *Poeciliopsis*. *Evolution (Lawrence, Kans.)* **26,** 294–306.

Cole, C. J. (1975). Evolution of parthenogenetic species of reptiles. In "Intersexuality in the Animal Kingdom" (R. Reinboth, ed.), pp. 340–355. Springer-Verlag, Berlin and New York.
Cole, C. J. (1978). Parthenogenetic lizards. *Science* **201,** 1154–1155.
Cole, C. J., Dessaurer, H. C., and Townsend, C. R. (1983). Isozymes reveal hybrid origin of neotropical unisexual lizards. *Isozyme Bull.* **16,** 74.
Comrie, L. (1938). Biological and cytological observations on tenthredinid parthenogenesis. *Nature (London)* **142,** 877–878.
Cuellar, O. (1971). Reproduction and the mechanism of meiotic restitution in the parthenogenetic lizard *Cnemidophorus uniparens. J. Morphol.* **133,** 1–28.
Cuellar, O. (1974). On the origin of parthenogenesis: The cytogenetic factors. *Am. Nat.* **108,** 625–648.
Cuellar, O. (1976). Cytology of meiosis in the triploid gynogenetic salamander *Ambystoma tremblayi. Chromosoma* **58,** 355–364.
Cuellar, O. (1984). Histocompatibility in Hawaiian and Polynesian populations of the parthenogenetic gecko *Lepidodactylus lugubris. Evolution* **38,** 176–185.
Cuellar, O., and Kluge, A. G. (1972). Natural parthenogenesis in the gekkonoid lizard *Lepidodactylus lugubris. J. Genetics* **61,** 14–28.
Cuellar, O., and McKinney, C. O. (1976). Natural hybridization between parthenogenetic and bisexual lizards: Detection of uniparental source by skin grafting. *J. Exp. Zool.* **196,** 341–350.
Cuellar, O., and Uyeno, T. (1972). Triploidy in rainbow trout. *Cytogenetics* **11,** 508–515.
Darevsky, I. S. (1966). Natural parthenogenesis in a polymorphic group of Caucasian rock lizards, *Lacerta saxicola* Eversmann. *J. Ohio Herpetol. Soc.* **5,** 115–152.
Darevsky, I. S., and Kulikova, V. N. (1961). Naturliche Parthenogenese in der polymorphen gruppe der Kaukasischen Felseidechse (*Lacerta saxicola* Eversmann). *Zool. Jahrb., Abt. Syst. Oekol. Geogr. Tiere* **89,** 119–176.
Darevsky, I. S., Kupriyanova, L. A., and Roshchin, V. V. (1984). A new all-female triploid species of gecko and karyological data on the bisexual *Hemidactylus frenatus* from Vietnam. *J. Herpetol.* **18,** 277–284.
Darlington, C. D. (1939). "The Evolution of Genetic Systems." Oliver & Boyd, Edinburgh and London.
Dessauer, H. C., and Cole, C. J. (1984). Influence of gene dosage on electrophoretic phenotypes of proteins from lizards of the genus *Cnemidophorus. Comp. Biochem. Physiol.* **77,** 181–189.
Dobzhansky, T. (1941). "Genetics and the Origin of Species," 2nd ed. Columbia Univ. Press, New York.
Doncaster, L. (1906). On the maturation of the unfertilized egg and the fate of the polar bodies in the Tenthredinidae (sawflies). *Q. J. Microsc. Sci.* **49,** 561–590.
Downs, F. L. (1978). Unisexual *Ambystoma* from the Bass Islands of Lake Erie. *Occas. Pape. Mus. Zool. Univ. Mich.* **685,** 1–36.
Echelle, A. A., and Mosier, D. T. (1981). All female fish: A cryptic species of *Menidia* (Atherinidae). *Science* **212,** 1411–1413.
Ernst, A. (1918). "Bastardierung als Ursache der Apogamie im Pflanzenreich." Jena.
Fisher, R. A. (1930). "The Genetical Theory of Natural Selection." Oxford Univ. Press (Clarendon), London and New York.
Fritts, T. H. (1969). The systematics of the parthenogenetic lizards of the *Cnemidophorus cozumela* complex. *Copeia,* pp. 519–535.
Good, D. A., and Wright, J. W. (1984). Allozymes and the hybrid origin of the parthenogenetic lizard *Cnemidophorus exsanguis. Experientia* **40,** 1012–1014.
Grant, V. (1971). "Plant Speciation." Columbia Univ. Press, New York.
Gustafsson, Å. (1936). Studies on the mechanism of parthenogenesis. *Hereditas* **21,** 1–112.

3. Evolution of Parthenogenesis

Gustafsson, Å. (1946). Apomixis in higher plants. Part I. The mechanisms of apomixis. *Lund Univ. Arsskr., Avd. 2* [N.S.] **42**, 1–72.
Gustafsson, Å. (1947a). Apomixis in higher plants. Part II. The causal aspects of apomixis. *Lund Univ. Arsskr., Avd. 2*]N.S.] **43**, 73–182.
Gustafsson, Å. (1947b). Apomixis in higher plants. Part III. Biotype and species formation. *Lund Univ. Arrskr., Avd. 2* [N.S.] **43**, 183–372.
Gustafsson, Å. (1948). Polyploidy, life form and vegetative reproduction. *Hereditas* **34**, 1–22.
Hall, W. P. (1970). Three probable cases of parthenogenesis in lizards (Agamidae, Chamaeleontidae, Gekkonidae). *Experientia* **26**, 1271–1273.
Hamilton, A. G. (1953). Thelytokous parthenogenesis for four generations in the desert locust (*Schistocerca gregaria* Forsk) (Acrididae). *Nature (London)* **172**, 1153–1154.
Harlan, J. R., and deWet, J. M. J. (1975). On Ö. Winge and a prayer: The origins of polyploidy. *Bot. Rev.* **41**, 361–390.
Harrison, J. W. H. (1920). The inheritance of melanism in the genus *Tephrosia* (*Ectropis*). *J. Genet.* **10** (No. 1).
Harrison, J. W. H. (1927). The inheritance of melanism in hybrids between continental *Tephrosia crepuscularia* and British *T. bistortata*, with some remarks on the origin of parthenogenesis in interspecific crosses. *Genetica (The Hague)* **9**, 467–479.
Hertwig, P. (1920). Abweichende Form der Parthenogenese bei einer Mutation von *Rhabditis pelliio*. Eine experimentelle cytologische Untersuchung. *Arch. Mikrosk. Anat. Entwicklungsmech.* **94**, 303–337.
Hewitt, G. M. (1975). A new hypothesis for the origin of the parthenogenetic grasshopper *Moraba virgo*. *Heredity* **34**, 117–123.
Hiesey, W. M., and Nobs, M. A. (1982). Experimental studies on the nature of species. VI. Interspecific hybrid derivatives between facultatively apomictic species of bluegrasses and their response to contrasting environments. *Carnegie Inst. Washington Publ.* **636**, 119pp.
Holmgren, I. (1919). Zytologische Studien uber die Fortpflanzung bei den Gattungen en *Erigeron* und *Eupatorium*. *K. Sven. Vetenskaps akad. Handl.* **59**, 1–118.
Hoogmoed, M. S. (1973). "Notes on the herpetofauna of Surinam. IV. The Lizards and Amphisbaenians of Surinam." Junk Publ., The Hague.
Hubbell, T. H., and Norton, R. M. (1978). The systematics and biology of the cave-crickets of the North American tribe Hadenoecini (Orthoptera: Saltatoria: Ensifera: Rhaphidophoridae: Dolichopodinae). *Misc. Publ. Mus. Zool. Univ. Mich.* **156**, 1–124.
Hubbs, C. L. (1955). Hybridization between fish species in nature. *Syst. Zool.* **4**, 1–20.
Hughes-Schrader, S., and Tremblay, E. (1966). *Gueriniella* and the cytotaxonomy of iceryne coccids (Coccoidea: Margarodidae). *Chromosoma* **19**, 1–13.
John, B., and Freeman, M. (1975). Causes and consequences of Robertsonian exchange. *Chromosoma* **52**, 123–130.
King, M. (1980). C-Banding studies on Australian hylid frogs: Secondary constriction structure and the concept of euchromatin transformation. *Chromosoma* **80**, 191–217.
Kluge, A. G. (1982). The status of the parthenogenetic lizard *Gehyra variegata ogasawarasimae*. *J. Herpetol.* **16**, 86–87.
Kluge, A. G., and Eckardt, M. J. (1969). *Hemidactylus garnoti*, a triploid all-female species of gekkonid lizard. *Copeia,* 651–664, 1969.
Lamb, R. Y., and Willey, R. B. (1975). The first parthenogenetic populations of Orthoptera Saltatoria to be reported from North America. *Ann. Entomol. Soc. Am.* **68**, 721–722.
Leider, U. (1955). Manchenmangel und naturaliche Parthenogenese bei der Silberkarausche *Carassus auratus gibelio*. *Naturwissenschaften* **42**, 590.
Leider, U. (1959). Uber die Eintwicklung bei mannchenlosen Stammen der Silberkarausche *Carassus auratus gibelio*. *Biol. Zentralbl.* **78**, 284–291.

Lepori, N. G. (1950). Ciclo cromosomico con polipoidia ginogenesi ed endomitosi in popolozioni Italiane di *Polycelis nigra* Ehrenberg. *Caryologia* **2**, 301–324.
Leslie, J. F., and Vrijenhoek, R. C. (1977). Genetic analysis of natural populations of *Poeciliopsis monacha*. *J. Hered.* **68**, 301–306.
Levan, A., Fredga, K., and Sandberg, A. A. (1964). Nomenclature for centromeric position on chromosomes. *Hereditas* **45**, 665–672.
Lokki, J., Suomaleinen, G., Saura, A., and Lankinen, P. (1975). Genetic polymorphism and evolution in parthenogenetic animals. VI. Diploid and triploid *Ploydrosus mollis* (Coleoptera: Curculionidae). *Hereditas* **82**, 209–216.
Lowe, C. H., and Wright, J. W. (1966). Chromosomes and karyotypes of *Cnemidophorus teiid* lizards. *Mamm. Chromosome Newsl.* **22**, 199–200.
Lowe, C. H., Wright, J. W., Cole, C. J., and Bezy, R. L. (1970). Chromosomes and evolution of the species groups of *Cnemidophorus* (Reptilia: Teiidae). *Syst. Zool.* **19**, 128–141.
Macgregor, H. C., and Uzzell, T. M. (1964). Gynogenesis in salamanders related to *Ambystoma jeffersonianum*. *Science* **135**, 1043–1045.
McKinney, C. O., Kay, F. R., and Anderson, R. A. (1973). A new all-female species of the genus *Cnemidophorus*. *Herpetologica* **29**, 361–366.
Makino, S., and Momma, E. (1949). An idiogram study of the chromosomes in some species of reptiles. *Cytologia* **15**, 96–108.
Maslin, T. P. (1971). Parthenogenesis in reptiles. *Am. Zool.* **11**, 361–380.
Matthey, R. (1945). Cytologie de la Parthénogénèse chez *Pycnoscelus surinamensis*. *Rev. Suisse Zool.* **52**, 1–109.
Maynard Smith, J. (1971). The origin and maintenance of sex. *In* "Group Selection" (G. C. Williams, ed.), pp. 163–175. Atherton, Chicago, Illinois.
Maynard Smith, J. (1978). "The Evolution of Sex." Cambridge Univ. Press, London and New York.
Mayr, E. (1969). "Population, Species and Evolution." Harvard Univ. Press, Cambridge, Massachusetts.
Monaco, P. J., Rasch, E. M., and Balsano, I. S. (1984). Apomictic reproduction in the Amazon molly *Poecilia formosa*, and its triploid hybrids. *In* "Evolutionary Genetics of Fishes" (B. J. Turner, ed.), pp. 311–328. Plenum, New York.
Moore, W. S. (1984). Evolutionary ecology of unisexual fishes. *In* "Evolutionary Genetics of Fishes" (B. J. Turner, ed.), pp. 329–398. Plenum, New York.
Moritz, C. (1983). Parthenogenesis in the endemic Australian lizard *Heteronotia binoei* (Gekkonidae). *Science* **220**, 735–737.
Moritz, C. (1984). The origin and evolution of parthenogenesis in *Heteronotia binoei* (Gekkonidae). *Chromosoma* **89**, 151–162.
Muldal, S. (1952). The chromosomes of the earthworm. *Heredity* **6**, 55–76.
Muller, H. J. (1932). Some genetic aspects of sex. *Am. Nat.* **66**, 118–138.
Muntzing, A. (1940). Further studies on apomixis and sexuality in *Poa*. *Hereditas* **26**, 115–190.
Murdy, W. H., and Carson, H. L. (1959). Parthenogenesis in *Drosophila mangabeirai*. *Am. Nat.* **93**, 355–363.
Nabours, R. F. (1919). Parthenogenesis and crossing over in the grouse locust *Apotettix*. *Am. Nat.* **53**, 131–142.
Nabours, R. K. (1925). Studies of inheritance and evolution in Orthoptera. V. The grouse locust, *Apotettix eurocephalus*. *Kans., Agric. Exp. Stn., Tech. Bull.* **17**, 1–231.
Nachtsheim, H. (1913). Cytologische Studien über die Geschlechtsbestimmung bei der Honigbiene (*Apis mellifica*). *Arch. Zellforsch.* **11**, 169–241.
Narbel, M. (1946). La cytologie de la parthénogénèse chez *Apterona helix* Sieb. (Lep. Psychidae). *Rev. Suisse Zool.* **53**, 625–681.

3. Evolution of Parthenogenesis

Narbel-Hofstetter, M. (1950). La cytologie de la parthénogénèse chez *Solenobia* sp. (*lichenella* L.) (Lep. Psych.). *Chromosoma* **4**, 56–90.
Narbel-Hofstetter, M. (1964). Les altérations de la méïose ches les animaux parthénogénètiques. *Protoplasmatologia* **6**, F2.
Neaves, W. B. (1969). Adenosine deaminase phenotypes among sexual and parthenogenetic lizards in the genus *Cnemidophorus* (Teiidae). *J. Exp. Zool.* **171**, 175–183.
Nur, U. (1963). Meiotic parthenogenesis and heterochromatinization in a soft scale, *Pulvinaria hydrangeae* (Coccidae: Homoptera). *Chromosoma* **14**, 123–139.
Okada, Y. (1930). Notes on the herpetology of the Chichijima, one of the Bonin islands. *Bull. Biol. Soc. Jpn.* **1**, 187–194.
Olsen, M. W. (1965). Twelve-year summary of selection for parthenogenesis in Beltsville small white turkeys. *Br. Poult. Sci.* **6**, 1–6.
Olsen, M. W., Wilson, S. P., and Marks, H. L. (1968). Genetic control of parthenogenesis in chickens. *J. Hered.* **59**, 41–42.
Omodeo, P. (1951). Il fenomeno della restituzione premeiotica in lombrichi parthenogenetici. *Boll. Soc. Ital. Biol. Sper.* **27**, 1292–1293.
Omodeo, P. (1952). Cariologia dei Lumbricidae. *Caryologia* **4**, 173–275.
Osawa, J. (1912). Studies on the cytology of some species of *Taraxacum*. *Arch. Zellforsch.* **10**, 45–469.
Ostenfeld, C. H. (1912). Experiments on the origin of species in the genus *Hieracium* (apogamy and hybridism). *New Phytol.* **11**, 347–353.
Ostenfeld, C. H., and Raunkiaer, C. (1903). Kastreringsforsög med *Hieracium* og andre Cichorieae (English abstract). *Bot. Tidsskr.*, **25** (No. 3).
Parker, E. D., Jr. (1979). Phenotypic consequences of parthenogenesis in *Cnemidophorus* lizards. I. Variability in parthenogenetic and sexual populations. *Evolution (Lawrence, Kans.)* **33**, 1150–1156.
Parker, E. D., Jr., and Selander, R. K. (1976). The organization of genetic diversity in the parthenogenetic lizard *Cnemidophorus tesselatus*. *Genetics* **84**, 791–805.
Peacock, A. D. (1925). Animal parthenogenesis in relation to chromosomes and species. *Am. Nat.* **59**, 218–224.
Peacock, A. D., and Harrison, J. W. H. (1926). Hybridity, parthenogenesis and segregation. *Nature (London)* **117**, 378–379.
Peacock, A. D., and Weidmann, U. (1961). Recent work on the cytology of animal parthenogenesis. *Przegl. Zool.* **5**, 5–27.
Peccinini, D. (1971). Chromosome variation in populations of *Cnemidophorus lemniscatus* in the Amazon Valley. *Cienc. Cult. (Sao Paulo)* **23**, 133–136.
Pennock, L. A. (1965). Triploidy in a parthenogenetic species of the teiid lizard genus *Cnemidophorus*. *Science* **149**, 593–540.
Peters, G. (1971). Die intragenerischen Gruppen und die Phylogenese der Schmetterlingsagamen (Agamidae: *Leiolepis*). *Zool. Jahrb.* **98**, 11–130.
Petrunkewitsch, A. (1901). Die richtungskorper und ihr Schicksal im befruchteten und undbrefuchteten Bienenei. *Zool. Jabrb., Anat. Anat. Ontog. Tiere* **14**, 573–608.
Pijnacker, L. P. (1969). Automictic parthenogenesis in the stick insect *Bacillus rossius* (Cheleutoptera, Phasmidae). *Genetica* **40**, 393–399.
Porter, D. L. (1971). Oogenesis and chromosomal heterozygosity in the thelytokous midge, *Lundstroemia parthenogenetica* (Diptera, Chironomidae). *Chromosoma* **32**, 333–342.
Rasch, E. M., Monaco, P. J., and Balsano, J. S. (1982). Cytophotometric and autoradiographic evidence for functional apomixis in a unisexual fish, *Poecilia formosa* and its related triploid unisexuals. *Histochemistry* **73**, 515–533.
Reisinger, E. (1940). Die cytologische Grundlage der parthenogenetischen Dioogonie. *Chromosoma* **1**, 531–553.

Richards, A. J. (1970). Eutriploid facultative agamospermy in *Taraxacum*. *New Phytol.* **69,** 761–774.
Richards, A. J. (1973). The origin of *Taraxacum* agamospecies. *Bot. J. Linn. Soc.* **66,** 189–211.
Rosenberg, O. (1907). Cytological studies on the apogamy in *Hieracium*. *Bot. Tidsskr.* **28,** 143–170.
Rosenberg, O. (1930). Apogamie und Parthenogenesis bei Pflanzen. *Handb. Vererbungs wiss.* **12,** 1–66.
Rostand, J. (1950). "La parthénogénèse animale." Presses Univ. de France, Paris.
Saura, A., Lokki, J., and Suomaleinen, E. (1977). Selection and genetic differentiation in parthenogenetic populations. In "Measuring Selection in Natural Populations" (F. B. Christiansen and T. M. Fenchel, eds.), pp. 381–402. Springer-Verlag, Berlin.
Scholl, H. (1960). Die Ooganese einiger parthenogenetischer Orthocladiinen (Diptera). *Chromosoma* **11,** 380–401.
Schultz, J. R. (1973). Unisexual fish: Laboratory synthesis of a "species." *Science* **179,** 180–181.
Scudday, J. F. (1973). A new species of lizard of the *Cnemidophorus tesselatus* group from Texas. *J. Herpetol.* **7,** 363–371.
Seiler, J. (1923). Geschlechtschromen-Untersuchungen in Psychiden. IV. Die Parthenogenese der Psychiden. Biologische und zytologische Beobachtungen. *Z. Abstammungsl.* **31,** 1–99.
Seiler, J. (1942). Resultate aus der kreuzung parthenogenetischer und zwei geschlechtlicher schmetterlinge. *Arch. Julius Klaus-Stift. Vererbungsforsch., Sozialanthropol. Rassenhyg.* **17,** 513–528.
Seiler, J. (1960). Untersuchungen uber die Enstehung der Parthenogenese bei *Solenobia triquetrella*. II. Mitt Analyse der diploid-parthenogenetischen *S. triquetrella*. Verhalten, Anfzuchtresultate und Zytologie. *Chromosoma* **11,** 29–102.
Seiler, J., and Schaeffer, K. (1941). Der Chromosomenzyklus einer diploid parthenogenetischen *Solenobia triqitrella*. *Rev. Suisse Zool.* **48,** 537–540.
Sessions, S. K. (1982). Cytogenetics of diploid and triploid salamanders of the *Ambystoma jeffersonianum* complex. *Chromosoma* **84,** 599–621.
Smith, S. G. (1951). A new form of spruce sawfly identified by means of its cytology and parthenogenesis. *Sci. Agic. (Ottawa)* **21,** 243–305.
Smith, S. G. (1971). Parthenogenesis and polyploidy in beetles. *Am. Zool.* **11,** 341–349.
Stalker, H. O. (1954). Parthenogenesis in *Drosophila*. *Genetics* **39,** 4–34.
Stalker, H. O. (1956). On the evolution of parthenogenesis in the *Lonchoptera* (Diptera). *Evolution (Lawrence, Kans.)* **10,** 345–59.
Stebbins, G. L., Jr. (1950). "Variation and Evolution in Plants." Columbia Univ. Press, New York.
Stebbins, G. L., Jr. (1971). "Chromosomal Evolution in Higher Plants." Edward Arnold, London.
Stebbins, G. L., Jr., and Jenkins, J. A. (1939). Aposporic development in the North American species of *Crepis*. *Genetica (The Hague)* **21,** 191–224.
Stefani, R. (1960). L'*Artemia salina* partenogenetica a Cagliari. *Riv. Biol.* **53,** 463–491.
Suomaleinen, E. (1940a). Polyploidy in parthenogenetic Curculionidae. *Hereditas* **26,** 51–64.
Suomaleinen, E. (1940b). Beiträge zur Zytologie der parthenogenetischen Insecten. II. *Lecanium hemisphaericum* (Coccidae). *Ann. Acad. Sci. Fenn.* **57,** 1–30.
Suomaleinen, E. (1947). Parthenogenese und polyploidie bei Russelkafern (Curculionidae). *Hereditas* **33,** 425–456.
Suomaleinen, E. (1950). Parthenogenesis in animals. *Adv. Genet.* **3,** 193–253.
Suomaleinen, E. (1961). On morphological differences and evolution of different polyploid parthenogenetic weevil populations. *Hereditas* **47,** 309–341.
Suomaleinen, E. (1962). Significance of parthenogenesis in the evolution of insects. *Entomology* **7,** 349–366.
Suomaleinen, E., and Saura, A. (1973). Genetic polymorphism and evolution in parthenogenetic animals. I. Polyploid Curculionidae. *Genetics* **74,** 489–508.

3. Evolution of Parthenogenesis 103

Suomaleinen, E., Lokki, J., and Saura, A. (1979). Evolution in parthenogenetic populations. *Aquilo, Ser. Zool.* **20,** 83–91.

Suomaleinen, E., Lokki, J., and Saura, A. (1981). Genetic polymorphism and evolution in parthenogenetic animals. X. *Solenobia* species (Lepidoptera: Psychidae). *Hereditas* **95,** 31–35.

Swanson, C. P., Merz, T., and Young, W. J. (1981). "Cytogenetics," 2nd ed. Prentice-Hall, Englewood Cliffs, New Jersey.

Takenouchi, Y., Okamoto, H., and Sugawara, H. (1981). A study of the influence of low temperature on the eggs of the tetraploid parthenogenetic *Catapionus gracilicornis* Roelofs (Curculionidae: Coleoptera). *J. Hokkaido Univ. Educ., Sect. IIB*, **32,** 1–15.

Takenouchi, Y., Suomaleinen, E., Saura, A., and Lokki, J. (1983). Genetic polymorphism and evolution in parthenogenetic animals. XII. Observations on Japanese polyploid Curculionidae (Coleoptera). *Jpn. J. Genet.* **58,** 153–157.

Templeton, A. R. (1982). The prophecies of parthenogenesis. *In* "Evolution and Genetics of Life Histories" (H. Dingle and J. P. Hegman, eds.), pp. 75–101. Springer-Verlag, Berlin and New York.

Thomas, R. (1965). The smaller teiid lizards (*Gymnophthalmus* and *Bachia*) of the southeastern Caribbean. *Proc. Biol. Soc. Wash.* **78,** 141–154.

Thomsen, M. (1927). Studien über die parthenogenese bei einigen Cocciden und Aleurodiden. *Z. Zellforsch. Mikrosk. Anat.* **5,** 1–116.

Turner, B. J. (1982). The evolutionary genetics of a unisexual fish, *Poecilia formosa. In* "Mechanisms of Speciation" (C. Barigozzi, ed.), pp. 265–305. A. R. Liss, New York.

Turner, B. J., Brett, B. H., and Miller, R. R. (1980). Interspecific hybridization and the evolutionary origin of a gynogenetic fish, *Poecilia formosa,* the Amazon molley. *Evolution (Lawrence, Kans.)* **34,** 246–258.

Uzzell, T. (1970). Meiotic mechanisms of naturally occurring unisexual vertebrates. *Am. Nat.* **104,** 433–445.

Uzzell, T. M., and Barry, J. C. (1971). *Leposoma pericarinatum,* a unisexual species related to *L. guianense,* and *L. Ioanna,* a new species from pacific coastal Columbia (Sauria, Teiidae). *Postilla* **154,** 1–39.

Uzzell, T. M., and Darevsky, I. S. (1975). Biochemical evidence for the hybrid origin of the parthenogenetic species of the *Lacerti saxicola* complex (Sauria Lacertidae), with a discussion of some ecological and evolutionary implications. *Copeia,* pp. 204–222.

Vandel, A. (1931). "La parthénogénèse Animal." Doin, Paris.

Vandel, A. (1936). L'évolution de la parthénogénèse naturelle. *C. R. Cogr. Int. Zool. 12th, 1935,* Vol. 1, pp. 51–64.

Vanzolini, P. E. (1970). Unisexual *Cnemidophorus lemniscatus* in the Amazonas Valley: A preliminary note (Sauria Teiidae). *Pap. Avulsos Dep. Zool., Secr. Agric., Ind. Comer. (Sao Paulo)* **34,** 63–68.

Vrijenhoek, R. C. (1979). Factors affecting clonal diversity and coexistence. *Am. Zool.* **19,** 787–797.

Vrijenhoek, R. C. (1984). The evolution of clonal diversity in *Poeciliopsis. In* "Evolutionary Genetics of Fishes" (B. J. Turner, ed.), pp. 399–429. Plenum, New York.

Webb, G. C., and White, M. J. D. (1975). Heterochromatin and timing of DNA replication in morabine grasshoppers. *In* "The Eukaryote Chromosome" (W. J. Peacock and R. D. Brock, eds.), pp. 395–408. Aust. N. Univ. Press, Canberra.

White, M. J. D. (1948). "Animal Cytology and Evolution." Cambridge Univ. Press, London and New York.

White, M. J. D. (1954). "Animal Cytology and Evolution." Cambridge Univ. Press, London and New York.

White, M. J. D. (1966). Further studies on the cytology and distribution of the Australian parthenogenetic grasshopper *Moraba virgo. Rev. Suisse Zool.* **73,** 383–398.

White, M. J. D. (1970). Heterozygosity and genetic polymorphism in parthenogenetic animals. In "Essays in Evolution and Genetics in Honor of Theodosius Dobzhansky" (M. K. Hecht and W. C. Steere, eds.), pp. 237–262. North-Holland Publ., Amsterdam.

White, M. J. D. (1973). "Animal Cytology and Evolution." Cambridge Univ. Press, London and New York.

White, M. J. D. (1974). Speciation in the Australian morabine grasshoppers. The cytogenetic evidence. In "Genetic Mechanisms of Speciation in Insects" (M. J. D. White, ed.), pp. 43–68. New Zealand Book Co., Sydney, Australia.

White, M. J. D. (1978). "Modes of Speciation." Freeman, San Francisco, California.

White, M. J. D., and Cheney, J. (1966). Cytogenetics of the *cultrata* group of morabine grasshoppers. I. A group of species with XY and X_1X_2Y sex chromosome mechanisms. *Aust. J. Zool.* **14**, 821–834.

White, M. J. D., and Webb, G. C. (1968). Origin and evolution of parthenogenetic reproduction in the grasshopper *Moraba virgo* (Emasticidae, Morabinae). *Aust. J. Zool.* **16**, 647–671.

White, M. J. D., Cheney, J., and Key, K. H. L. (1963). A parthenogenetic species of grasshopper with complex structural heterozygosity (Orthoptera Acridoidea). *Aust. J. Zool.* **11**, 1–19.

White, M. J. D., Key, K. H. L., Andre, M., and Cheney, J. (1969). Cytogenetics of the *viatica* group of morabine grasshoppers. II. Kangaroo Island populations. *Aust. J. Zool.* **17**, 313–328.

White, M. J. D., Webb, G. C., and Cheney, J. (1973). Cytogenetics of the parthenogenetic grasshopper *Moraba virgo* and its bisexual relatives. I. A new species of the *virgo* group with a unique sex chromosome mechanism. *Chromosoma* **40**, 199–212.

White, M. J. D., Contreras, N., Cheney, J., and Webb, G. C. (1977). Cytogenetics of the parthenogenetic grasshopper *Warramaba* (formerly *Moraba*) *virgo* and its bisexual relatives. *Chromosoma* **61**, 127–148.

Whiting, P. W. (1945). The evolution of male haploidy. *Q. Rev. Biol.* **20**, 231–260.

Williams, G. C. (1975). "Sex and Evolution." Princeton Univ. Press, Princeton, New Jersey.

Winge, Ö. (1917). The chromosomes. Their numbers and general importance. *C. R. Trav. Lab. Carlsberg* **13**. 131–275.

Winkler, H. (1920). "Verbreitung und Ursache der Parthenogenesis im Pflanzen-un Tierreiche." Jena.

Wright, J. W. (1967). A new uniparental whiptail lizard (*Cnemidophorus*) from Sonora, Mexico. *J. Ariz. Acad. Sci.* **4**, 185–193.

Wright, J. W. (1978). Parthenogenetic lizards. *Science* **201**, 1152–1154.

Wright, J. W., and Lowe, C. H. (1967). Evolution of the allodiploid parthenospecies *Cnemidophorus tesselatus* (say). *Mamm. Chromosome Newsl.* **8**, 95–96.

Wright, J. W., Spolsky, C., and Brown, W. M. (1983). The origin of the parthenogenetic lizard *Cnemidophorus laredoensis* inferred from mitochondrial DNA analysis. *Herpetologica* **39**, 410–416.

II
Recombination

4

Meiotic Recombination Interpreted as Heteroduplex Correction

P. J. HASTINGS

Department of Genetics
University of Alberta
Edmonton, Alberta, Canada T6G 2E9

I. INTRODUCTION

Where genes are situated on different chromosomes, characters recombine by independent assortment. Characters determined by genes on the same chromosome recombine predominantly by crossing-over, a reciprocal process, while alleles of the same gene recombine predominantly by processes called conversion, which are not reciprocal.

Reciprocal and nonreciprocal recombination appear to be related processes, since their occurrence is strongly correlated. For this reason, models of recombination seek to explain them together.

First, I shall describe the observations and parameters of meiotic recombination in fungi as revealed by genetic methods, trying to avoid model-dependent interpretation. Then the observations will be interpreted in terms of current ideas on mechanisms. This is discussed further in the next chapter. Finally, I shall show where current models are not fully satisfying and shall try to show the direction in which they are changing.

II. THE OBSERVATIONS

A. Crossing-over

Crossing-over is a breakage and rejoining process which leads to reciprocal exchange of material at homologous positions between two chromatids. Thus, a chromatid emerging from meiosis consists of lengths of material from one homologue joined end-to-end with material from the other homologue. The positions of crossovers will be seen cytologically at diplotene as chiasmata. The history of and justification for this view of crossing-over has been reviewed recently by Whitehouse (1982).

In most organisms, crossovers tend not to occur close to each other within chromosome arms. This phenomenon is called positive crossover position interference. It is, apparently, not a necessary effect of the mechanism of crossing-over, since *Aspergillus nidulans* (Strickland, 1958) and *Ascobolus immersus* (Nicolas, 1978) do not show it.

Maps of the positions of genes based on meiotic recombination are based on the frequency of crossing-over. The standard map unit is percent recombination, which is half the percentage of crossing-over, since one crossover causes only two of the four chromatids to be recombinant.

B. Intragenic Recombination

1. Selected Prototroph Frequencies

The ability to map genetic markers by the frequency of recombination is found, in many systems, to apply to very close markers. It is common to use selective techniques for frequencies below 10^{-2}. Extensive data are available for crosses between auxotrophic alleles in fungi, especially *Neurospora crassa*. Selection of prototrophic ascospores yields data such as those presented in Table I, which are taken from the *methionine-2* data of Murray (1963), and will be used here to demonstrate several characteristics of intragenic recombination. All crosses were set up with flanking markers, that is, the crosses were segregating for markers at a distance of a few map units from the locus under study. The data give the frequency with which methionine prototrophs occur in ascospores and the flanking marker configurations found in the prototrophs.

The first point to make about prototroph frequency data is that the repeatability of a result is not good. This is seen when the same allelic cross has been made with reciprocal flanking marker configurations. At least a part of the variation can be attributed to uncontrolled genetic variation. Next we observe that a map can be constructed from the prototroph frequencies using the principle that the higher value defines the longer length. The map shown in Fig. 1 has some inconsistencies in the sequence of alleles and does not show good additivity. I shall return to the subject of the deviation from additivity. Murray (1963) also found that distribution of alleles on the map of *me-2* was not random, in that alleles tend to be clustered. Thus the map shows regions of low recombination and regions of high recombination. It has been shown for *cyc1* in *Saccharomyces cerevisiae* that some of the clustering of alleles is related to the initiation codon

4. Meiotic Recombination as Heteroduplex Correction

TABLE I

Data for Selected Prototrophic Spores from Crosses between Alleles of the *me-2* Locus of *Neurospora crassa*[a]

Cross[b]	Frequency per 10^5 viable spores	Classification of methionine prototrophs (%)			
		Parental		Recombinant	
		$tryp^-\ pan^+$	$tryp^+\ pan^-$	$tryp^-\ pan^-$	$tryp^+\ pan^+$
−K44 + × +P2 −	8	17	38	10	36
−P2 + × +K44 −	5	40	11	42	7
−K23 + × +P2 −	4	5	60	10	26
−P2 + × +K23 −	7	48	11	28	13
−K44 + × +H98 −	15	11	44	9	36
−H98 + × +K44 −	45	43	7	44	7
−K44 + × +K5 −	19	10	44	11	36
−K5 + × +K44 −	17	45	8	39	9
−K23 + × +H98 −	43	13	41	17	29
−H98 + × +K23 −	33	35	16	39	16
−P133 + × +H98 −	26	10	43	11	36
−K44 + × +P24 −	48	12	55	12	21
−P24 + × +K44 −	33	61	9	19	11
−K23 + × +P24 −	10	9	51	14	26
−P24 + × +K23 −	32	51	8	27	14
−P133 + × +P159 −	29	11	52	13	24
−P2 + × +H98 −	12	22	35	14	29
−H98 + × +P2 −	10	32	22	32	13
−P2 + × +K5 −	9	19	32	18	31
−K5 + × +P2 −	9	38	22	28	12
−P2 + × +P24 −	30	20	43	16	22
−P24 + × +P2 −	31	51	18	17	13
−P2 + × +P159 −	18	14	45	17	25
−P159 + × +P2 −	32	43	19	28	9
−H98 + × +P24 −	10	23	52	11	14
−P24 + × +H98 −	14	44	17	23	16
−P159 + × +H98 −	37	52	17	17	14
−K5 + × +P24 −	16	18	44	13	25
−P24 + × +K5 −	13	54	16	19	11
−P159 + × +K5 −	24	49	20	18	13
−P133 + × +P24 −	110	13	40	19	28

[a] Abridged from the data presented by Murray (1963), with permission.
[b] Crosses were made in the form $tryp^-\ me^1\ pan^+ \times tryp^+\ me^2\ pan^-$.

and to the heme binding site of the gene product, iso-1-cytochrome *c* (Moore and Sherman, 1975). Since, in this case, clustering clearly has a functional basis, we cannot say whether any of the clustering relates to the nature of the recombination process. It is also possible that large spaces in a recombination map are caused by intervening sequences in genes.

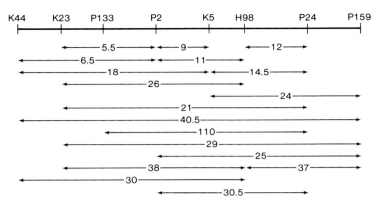

Fig. 1. A partial map of *methionine-2* in *Neurospora crassa* derived from the data of Murray (1963) shown in Table I. The map is oriented so that the flanking marker *tryp* is to the left and *pan* is to the right.

2. Flanking Marker Combinations

Table I shows that all four possible combinations of flanking markers occur commonly among selected intragenic recombinant products. Two of these combinations ($tryp^-$ pan^+ and $tryp^+$ pan^-) are parental, or non-crossover combinations. The other two classes are recombinant and show a crossover in the interval between *tryp* and *pan*. Crossover and non-crossover prototrophs are approximately equal. Extensive data show that they are rarely actually equal. In an unselected sample of ascospores, only 10% would be recombinant in this interval, since the map distance from *tryp* to *pan* is 10%. Thus, in selecting for methionine prototrophs, Murray (1963) selected for crossovers between the flanking markers.

There are two opposing aspects to this finding. On the one hand, intragenic recombination is strongly correlated with crossing-over. On the other hand, a very high proportion of intragenic recombinants (about 50%) are not accompanied by a crossover. So we may say that two different types of event give rise to intragenic recombination—crossover events and non-crossover events.

This frequent association of the selected exchange (giving rise to the prototroph) with unselected exchanges has been defined as intensive negative interference (Pritchard, 1955). It will be seen later, however, that much of this association results from a single event of nonreciprocal information transfer, in other words, from conversion.

3. Mapping by Flanking Marker Combinations

The two classes of prototrophic products which show recombination of flanking markers are often unequal. The more common class is called R_1 and the less

4. Meiotic Recombination as Heteroduplex Correction

common R_2. It can be seen in Table I that, when the position of the flanking markers with respect to *me-2* alleles is reversed, the genotypes of R_1 and R_2 are reversed. One of these classes can be described as a single exchange, occurring between the alleles, while the other class can be described as resulting from three exchanges. If one assumes that it is easier to obtain one exchange than three, then the R_1 class can be taken as the class arising from a single exchange. Hence, from a cross *tryp*$^-$ K44 *pan*$^+$ × *tryp*$^+$ P_2 *pan*$^-$, we see in Table I that the prototroph $+ + + +$ is more common (36%) than $- + + -$ (10%). For $+ + + +$ to be obtained by a single exchange, it is necessary for K44 to be to the left of P_2. If this method is applied to the data in Table I, it will be found that a consistent map order is obtained and that the order is also consistent with the map shown in Fig. 1 produced from prototroph frequencies. By this method, the map is also oriented with respect to flanking markers and, hence, to the whole linkage map.

4. Interference Properties of Intragenic Recombinants

It has been found that crossover position interference is a property of crossover recombination events. Non-crossover recombination events do not interfere with the occurrence of crossovers nearby. This was first shown by Stadler (1959) for *Neurospora*, and it was later reported for yeast (Mortimer and Fogel, 1974). The method used is to measure the length of a region adjacent to the region flanking the locus under study. The observation is that this further region has the same length in a sample selected for the presence of non-crossover events as it has in a nonselected sample, while in progeny showing crossover recombination events, the length of the further region is reduced.

Further data relevant to this point can be found in Savage (1979). These data include tetrads containing a product which does not require histidine, arising by recombination between alleles of *his1*. Crossovers were seen in the flanking regions: *hom3*, 2.5% recombination to *his1*, 6.8% r to *arg6*, 20.0% r to *trp2*. These distances are based on 8679 unselected tetrads, which showed 3289 tetratype (T) tetrads and 29 nonparental ditype (NPD) tetrads for *arg6* and *trp2*, where % r = $[\frac{1}{2}(T + 2 NPD)/Total] \times 100$. The length of this region was not reduced in tetrads selected for non-crossover events at *his1* (259 tetratype tetrads for *arg6* and *trp2*/579 non-crossover events at *his1* gives 22.4% r), but was reduced to 60% in events showing crossovers at *his1* (131/531 gives 12.3% r).

C. Ratios in Tetrads

A crossover per se by definition does not disturb the Mendelian segregation of the markers flanking the crossover; they still segregate 2:2 in meiotic tetrads. It is found, however, that a low frequency of tetrads (often about 1%) shows non-Mendelian segregation of a marker. When one looks at a tetrad of meiotic

TABLE II

Genotypes of Asci Producing Only Homokaryotic Cultures in Which There Is a Recombination Event at the *pan-2* Locus of *Neurospora crassa*[a,b]

Ascus 43	Ascus 78	Ascus 98
ad 23 + 36 +	+ 23 + + +	+ + 72 + tryp
ad 23 + 36 +	+ 23 + + +	+ + 72 + tryp
ad + 72 + +	+ + 72 + tryp	+ + + 36 +
ad + + 36 +	+ + 72 + tryp	+ + + 36 +
+ + 72 + tryp	ad 23 + 36 +	ad 23 72 + tryp
+ + 72 36 tryp	ad 23 + 36 +	ad 23 72 + tryp
+ 23 72 + tryp	ad 23 + 36 tryp	ad 23 + 36 +
+ 23 72 + tryp	ad 23 + 36 tryp	ad 23 + 36 +

Ascus 153	Ascus 497	Ascus 529
+ + 72 + tryp	+ + 72 + tryp	ad 23 + 36 tryp
+ + 72 + tryp	+ + 72 + tryp	ad 23 + 36 tryp
+ + 72 + tryp	+ + 72 + tryp	ad + + 36 +
+ + 72 + tryp	+ + 72 + tryp	ad + + 36 +
ad 23 + 36 +	ad + + 36 +	+ + + 36 tryp
ad 23 + 36 +	ad + + 36 +	+ + 72 + tryp
ad + + 36 +	ad 23 + 36 +	+ + 72 + +
ad + + 36 +	ad 23 + 36 +	+ + 72 + +

Ascus 581	Ascus 583	Ascus 614
ad + 72 + +	+ + 72 + tryp	+ + 72 + tryp
ad + 72 + +	+ + 72 + tryp	+ + 72 + tryp
ad 23 + 36 +	+ 23 + 36 tryp	+ + 72 36 +
ad 23 + 36 +	+ 23 + 36 tryp	+ + 72 36 +
+ + 72 + tryp	ad 23 + 36 +	ad 23 + 36 +
+ + 72 + tryp	ad 23 + 36 +	ad 23 + 36 +
+ + 72 + tryp	ad 23 72 + +	ad 23 + 36 tryp
+ + 72 + tryp	ad 23 72 + +	ad 23 + 36 tryp

Ascus 1021	Ascus 545	Ascus 565
ad 23 + 36 +	ad + 72 + tryp	ad 23 + 36 +
ad 23 + 36 +	ad + 72 + tryp	ad 23 + 36 +
ad + + 36 +	ad 23 + 36 +	ad + 72 + tryp
ad + + 36 +	ad 23 + 36 +	ad + 72 + tryp
+ + 72 + tryp	. . . .	+ + 72 + tryp
+ + 72 + tryp	+ + 72 + tryp	+ + 72 + tryp
+ + 72 + tryp	. . . .	+ + 72 + +
+ + 72 + tryp	+ + + 36 +	+ + 72 + +

4. Meiotic Recombination as Heteroduplex Correction 113

TABLE II (*Continued*)

Ascus 710
+ + 72 + *tryp*
+ + 72 + *tryp*
+ + + 36 +
+ + + 36 +
. . . .
ad 23 72 + *tryp*
ad 23 + 36 +
ad 23 + 36 +

^a Modified from Case and Giles (1964), with permission.
^b These asci were found among 1457 unselected asci dissected from crosses of the genotype *ad* + 72 + + / + 23 + 36 *tryp*, where *ad*/+ and +/*tryp* are flanking markers and 72, 23, and 36 are alleles of *pan-2*. Dotted lines indicate spores which did not germinate.

products in which intragenic recombination has occurred, one almost always finds that an allele has segregated with a non-Mendelian ratio.

The most common non-Mendelian ratios are 3 wild type:1 mutant spore (3:1), or 1:3 (6:2 and 2:6 in eight-spored tetrads as seen in *Neurospora crassa* or *Ascobolus immersus*). These ratios are said to show conversion. It appears that one chromatid has acquired, for this marker, the genotype of the homologue. The word "conversion" will be used here for this phenomenon and for the processes which give rise to it.

Odd ratios, 5:3 and 3:5, show postmeiotic segregation (PMS). The alleles segregated from each other at the first postmeiotic mitosis in one of the meiotic products. This gives rise to a mixed-spore pair in an eight-spored ascus. In four-spored yeast tetrads, PMS produces a mixed colony derived from a single ascospore. Postmeiotic segregation of a marker can occur simultaneously in two chromatids, giving an aberrant 4:4 ratio in a tetrad.

Wider ratios such as 7:1, 1:7, 8:0, and 0:8 occur. These events represent PMS and/or conversion occurring simultaneously on several chromatids. Such ratios have a frequency predicted for the simultaneous occurrence of two independent events.

Table II shows tetrads obtained by nonselective analysis by Case and Giles (1964). These are the tetrads which show evidence of a recombination event at the *pan-2* locus of *Neurospora crassa* found among 1457 ordered asci dissected. The presence of three alleles of *pan-2* and of flanking markers gives a picture of the recombination event which has since been confirmed repeatedly by more extensive data in more favorable systems. It is a useful exercise to study such tetrads, attempting to determine the pattern of events which gave rise to them. The following generalizations are widely valid.

1. It is often impossible to name a unique series of events as the origin of any particular tetrad. This is especially so if one allows that conversions, even though individually nonreciprocal, may occur reciprocally.
2. Conversion and PMS may occur in the same tetrad.
3. It is often possible to see that at least two chromatids have been involved.
4. The position of the crossover usually cannot be determined.
5. The crossover (if any) usually involves the chromatids on which conversion and PMS can be seen.
6. It is common to see two or three markers in a row being converted together to the genotype of the homologue. From this observation, it is taken that the conversion process involves a length of a chromatid, rather than the point which is the site of an allele. When two or more markers are converted together in such a length, the phenomenon is called "co-conversion."
7. PMS can also be seen to involve a length of a chromatid. When this occurs, it is seen that sub-chromatids retain parental linkage relationships.
8. Lengths of conversion or PMS tend to be related to one or other end of the gene, seldom confined to the middle.
9. Almost all recombination between alleles occurs at the end of a length of conversion or PMS.

D. Variation in Conversion Spectrum

Leblon (1972a,b) reported an important experiment with *Ascobolus*. He induced ascospore color mutants with several mutagens and analyzed the proportion of different non-Mendelian ratios occurring for each mutant allele. It was found that the frequency with which non-Mendelian ratios occur was more locus-specific than allele-specific or mutagen-specific. But the spectrum of aberrant ratios, that is, the relative frequency of different ratios, related quite well to the mutational origin of the marker.

Alleles which appeared to be frameshift mutations rarely showed PMS and had a strong disparity in the direction of conversion. For one sign of frameshift mutation, there was more conversion to wild type (6:2) than to mutant (2:6), while the other sign of frameshift mutation showed the opposite disparity. Alleles which behaved as substitution mutations showed a considerable frequency of PMS with, usually, only slight disparity in the direction of conversion. The experiment of Leblon (1972a,b) has been repeated, using *Sordaria brevicollis*, by Yu-Sun et al. (1977). The results were very similar to those reported by Leblon.

The conclusion that the conversion spectrum is a response to the nature of the mutation at the level of base sequence is very important to recombination theory, and will be considered again.

Yeasts, however, are different. The most complete listing of conversion spectrum of alleles in *Saccharomyces cerevisiae* is to be found in Fogel *et al.* (1981). The list reveals that all markers show a close approximation to parity in direction of conversion (3:1 = 1:3) and that, for most alleles, there is a very similar level of PMS. A few alleles are conspicuous for their abnormally high relative frequency of PMS. The more limited data for the fission yeast *Schizosaccharomyces pombe* (Thuriaux *et al.*, 1980) show the same general characteristics.

E. Polarity

In many systems it can be seen that meiotic recombination is a polarized process. Polarity can be seen in three ways:

1. Polarity was first described in *Ascobolus* (Lissouba *et al.*, 1962), where it was found that, in the generation of a wild-type recombinant from a two-point allelic cross, the same allele was always, or almost always, converted. When data for a series of crosses are seen, it emerges that in each cross the allele toward one end of the locus is the one to be converted. Unselected tetrad analysis reveals that the other allele is still being converted, but that such conversion does not lead to recombination because the two alleles are co-converted (Rizet and Rossignol, 1966).

2. A second way to see polarity is in a gradient of the frequency of aberrant ratio tetrads, with high frequency at one or both ends of the polaron. One of the best documented cases is the *b2* locus of *Ascobolus* (Paquette and Rossignol, 1978).

3. The third expression of polarity was described by Murray (1963) from the data presented in Table I. It was seen above that intragenic recombination usually results from conversion of one allele and not the other. If this conversion occurs without crossing-over, then the more common class of prototrophic spore will have the flanking marker configuration which entered the cross with the more frequently converted allele. It is conventional to name the class of prototroph having the flanking marker configuration which entered the cross with the centromere proximal allele "P_1" and that which entered with the distal allele "P_2." Reference to the map order of *me-2* given in Fig. 1 will reveal that, in the data in Table I, P_2 is always in excess of P_1. Thus, from the excess of P_2 over P_1, one can see that *me-2* is polarized, conversion being higher to the right-hand (centromere distal) end of the locus and relatively lower to the left, throughout the data presented here.

The three expressions of polarity are equatable if we understand that there is a tendency for conversion lengths to begin (or end) outside the gene and extend for a variable distance into the gene (Hastings and Whitehouse, 1964). The idea that the unit of polarity (the polaron) corresponds in some way to the unit of function is based on the observation that there are no known peaks of conversion frequency within genes, but many genes contain a trough, the frequency of conversion being higher at both ends of a gene than it is in the middle. Two cases of peaks within a locus in *Ascobolus* are reported by Lissouba et al. (1962) but neither case stands up to scrutiny or later work.

F. Symmetry and Asymmetry

It can be seen in the tetrads of Case and Giles (1964) in Table II that the lengths of conversion and PMS may be on one chromatid or on two. There are several ways to tell that these events do not uniformly involve two chromatids over equal lengths. For example, if conversion of an allele occurred on two homologous (i.e., non-sister) chromatids, the conversions would occur reciprocally, giving the appearance of an intragenic crossover. Reciprocal intragenic recombination is much less common than nonreciprocal.

The most widely used criterion for the number of chromatids involved in an event is the frequency of aberrant 4:4 segregation, showing PMS on two homologous chromatids. For markers which show significant levels of PMS, seen by 5:3 and 3:5 segregations, one may find very few aberrant 4:4 segregations. This is interpreted to mean that the recombination event was usually asymmetrical (involving only one chromatid) at the site of those markers.

Paquette and Rossignol (1978) found evidence of a gradient in the frequency of aberrant 4:4 segregation, relative to other non-Mendelian ratios, for alleles of the *b2* locus of *Ascobolus* which give a high frequency of all non-Mendelian segregations. So it seems that the event is more likely to be symmetrical in the low conversion part of the gene. This point will be discussed further when considering models of recombination.

G. Where Is the Crossover?

The question can be asked at different levels. First, it has already been noted that the crossover (if any) involves the chromatid or chromatids which show conversion (in those cases in which the chromatid which was converted can be identified) or which show PMS. The number of crossovers found between the flanking markers which do not involve a conversion chromatid corresponds to the number expected for unrelated crossovers occurring in conjunction with non-crossover conversion events. It was noted above that non-crossover events do not show interference.

4. Meiotic Recombination as Heteroduplex Correction

The question can also be asked where in the map length between the flanking markers the crossover occurs. It is found that the proportion of crossover to noncrossover events is independent of the length of the flanking intervals, except that longer intervals show more unrelated crossovers. This rule applies when the flanking markers are very close to the locus under study (less than 1% recombination) so it is apparent that the crossover occurs in close proximity to the conversion event.

Finally, it can be asked where the crossover is in respect to the lengths of conversion and/or PMS. It was discussed in Section II,E that many conversion or PMS lengths can be seen to have one end outside the region under study, and apparently outside the gene at the high conversion end, while the other end of the conversion or PMS length is apparently within the gene at a variable position. So the question can be rephrased to ask whether the crossover is outside the gene at its high conversion end or within the gene at the variable end of the conversion length. It is not possible to say on which side of a conversion the crossover lies. But one can see the relative position of a crossover and a site showing PMS, as illustrated in Fig. 2. For this method, it is necessary to know that the marker has not been converted on another chromatid, so the method is valid only where there is strong evidence of a preponderance of asymmetrical events. Fogel *et al.* (1979) used this method with *arg4* in yeast, which has strong polarity and strong asymmetry. Their data show that the crossover occurs on either side of the PMS site, so it apparently has an indeterminate position with respect to a length of chromatid in which PMS is occurring.

This result shows, at least, that the crossover is not uniformly outside the gene at the high conversion end. Any further discussion of this problem is model-dependent and, for this reason, will not be included until some theoretical considerations have been introduced. We shall, therefore, return to this matter later.

H. Variation between Loci

Some of the properties of recombination which I have been describing are not universal in occurrence.

1. Mappability

It was pointed out by Hastings (1975) that, although some loci give a very clear map based on recombination frequency, others (as seen in Table I) give a rather poor map, and some loci can hardly be mapped at all. Examples of the last class are *mtr* in *Neurospora* (Stadler and Kariya, 1969) and *SUP6* in yeast (DiCaprio and Hastings, 1976).

Loci which are immappable by recombination frequency are also difficult or

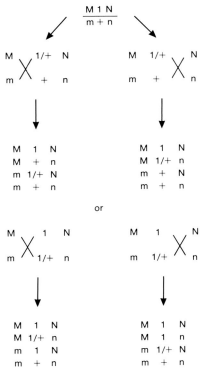

Fig. 2. Determination of crossover position in a tetrad showing PMS. The cross was M 1 N / m + n, where M/m and N/n are flanking markers. The diagram shows various positions of the crossover and the PMS for the two chromatids involved and the compositions of the resultant tetrads. When a chromatid showing PMS is involved in a crossover, the position of the crossover is revealed by the genotype of the other crossover chromatid.

impossible to map by the relative numbers of R_1 and R_2 recombinants. The two crossover combinations of recombinant products are equally common, or vary relative to each other in an inconsistent way.

2. Additivity

For a map sequence a-b-c, when the map is additive, the amount of recombination between a and c is equal to the sum of recombination between a and b and between b and c ($r_{a \text{ to } c} = r_{a \text{ to } b} + r_{b \text{ to } c}$). Maps of intergenic distances show map contraction ($r_{a \text{ to } c} < r_{a \text{ to } b} + r_{b \text{ to } c}$). This is caused by multiple recombination events canceling out the effects of each other. On an intragenic scale one often finds a phenomenon described by Holliday (1968) and called map expansion. When expansion occurs, $r_{a \text{ to } c} > r_{a \text{ to } b} + r_{b \text{ to } c}$. Expansion is sometimes very marked, sometimes weak, and sometimes absent.

3. Polarity

Some of the spore color loci in *Ascobolus* show absolute polarity, that is, recombination always (or almost always) results from conversion of the allele on one side, never conversion of the allele on the other side, in a two-point cross. The data for *me-2* in *Neurospora* in Table I also show polarity, but it is certainly not absolute. There is a tendency for recombination to arise from conversion of the right-hand allele more often than from conversion of the left-hand allele. Further, there are many loci which do not show polarity at all.

4. Frequency of Recombination Events at a Locus

Another very important locus-specific variable is the frequency with which recombination events occur, although the frequency with which they are detected also varies according to the position of a site within the locus, as described in Section II,E.

I shall show in further sections that the range of variation in properties between loci can be explained by very few variables.

I. General Description of a Recombination Event

The observations in Section II can be summed up by a description of what we visualize as a recombination event. A recombination event is any event which is detected by intragenic recombination or by disturbance of the Mendelian segregation of markers. It is found to consist of lengths of chromatid within which lengths of conversion and/or PMS occur, sometimes interspersed with lengths in which no change is apparent. These lengths tend to have one end outside the gene. There may be one or two chromatids involved as recipients of information, and the lengths on the two chromatids are often unequal. In at least one locus it can be seen that it is common for the event to be asymmetrical (involving one chromatid) at the high conversion end of the gene and symmetrical at the low conversion end.

A crossover is often found to occur between these same two chromatids within the gene where the recombination event is occurring. Many events, however, have no associated crossover. The relative amount of conversion to mutant or to wild type or of PMS is an allele-specific parameter.

This description of a recombination event is clearly not based on direct observation, but is an interpretation derived from a vast body of observation. Howev-

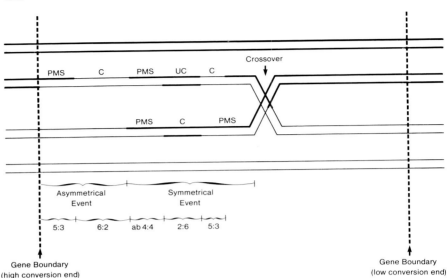

Fig. 3. The distribution of the genetic contribution of the two parents in a crossover recombination event as it might be seen in the *b2* locus of *Ascobolus immersus*. All four chromatids are shown, the two middle chromatids being the ones involved in the event. Each chromatid is shown as two subchromatids. One parental contribution is shown by thick lines and the other by thin lines. The positions of lengths of conversion and of postmeiotic segregation will vary. There will also be variation in the position of the transition from an asymmetrical to a symmetrical event, and in the position of the right-hand end of the symmetrical part of the event. The position of the crossover is still open to debate. We do not know what occurred outside the gene. The figure indicates the ratios of alleles which would be seen in various parts of this event. PMS = postmeiotic segregation; C = conversion; UC = unconverted segment.

er, I think that it is not dependent on any preconceived model of the mechanism of recombination. It is, rather, the situation which any model must seek to explain.

The imaginative drawing of a recombination event shown in Fig. 3 can be derived descriptively from the data concerning recombination in the *b2* locus of *Ascobolus* in every point except the position of the crossover.

III. HETERODUPLEX REPAIR MODELS OF CONVERSION

The present style of recombination model originated in the early 1960s with the suggestion that PMS represented a precursor to conversion in that it showed the survival through meiosis of a heteroduplex molecule formed by the hydrogen bonding of a DNA nucleotide chain from one parent to a DNA chain from the

4. Meiotic Recombination as Heteroduplex Correction

other parent. Mismatches in the heteroduplex would be corrected by an excision repair process, one outcome of which would be conversion (Holliday, 1962, 1964; Whitehouse, 1963; Meselson, 1965).

The validity and universality of this hypothesis, and its usefulness in explaining observations, is a subject which can stand apart from ideas on how the heteroduplex is produced and how it is associated with, or leads to, crossing-over.

A. The Relative Proportions of Aberrant Ratio Tetrads

The heteroduplex correction model of conversion is well supported by the findings (e.g., Leblon, 1972a) that the conversion spectrum is allele-specific and that this is related to the mutagenic origin of the allele. This implies that whether or not correction occurs, as well as the direction of correction, is dictated by the nature of the mismatch at the level of specific base pairs. It follows from this understanding that the quantitative relationships of the various classes of aberrant tetrad can be defined on the basis of very few parameters.

As an example of how this works, we can see that a 5 wild type to 3 mutant segregation will result from any one of three situations. These are shown in Fig. 5 of Chapter 5. They are (1) the heteroduplex was asymmetrical on a chromatid which was originally mutant, and was not corrected; (2) the heteroduplex was formed symmetrically, was uncorrected on the chromatid which was originally mutant, and was corrected to wild type on the chromatid which was originally wild type; or (3) the heteroduplex was formed symmetrically, remained uncorrected on the chromatid which was originally wild type, but was corrected (converted) to wild type on the chromatid which was originally mutant. This last class can be distinguished from the other two in non-crossover events if flanking markers are present. Similar lists of origins can be derived for other aberrant segregations. Some of these are illustrated in Fig. 3.

Whitehouse and Hastings (1965) attempted to analyze the conversion spectrum in this way. They used three parameters: whether correction occurs, the direction of correction, and whether heteroduplex occurs on a chromatid. The method had only moderate success in explaining the spectrum of the *gray* mutation in *Sordaria fimicola*. The method was criticized (Emerson, 1966, 1969) because it made the assumption that the mismatch was the same, and therefore behaved in the same way, on both chromatids.

Further complications have been introduced by other authors (see Gutz, 1971), but the most successful analysis of this sort is as simple as that of Whitehouse and Hastings (1965). It is the analysis of the *b2* locus of *Ascobolus*. The algebra used is in the appendix of a paper by Paquette and Rossignol (1978). It is found that conversion spectra in *b2* can be described by four parameters, with very few cases of significant deviation from a fit of the data to the equations. The parameters are (1) recombination event or not, (2) asymmetrical or not, (3) heteroduplex corrected or not, and (4) correction to wild type or not. The success of the method for the *b2* locus is well documented by Rossignol *et al.* (1979) and leaves very little room for doubt that the theoretical basis of the analysis is on the right track.

The reader will have noticed that a new phenomenon has been introduced to this discussion. It is proposed that heteroduplex correction may be to the genotype of either parent of the heteroduplex. Thus, the correction will sometimes convert a chromatid to the genotype of the homologue and sometimes restore the original genotype of the chromatid. Restoration will be discussed in more detail in Section III,C.

B. Co-correction of High PMS Alleles

We have noted that adjacent markers are often converted together. This is interpreted to mean that the conversion process involves a length of a chromatid rather than a point. On a heteroduplex repair model of conversion, a length of co-conversion is interpreted as a length of co-correction, that is, the mismatch correction process is an excision process in which a contiguous length of a single nucleotide chain is excised and replaced by replication using the other chain as a template. A consequence of this is that one expects the conversion spectra of alleles to interact, since a correction tract may originate at the mismatch for one allele and overlap the site of a second allele, thus imposing the characteristics of conversion of one marker upon another. That this is, in fact, what is seen provides the best evidence we have that conversion occurs by excision correction of mismatches in heteroduplex DNA.

The experiments were performed with markers in the *b2* locus of *Ascobolus* and reported by Leblon and Rossignol (1973). The observations are as follows: When a cross was made between two alleles both of which show low PMS, one being converted predominantly to wild type (A) and the other predominantly to mutant (B), there was a reciprocal interaction in the co-conversion events, such that they showed parity in conversion in the two directions. However, when an allele which shows a high frequency of PMS (C) is placed beside an allele showing low PMS (B), the C allele acquires the conversion spectrum of the B allele in co-conversion events, but not when it is converted alone. Conversely, the high PMS characteristic of the C allele is not acquired by the B allele.

These observations seem to show that conversion is an expression of a later stage in a process than the condition which has postmeiotic segregation as its final outcome. If we assume the obvious explanation for PMS—that it shows persistence through meiosis of heterozygosity within a DNA molecule, followed by replicative resolution of the heterozygosity prior to the first postmeiotic mitosis—then it follows that conversion is a nonreplicative resolution of the same structure. Thus, we see that conversion in the *b2* locus of *Ascobolus* is a process of excision correction of mismatches in heteroduplex DNA, induced by the presence and nature of the mismatches themselves. The existence throughout the map of the *b2* locus of markers which give much more PMS than conversion shows that the heteroduplex is widespread and is approximately as common as conversion of the markers which show little or no PMS. This is consistent with the belief that all, or almost all, conversion in the *b2* locus is of this type.

Alleles which show high PMS are not very common in yeast. However, an experiment equivalent to that of Leblon and Rossignol (1973) has been reported in *Saccharomyces cerevisiae* using the *arg4* locus by Fogel *et al.* (1979) and using the *his1* locus by Hastings (1984). In both cases, the result was the same as that seen in *Ascobolus*: a high PMS allele is co-converted by a low PMS allele so that, in the presence of nearby heterozygosity, it shows PMS much less often than it does when alone. It has been argued by Hastings (1984) that the existence of high PMS alleles, and the co-conversion of them with nearly normal alleles, must have the same interpretation in yeast as in *Ascobolus*: that conversion is a process of correction of mismatch in heteroduplex DNA.

C. Restoration of Parental Genotype from Heteroduplex

1. Restoration in Ascobolus

It is a prediction of the heteroduplex repair hypothesis that correction of a mismatch will restore the parental genotype of a chromatid as often as it will convert the chromatid to the genotype of the homologue. The reason for this prediction is shown in Fig. 4. It results from the postulate that the mismatch repair system will exercise a preference for excision of one base rather than the other at any mismatch without reference to which information was the recipient and which the donor in the formation of heteroduplex. Conversion is seen only when the recipient information is excised so that the donor information becomes homozygous within the DNA duplex. If there is a disparity in the direction of conversion with, for example, the mutant information being excised more often than the wild-type information, this will be seen as an excess of 6:2 segregations over 2:6 segregations. The hypothesis predicts that there will be, unseen, an equal disparity in restoration to wild type exceeding restoration to mutant. The rule that conversion will equal restoration in frequency applies for any relative frequency of excision of the two bases.

This prediction has been tested for a pair of frameshift mutations in the *b2* locus of *Ascobolus*, using the scheme shown in Fig. 5 (Hastings *et al.*, 1980).

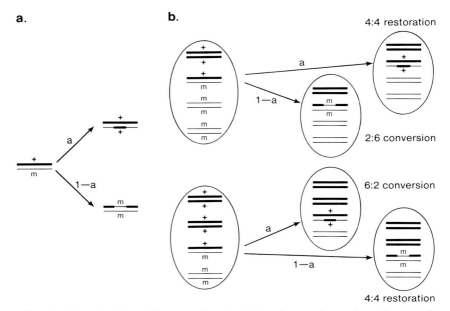

Fig. 4. The mismatch excision correction hypothesis of conversion. a shows the correction mechanism in one chromatid. b shows the consequences of correction as seen in a tetrad, revealing that conversion and restoration should be equally common. Reprinted from Hastings (1984).

Two well-spaced alleles (*17* and *A4*) which give high PMS were used to select tetrads in which heteroduplex covered the locus on one chromatid. Between them were two frameshift mutations of opposite sign (*E1* and *E2*) which gave incomplete mutual suppression, resulting in a phenotype intermediate between the wild-type spores (black) and the mutant spores (white). The alleles *17* and *A4* have the white phenotype. The *E1E2* pair of mutations does not show PMS. Figure 5 shows how conversion and restoration in coupling and repulsion crosses give asci of different composition, allowing determination of the parameters of heteroduplex correction. The values shown on the arrows show a close approximation to the prediction from Fig. 4. The cause of the small deviation from expectations was identified as co-conversion of *17* with *E1E2* being included in the data, while co-restoration was not detected.

This result seems to leave very little room to doubt that the heteroduplex repair hypothesis is a good description of what happens in the *b2* locus with these mutations. However, we already know that it is not the whole story, even for the *b2* locus. When the experiment was repeated using a large heterology (believed to be a deletion of a few hundred base pairs) in the middle of the locus, it was found that, for the heterology, conversion was 10 times more common than restoration (Rossignol *et al.*, 1984).

4. Meiotic Recombination as Heteroduplex Correction

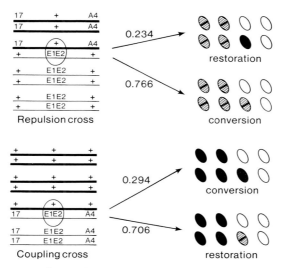

Fig. 5. Measurement of restoration in the *b2* locus of *Ascobolus immersus*. The diagram is the equivalent of Fig. 4b and shows how conversion and restoration on chromatids of each of the two homologues lead to different phenotypic consequences in tetrads. Black spores are wild type, white spores are *17*, *A4*, or *17,A4*, and belted spores are *E1E2*. White is epistatic to belted. The values on the four arrows are those determined for the correction of *b2-E1E2* by Hastings *et al.* (1980). Reprinted from Hastings (1984).

2. Crossover Position Revisited

Now that the concept of restoration as a correction process from heteroduplex has been established, we can return to the question of the position of the crossover. In turn, the measurement of restoration in yeast which follows depends upon an assumption about the position of the crossover. It has been shown for the *b2* locus in *Ascobolus* that more than one conversion tract can occur in the same length of heteroduplex. Kalogeropoulos and Rossignol (1980) saw conversion of two markers, separated by PMS of a marker between them. This, with the demonstration of restoration described above, makes it reasonable to assume that we may see independent correction giving both restoration and conversion within one length of heteroduplex. Such independent correction would be a source of intragenic recombination.

If we consider the origin of various classes of tetrad containing prototrophs from heteroallelic crosses in this context, we may gain insight on the distribution of various aspects of a recombination event from the relative strength of different tetrad classes. This was done with data for the *his1* locus of yeast by Savage and Hastings (1981).

In these data, tetrads which included a prototrophic spore were selected and the whole tetrad was analyzed. The results were complemented by nonselective analysis. The classification of such tetrads is shown, for crossover events, in Fig. 6. The designations R_1 and R_2 refer to the flanking marker combination of the prototrophic product, as described above. The rest of the tetrad is described by p, meaning a 3:1 segregation of the left-hand allele; q, meaning 3:1 at the right-hand allele; and r, meaning that the tetrad included the double-mutant product reciprocal to the prototroph. Unselected data have shown (1) the locus is strongly polarized, conversion being high to the left; (2) essentially all events are asymmetrical; and (3) events occur equally often on chromatids of the two homologues.

Figure 7 shows how the various types of tetrad would arise in the situation described for the *his1* locus if (a) the crossover were outside the gene or if (b) the crossover were within the gene at the variable, low conversion end of the heteroduplex. In Fig. 7a we see that a crossover outside the gene to the left (the high conversion end) would give mainly R_1p tetrads, with R_2q as a minor class. R_1q tetrads would arise only from events entering from the right, and these would

From the cross $\dfrac{M\ 1\ +\ N}{m\ +\ 2\ n}$:

where R_1 = m + + N
       R_2 = M + + n

p =   1 +   q =   1 +   r =   1 +
      + +   1 +   1 2
      + 2   + +   + +
      + 2   + 2   + 2

we get R_1p M 1 + N R_2p M 1 + N
                   M + 2 n   M + + n
                   m + + N   m + 2 N
                   m + 2 n   m + 2 n

         R_1q M 1 + N R_2q M 1 + N
                   M 1 + n   M + + n
                   m + + N   m 1 + N
                   m + 2 n   m + 2 n

         R_1r M 1 + N
                   M 1 2 n
                   m + + N
                   m + 2 n

Fig. 6. Classification of prototroph-containing crossover tetrads. This is a diagram of the scheme described in the text. *1* and *2* are alleles of the gene under study and M/m and N/n are flanking markers.

4. Meiotic Recombination as Heteroduplex Correction

(a) Crossover outside the gene

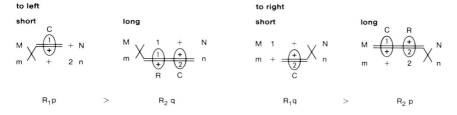

(b) Crossover at variable end of heteroduplex

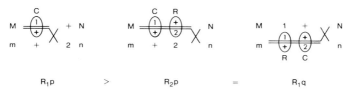

Fig. 7. Position of the crossover. Two hypotheses on the position of the crossover in a recombination event are compared using a model of mismatch repair of heteroduplex. The double line represents heteroduplex in a gene represented by alleles *1* and *2*. M/m and N/n are flanking markers. C means conversion and R means restoration. The resultant tetrads are classified according to the scheme in Fig. 6. The numbers reported in the text fit hypothesis b much better than hypothesis a.

give R_2p as a minor class. Reciprocal recombinants would not occur, since the crossover is not within the gene, and reciprocal conversion cannot occur without symmetrical heteroduplex.

Figure 7b shows that the consequence of placing the crossover at the variable end of the heteroduplex would be that when R_1p is common, resulting from events entering from the left, R_2p and R_1q would be equally common minority classes. On this configuration, R_2q does not occur, though it can occur if a few events enter from the right. R_1r tetrads are expected on this configuration, having the same origin as R_1p, but with restoration instead of conversion. The observations for the cross of *his1-315* × *his1-1*, corrected for unrelated crossovers, are $R_1p = 56$, $R_1q = 16$, $R_2p = 16$, $R_2q = 0$, and $R_1r = 13$. Other crosses give similar results (Savage and Hastings, 1981; Hastings and Savage, 1984). This looks like a very good fit to the expectation of a crossover within the gene and does not seem to conform at all to the hypothesis of a crossover outside the gene, since R_2p and R_1q are approximately equal, while R_2q is a very rare class and R_1r is reasonably common. The ability to provide a simple explanation of the occurrence of R_1r tetrads—intragenic reciprocal events with a crossover apparently between the alleles—is a major advantage of the view taken here that the crossover is within the gene.

3. Restoration in Yeast

For yeast, as for the heterologies in the *b2* locus, there appears to be less restoration than conversion. In this case, however, the evidence is indirect and is dependent on the interpretation of observations which gave the description of a recombination event shown in Fig. 3 and the position of the crossover, discussed in the preceding section. Figure 8 shows that there are cases where the same basic event has a different outcome in a tetrad depending on whether mismatch correction is by conversion or by restoration. These comparisons give rise to two methods for calculation of the relative amount of restoration and conversion. In method 1 (Fig. 8), the short events with heteroduplex covering the left-hand allele and ending in a crossover within the gene lead to a prototrophic product with 3:1 segregation of the left-hand allele when the left-hand allele is converted. Prototrophic products can be obtained in two ways when the left-hand allele is restored, and this shows up as reciprocal intragenic recombination. In method 2, two classes of crossover events with long heteroduplex are compared. These classes of tetrad include prototrophic meiotic products because independent correction tracts have occurred in the same length of heteroduplex. If one of the alleles is held common to all crosses, the ratio of the two classes varies as the ratio of conversion to restoration for the other allele.

Values obtained by the two methods are significantly correlated and show, for alleles of *his1*, an apparently allele-specific excess of conversion over restoration, with values ranging from a 3-fold excess to a 14-fold excess of conversion

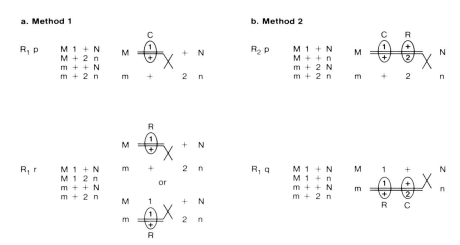

Fig. 8. Two methods for the measurement of restoration in yeast. Tetrads are classified according to the scheme in Fig. 6, and the diagrams use the conventions described in Fig. 7. See text for explanation. Reprinted from Hastings (1984).

(Hastings, 1984). A system for direct measurement of restoration in yeast, comparable to the system in *Ascobolus*, is now being developed. Direct evidence on the amount of restoration in yeast is urgently needed because of the theoretical significance of the phenomenon.

D. Mappability and Map Expansion

The theory of intragenic mapping on the heteroduplex repair model of conversion was developed by Fincham and Holliday (1970) to explain fine-structure map expansion. Slight modification to their postulates shows that the heteroduplex repair model can account for the variability found between loci in recombination frequency, mappability, flanking marker (R_1:R_2) mappability, map expansion, and polarity (Hastings, 1975).

Fincham and Holliday (1970) postulated that excision tracts of constant length (x) were substantially shorter than heteroduplex tracts of constant length (h), but of randomly variable position. Recombination between alleles will occur when the length of heteroduplex ends between them. This will have a probability which will depend on the distance between the markers, generating a strictly additive map. Recombination for alleles which are farther apart than the length of excision, x, will also occur when heteroduplex correction occurs in opposite directions. Thus, the recombination frequency will be expanded for distances exceeding x relative to distances less than x. This provides a reasonable description of the pattern of map additivity and expansion observed in several loci from diverse organisms.

The modification proposed by Hastings (1975) was the suggestion that the ratio of excision lengths to heteroduplex lengths is a locus-specific parameter. This explains the secondary parameters as follows: (1) Recombination frequency mapping will give an additive map when only h endings occur, an expanding map when both h and x endings occur, and little or no mappability when all recombination results from x endings, assuming that the occurrence of x endings has little distance dependence. (2) Flanking marker mappability also depends on h endings. Independent correction tracts in the same length of heteroduplex will give R_2 equal to R_1 (see, for example, Fig. 8). Heteroduplex endings between the alleles give only R_1 recombinants, causing R_1 to exceed R_2 when h endings are occurring. (3) If polarity is explained by polarized formation of heteroduplex, having one end outside the gene at the high conversion end and the other end of the heteroduplex at a variable position within the gene (Hastings and Whitehouse, 1964), then polarity is also dependent on h endings. Absolute polarity will be seen when all recombination results from h endings, less than absolute polarity when both h and x endings occur, and no polarity when there are only x endings occurring in a locus, since all markers will have the same opportunity to be converted.

TABLE III

Explanation of Variation in Recombination Parameters between Genes in Terms of the Relative Lengths of Heteroduplex, Excision Tracts, and the Distance between Markers

	Relationship of lengths		
	$x > h < d$ [a]	$h > d$	$x \leqq h$
Recombination results from:	h endings only	x endings only	Both h and x endings
Locus mappable by R_1/R_2:	Yes (no R_2)	No ($R_1 = R_2$)	Yes ($R_1 > R_2$)
Locus mappable by recombination frequency:	Yes	No	Yes
Additivity:	Additive map	No map	Map expansion
Polarity:	Yes (absolute)	No	Yes (relative)
Examples:	*Ascobolus*, Series 19A[b] *Ascobolus*, Series 46[b]	Yeast, *SUP6*[c] *Neurospora*, *mtr*[d]	*Ascobolus*, *b2*[e] *Neurospora*, *me-2*[f] *Neurospora*, *pan-2*[g] Yeast, *his1*[h]

[a] x = length of excision tract; h = length of heteroduplex; d = physical length of DNA between markers.
[b] Lissouba *et al.* (1962).
[c] DiCaprio and Hastings (1976).
[d] Stadler and Kariya (1969).
[e] Rossignol *et al.* (1984a).
[f] Murray (1963).
[g] Case and Giles (1959, 1964).
[h] Fogel and Hurst (1967); Savage and Hastings (1981).

Table III summarizes the properties of loci with different ratios of h to x and to d (the distance between markers) and gives examples of loci which fit the descriptions. Only one locus, *am-1* in *Neurospora* (Fincham, 1974), is in conflict with this scheme. This locus appears to have long heteroduplex in crossover events, giving difficulty with mappability, while it has short heteroduplex in non-crossover events, giving polarity by the $P_1:P_2$ method.

E. The Status of the Heteroduplex Repair Hypothesis of Conversion

The preceding sections describe results which seem to leave little room for doubt that conversion, at least in *Ascobolus*, occurs by excision repair of heterozygosity in heteroduplex. The evidence says that there was a heteroduplex intermediate which is resolved to give conversion. Many more detailed points leading to the same general conclusion have been reviewed by Rossignol *et al.* (1986). The discussion above is not, however, limited to data from *Ascobolus*. It seems

4. Meiotic Recombination as Heteroduplex Correction

likely that what we understand for *Ascobolus* can be applied quite easily to *Neurospora* and *Sordaria*.

For yeast, the evidence for a heteroduplex intermediate appears to be the same as the evidence for a heteroduplex intermediate in *Ascobolus*. However, the evidence that the heteroduplex is resolved by excision repair to give conversion does not fit in the expected way. The problems, as described above, are (1) parity in conversion to mutant and to wild type, (2) lack of evidence that the spectrum of aberrant ratios responds to the nature of the heterozygosity, and (3) evidence that conversion may be substantially in excess of restoration.

Rossignol *et al.* (1986) present evidence that conversion of deletions in *Ascobolus* is like the conversion of normal alleles of yeast. Deletions in *Ascobolus* are converted with a frequency which is about two-thirds the frequency of conversion of point mutations in the same part of the map. The characteristics of the conversion spectra of deletions in *Ascobolus* are, like point mutations in yeast, that (1) they show parity, (2) PMS is rare, (3) events appear to be asymmetrical, and (4) there is much less restoration than conversion.

Those authors suggest that yeast systems react to almost all heterozygosity in the way that *Ascobolus* systems react to heterology. They conclude that there are at least two mechanisms of conversion. The hypothesis of excision repair of mismatches in heteroduplex DNA is satisfactory for most markers in *Ascobolus*, but unsatisfactory for conversion of heterologies in *Ascobolus* and for most or all conversion in yeast. We must, therefore, seek an alternative hypothesis.

IV. ALTERNATIVES TO EXCISION REPAIR OF HETERODUPLEX

There have almost always been available to us models of recombination which achieve conversion in meiosis without involving extensive heteroduplex. In these models, various devices are used to give the net effect that a segment of one homologue was replicated twice and the other homologue not at all. The copy-choice models which preceded heteroduplex models did this by supposing that replication was conservative and could switch templates.

The models of Stahl (1969) and Paszewski (1970) are interesting to study as examples of different ways to achieve net overreplication. Both authors appear to have offered their model, in part, to show us that we could not assume that heteroduplex repair was the only mechanism of conversion. More recently, Szostak *et al.* (1983) proposed a model based on the mechanism of repair of double-strand gaps in DNA. Also in this category, we should include the use of overreplication for special purposes such as the conversion of a heterology. Such a mechanism was offered recently by Rossignol *et al.* (1986).

A. Conversion by Double-Strand Gap Filling

Saccharomyces cerevisiae has an efficient mechanism for the repair of double-strand breaks (Resnick, 1975; Ho and Mortimer, 1975). We now know that this double-strand break repair mechanism can fill in the missing information in a double-strand gap and that it does so by putting in a copy of homologous information. Here, we are concerned with the implications of this mechanism for the understanding of meiotic conversion.

The key observations, reported by Orr-Weaver *et al.* (1981), are that a plasmid with a portion of a gene removed by restriction cuts can be used to transform yeast with the same efficiency as a plasmid which was cut only once in the same gene, and that, in this process of transformation, the gap is filled with a copy of the homologous information which was present in the transformed cell. This process is clearly analogous to meiotic conversion as defined here; the plasmid has acquired the genotype of a homologue. The crucial point is, of course, that there cannot have been a heteroduplex precursor to this conversion in the region where the plasmid was gapped before transformation.

Szostak *et al.* (1983) suggested that meiotic recombination is initiated by the formation of a double-strand cut. The cut is enlarged to give a double-strand gap. The gap is then filled by the 3' nucleotide chain ends at either end of the gap invading the homologue, giving short lengths of heteroduplex. Repair replication using both 3' ends as primers, and both chains of the homologue as templates, replaces the length of DNA lost when the gap was formed. This will produce a non-Mendelian ratio for any marker within the gap, and so be seen as conversion.

Those authors point out that such a model accounts for the peculiarities of yeast meiotic recombination listed above. Such conversion is not a response to heterozygosity and is, therefore, not expected to vary according to the nature of the heterozygosity. Parity in conversion in the two directions simply reflects an equal probability of forming the primary gap on chromatids of the two homologues. The model requires that we find an excess of conversion over restoration, since the gap-filling mechanism gives only conversion. The low levels of PMS and of restoration which are seen reflect the frequency with which a marker falls into the heteroduplex regions flanking the gap. A further point is that the gap-filling mechanism is necessarily asymmetrical, thus explaining why we find no evidence of symmetrical heteroduplex in yeast.

On the other hand, the gap-filling model of Szostak *et al.* (1983) does not provide a satisfactory explanation for the existence in yeast of alleles which give a high frequency of PMS. Nor can it explain the experiments cited above which show high PMS alleles being co-converted with normal markers nearby. Both observations seem to indicate that most conversion in yeast had a heteroduplex precursor (Hastings, 1984).

B. Modified Excision Processes

We are saying, then, that there is a heteroduplex precursor to conversion, but that, in yeast especially, the data do not seem to accord with the classical style of excision repair of mismatches in heteroduplex. Two suggestions are now available of ways in which the excision process may be modified to explain the discrepancies.

The first suggestion (Savage and Hastings, 1981) is that the excision repair system recognizes which is the donor material and which is the recipient material in the formation of a heteroduplex molecule. If there is then a preference for excision of the recipient information, the donor chain will most frequently act as the template, giving conversion, so that conversion will be seen to exceed restoration. In these circumstances there is little opportunity for the nature of the mismatch to influence the conversion spectrum, since a disparity in direction of conversion is not expected. The mismatch would still be expected to influence the relative amount of PMS to conversion. This hypothesis suffers from lack of a known mechanism to allow this discrimination, although the converse situation has been shown to pertain in *Streptococcus* transformation; the donor information is preferentially excised, giving a low efficiency of transformation when the mismatch is recognized with high efficiency (Claverys *et al.*, 1983).

The second suggestion (Hastings, 1984) is developed from the idea of Szostak *et al.* (1983) that the double-strand gap-filling process described by Orr-Weaver *et al.* (1981) may play a role in meiotic conversion. It is postulated that yeast may have an excision system which may cut one chain at a mismatch or it may cut both nucleotide chains. If one chain is cut, the other chain acts as a template to fill the single-strand gap, giving an equal probability of restoration and conversion. If both chains are cut, the double-strand cut will be enlarged to form a double-strand gap. This gap will be filled by replication of both chains of a homologue. This gap-filling process will give only conversion. It cannot lead to restoration, since all resident information has been removed in the region of the gap. If heteroduplex forms equally on chromatids of the two homologues, conversion by the gap-filling process will show parity in conversion to wild type and conversion to mutant. The net result of the mixture of single- and double-strand cuts will be an excess of conversion over restoration. The postulate that the gap-filling mechanism of Orr-Weaver *et al.* (1981) may be used in this way has most of the advantages for explaining meiotic conversion in yeast that the Szostak *et al.* (1983) model has. But this suggestion has the further advantage that it explains all the evidence that conversion had a heteroduplex precursor. However, the hypothesis has some difficulties. These may be caused by lack of information about the properties of enzymes which could be involved.

One difficulty is that the double-strand-cutting hypothesis of mismatch repair appears to predict that the ratio of PMS to conversion will be allele-specific, as it would be if only one chain were excised, since this ratio depends on the probability of the mismatch being recognized, which should depend on the nature of the mismatch. The observation in yeast is that there is a fairly constant ratio of PMS to conversion, except for the few alleles showing a high relative frequency of PMS.

A second difficulty stems from the nature of the eukaryotic enzyme which has been reported to cut DNA at a mismatch (Pukkila, 1978; Holliday *et al.*, 1979). These authors saw that nuclease α from *Ustilago maydis* made a mixture of single- and double-strand cuts at some, but not all, mismatches in bacteriophage φ80 DNA. The kinetics of the reaction revealed that the double-strand cuts were made sequentially. Nuclease α is a single-strand binding endo-exonuclease and, if this is the class of enzyme which is responsible for mismatch repair in eukaryotes, such enzymes are presumably responding to local melting caused by reduced hydrogen bonding at the mismatch. One might then expect both DNA chains to be cut, either because both chains are single stranded in the melted region or because one excision tract exposes a single-stranded region on the other chain. The difficulty is that neither situation seems to leave room for the proportion of single- to double-strand cuts to be allele-specific. This is true especially with sequential cutting, since the mismatch is no longer present once excision has occurred on one chain. So the hypothesis of a mixture of single- and double-strand cuts does not seem to offer an explanation of an allele-specific ratio of conversion to restoration. The limited data implying that this ratio is allele-specific were reported above. It is too early to try to resolve these problems because biochemical work on mismatch repair is only just beginning, and because there is still room for doubt about the interpretation of genetic data on recombination in yeast.

C. Is There Any Restoration at All?

An interesting point was made recently by A. Campbell (personal communication). He pointed out that it is difficult to justify an assumption underlying the foregoing discussion—that only two chromatids are involved in a meiotic recombination event. If two non-sister chromatids are involved in an asymmetrical heteroduplex formation, one as donor and the other as recipient, the heteroduplex chromatid may then be enzymatically gapped, as discussed above. The DNA molecule which is copied to fill the gap may not be the original donor in the heteroduplex formation and, in fact, it could be the sister to the chromatid on which the heteroduplex formed. In this case we would see the parental genotype of that chromatid restored, but it would occur by conversion to the genotype of the sister rather than by restoration of the recipient chromatid genotype.

4. Meiotic Recombination as Heteroduplex Correction

By this argument, there may be no such thing as restoration as described here. If the gap were to be filled from one of the other three chromatids chosen at random, we would expect a 2:1 ratio of conversion to apparent restoration. A corollary of this is that an organism which has symmetrical heteroduplex (such as *Ascobolus*) would have only two chromatids left to act as donors in gap filling, one homologue and one sister, so that a random choice would give equality of conversion to apparent restoration.

These thoughts are introduced to illustrate the extent of our lack of hard information and the extent to which the whole story being related here depends on presumption and interpretation. Although the story is coherent, it may not necessarily be closely related to what is happening.

ACKNOWLEDGMENTS

I am grateful to the many people who have read and commented on the manuscript for me. I also wish to thank Anne Galloway for assistance with the figures and Dorothy Woslyng for the typing. This work is supported by an operating grant from the Natural Sciences and Engineering Research Council of Canada.

REFERENCES

Case, M. E., and Giles, N. H. (1959). Recombination mechanisms at the *pan-2* locus in *Neurospora crassa*. *Cold Spring Harbor Symp. Quant. Biol.* **23,** 119–135.

Case, M. E., and Giles, N. H. (1964). Allelic recombination in *Neurospora:* Tetrad analysis of a three-point cross within the *pan-2* locus. *Genetics* **49,** 529–540.

Claverys, J.-P., Mejean, V., Gasc, A.-M., and Sicard, A. M. (1983). Mismatch repair in *Streptococcus pneumoniae:* Relationship between base mismatches and transformation efficiencies. *Proc. Natl. Acad. Sci. U.S.A.* **80,** 5956–5960.

DiCaprio, L., and Hastings, P. J. (1976). Gene conversion and intragenic recombination at the *SUP6* locus and the surrounding region in *Saccharomyces cerevisiae*. *Genetics* **84,** 697–721.

Emerson, S. (1966). Quantitative implications of the DNA-repair model of gene conversion. *Genetics* **53,** 475–485.

Emerson, S. (1969). Linkage and recombination at the chromosome level. *In* "Genetic Organization" (E. W. Caspari and A. W. Ravin, eds.), Vol. 1, pp. 267–360. Academic Press, New York.

Fincham, J. R. S. (1974). Negative interference and the use of flanking markers in fine-structure mapping in fungi. *Heredity* **33,** 116–121.

Fincham, J. R. S., and Holliday, R. (1970). An explanation of fine structure map expansion in terms of excision repair. *Mol. Gen. Genet.* **109,** 309–322.

Fogel, S., and Hurst, D. D. (1967). Meiotic gene conversion in yeast tetrads and the theory of recombination. *Genetics* **57,** 455–481.

Fogel, S., Mortimer, R., Lusnak, K., and Tavares, F. (1979). Meiotic gene conversion: A signal of the basic recombination event in yeast. *Cold Spring Harbor Symp. Quant. Biol.* **43,** 1325–1341.

Fogel, S., Mortimer, R. K., and Lusnak, K. (1981). Mechanisms of meiotic gene conversion or wanderings on a foreign strand. In "Molecular Biology of the Yeast *Saccharomyces*: Life Cycle and Inheritance" (J. N. Strathearn, E. W. Jones, and J. R. Broach, eds.), pp. 289–339. Cold Spring Harbor Lab., Cold Spring Harbor, New York.

Gutz, H. (1971). Gene conversion: Remarks on the quantitative implications of hybrid DNA models. *Genet. Res.* **17**, 45–52.

Hastings, P. J. (1975). Some aspects of recombination in eukaryotic organisms. *Annu. Rev. Genet.* **9**, 129–144.

Hastings, P. J. (1984). Measurement of restoration and conversion: Its meaning for the mismatch repair hypothesis of conversion. *Cold Spring Harbor Symp. Quant. Biol.* **49**, 49–53.

Hastings, P. J., and Savage, E. A. (1984). Further evidence of a disparity between conversion and restoration in the *his1* locus of *Saccharomyces cerevisiae*. *Curr. Genet.* **8**, 23–28.

Hastings, P. J., and Whitehouse, H. L. K. (1964). A polaron model of genetic recombination by the formation of hybrid deoxyribonucleic acid. *Nature (London)* **201**, 1052–1054.

Hastings, P. J., Kalogeropoulos, A., and Rossignol, J.-L. (1980). Restoration to the parental genotype of mismatches formed in recombinant DNA heteroduplex. *Curr. Genet.* **2**, 169–174.

Ho, K., and Mortimer, R. K. (1975). X-ray induced dominant lethality and chromosome breakage and repair in a radiosensitive strain of yeast. In "Molecular Mechanisms for Repair of DNA" (P. Hanawalt and R. B. Setlow, eds.), Part B, pp. 545–547. Plenum, New York.

Holliday, R. (1962). Mutation and replication in *Ustilago maydis*. *Genet. Res.* **3**, 472–486.

Holliday, R. (1964). A mechanism for gene conversion in fungi. *Genet. Res.* **5**, 282–304.

Holliday, R. (1968). Genetic recombination in fungi. In "Replication and Recombination of Genetic Material" (W. J. Peacock and R. B. Brock, eds.), pp. 157–174. Aust. Acad. Sci., Canberra.

Holliday, R., Pukkila, P. J., Dickson, J. M., Spanos, A., and Murray, V. (1979). Relationships between the correction of mismatched bases in DNA and mutability. *Cold Spring Harb. Symp. Quant. Biol.* **43**, 1317–1323.

Kalogeropoulos, A., and Rossignol, L.-L. (1980). Evidence for independent mismatch corrections along the same hybrid DNA tract during meiotic recombination in *Ascobolus*. *Heredity* **45**, 263–270.

Leblon, G. (1972a). Mechanism of gene conversion in *Ascobolus immersus*. I. Existence of a correlation between the origin of mutants induced by different mutagens and their conversion spectrum. *Mol. Gen. Genet.* **115**, 36–48.

Leblon, G. (1972b). Mechanism of gene conversion in *Ascobolus immersus*. II. The relationships between the genetic alterations in b_1 or b_2 mutants and their conversion spectrum. *Mol. Gen. Genet.* **116**, 322–335.

Leblon, G., and Rossignol, J.-L. (1973). Mechanism of gene conversion in *Ascobolus immersus*. III. The interaction of heteroalleles in the conversion process. *Mol. Gen. Genet.* **122**, 165–182.

Lissouba, P., Mousseau, J., Rizet, G., and Rossignol, J.-L. (1962). Fine structure of genes in the ascomycete *Ascobolus immersus*. *Adv. Genet.* **11**, 343–380.

Meselson, M. S. (1965). The duplication and recombination of genes. In "New Ideas in Biology" (J. A. Moore, ed.), pp. 3–16. Natural History Press, New York.

Moore, C. W., and Sherman, F. (1975). Role of DNA sequences in genetic recombination in the iso-1-cytochrome *c* gene of yeast. I. Discrepancies between physical distances and genetic distances determined by five mapping procedures. *Genetics* **79**, 397–418.

Mortimer, R. K., and Fogel, S. (1974). Genetic interference and gene conversion. *Mech. Recomb.* [*Proc. Biol. Div. Res. Conf.*], 27th, 1974, pp. 263–275.

Murray, N. E. (1963). Polarized recombination and fine structure within the *me-2* gene of *Neurospora crass*. *Genetics* **48**, 1163–1183.

Nicolas, A. (1978). Carte génétique et fréquence de conversion dans le génome d'*Ascobolus immersus*. Thése Doctorat, Université Paris-Sud, Centre d'Orsay, France.

Orr-Weaver, T. L., Szostak, J. W., and Rothstein, R. J. (1981). Yeast transformation: A model system for the study of recombination. *Proc. Natl. Acad. Sci. U.S.A.* **78,** 6354–6358.

Paquette, N., and Rossignol, J.-L. (1978). Gene conversion spectrum of 15 mutants giving postmeiotic segregation in the *b2* locus of *Ascobolus immersus*. *Mol. Gen. Genet.* **163,** 313–326.

Paszewski, A. (1970). Gene conversion: Observations on the DNA hybrid models. *Genet. Res.* **15,** 55–64.

Pritchard, R. H. (1955). The linear arrangement of a series of alleles of *Aspergillus nidulans*. *Heredity* **9,** 343–371.

Pukkila, P. J. (1978). The recognition of mismatched base pairs in DNA by DNase I from *Ustilago maydis*. *Mol. Gen. Genet.* **161,** 245–250.

Resnick, M. A. (1975). The repair of double-strand breaks in chromosomal DNA of yeast. *In* "Molecular Mechanisms for Repair of DNA" (P. Hanawalt and R. B. Setlow, eds.), Part B, pp. 549–556. Plenum, New York.

Rizet, G., and Rossignol, J.-L. (1966). Sur la dimension probable des échanges réciproques au sein d'un locus complexe d'*Ascobolus immersus*. *C. R. Hebd. Seances Acad. Sci., Ser. D* **262,** 1250–1253.

Rossignol, J.-L. Paquette, N., and Nicolas, A. (1979). Aberrant 4:4 asci, disparity in the direction of conversion, and frequencies of conversion in *Ascobolus immersus*. *Cold Spring Harbor Symp. Quant. Biol.* **43,** 1343–1352.

Rossignol, J.-L., Nicolas, A., Hamza, H., and Langin, T. (1984). Origins of gene conversion and reciprocal exchange in *Ascobolus*. *Cold Spring Harbor Symp. Quant. Biol.* **49,** 13–21.

Rossignol, J.-L., Nicolas, A., Hamza, H., and Kalogeropoulos, A. (1986). Recombination and gene conversion in *Ascobolus*. *In* "The Recombination of Genetic Material" (B. Low, ed.), Academic Press, New York. (In press).

Savage, E. A. (1979). A comparative analysis of recombination at the *his1* locus among five related diploid strains of *Saccharomyces cerevisiae*. Ph.D. Thesis, University of Alberta, Edmonton, Alberta, Canada.

Savage, E. A., and Hastings, P. J. (1981). Marker effects and the nature of the recombination event at the *his1* locus of *Saccharomyces cerevisiae*. *Curr. Genet.* **3,** 37–47.

Stadler, D. R. (1959). The relationship of gene conversion to crossing-over in *Neurospora*. *Proc. Natl. Acad. Sci. U.S.A.* **45,** 1625–1629.

Stadler, D. R., and Kariya, B. (1969). Intragenic recombination at the *mtr* locus of *Neurospora* with segregation at an unselected site. *Genetics* **63,** 291–316.

Stahl, F. W. (1969). One way to think about gene conversion. *Genetics* **61,** Suppl., 1–13.

Strickland, W. N. (1958). An analysis of interference in *Aspergillus nidulans*. *Proc. R. Soc. London, Sec. B* **149,** 82–101.

Szostak, J. W., Orr-Weaver, T. L., Rothstein, R. J., and Stahl, F. W. (1983). The double-strand-break repair model for recombination. *Cell* **33,** 25–35.

Thuriaux, P., Minet, M., Munz, P., Ahmad, A., Zbaeren, D., and Leupold, U. (1980). Gene conversion in nonsense suppressors of *Schizosaccharomyces pombe*. II. Specific marker effects. *Curr. Genet.* **1,** 89–95.

Whitehouse, H. L. K. (1963). A theory of crossing-over by means of hybrid deoxyribonucleic acid. *Nature (London)* **199,** 1034–1040.

Whitehouse, H. L. K. (1982). "Genetic Recombination: Understanding the Mechanisms." Wiley, New York.

Whitehouse, H. L. K., and Hastings, P. J. (1965). The analysis of genetic recombination on the polaron hybrid DNA model. *Genet. Res.* **6,** 27–92.

Yu-Sun, C. C., Wickramaratne, M. R. T., and Whitehouse, H. L. K. (1977). Mutagen specificity of conversion pattern in *Sordaria brevicollis*. *Genet. Res.* **29,** 65–81.

5

Models of Heteroduplex Formation

P. J. HASTINGS

Department of Genetics
University of Alberta
Edmonton, Alberta, Canada T6G 2E9

I. INTRODUCTION

This chapter provides a very partial coverage of the wealth of schemes available for achieving a recombination event involving heteroduplex. In fact, very few models describe the distribution of heteroduplex as it is seen in the $b2$ locus of *Ascobolus*—asymmetrical at the high conversion end and symmetrical at the low conversion end in the same event. The possibilities can be narrowed even more by applying the knowledge that the event moves from the high conversion end toward the low conversion end (i.e., the asymmetrical phase comes first) and that the asymmetrical phase and the symmetrical phase are propagated differently. But we must remember that the evidence seems to be telling us that we are looking at only a part of a recombination event, since part of it—conceivably a large part—is outside the gene where we cannot study it.

I shall review the limited evidence on the direction of travel and then describe only the main stream of recombination models leading to the style of scheme which is actively considered today. Then I shall attempt to reconcile the models with some outstanding problems, notably the crossover/non-crossover decision and crossover position interference. Finally, there is a discussion of the mounting evidence that crossovers can occur alone, without a heteroduplex tract.

II. DIRECTION OF TRAVEL OF THE EVENT

We first learned that a recombination event travels on one chromatid before involving two chromatids from the work of Catcheside and Angel (1974). Their study involved the selection of prototrophic ascospores from crosses in the *his-3* locus of *Neurospora* where one of the *his-3* mutants in each cross was the breakpoint of a translocation. Consequently, one chromatid in each event was discontinuous. The work also used a system of genetic control of recombination which apparently allowed the authors to control whether any events were initiated on the continuous chromatid. When initiation occurred only on the discontinuous chromatid, no events were found which had passed the break-point. When initiation was allowed on the continuous chromatid, events, including crossover events, were found to involve the region beyond the break-point. The interpretation is that a recombination event is initiated on one chromatid and can travel on that chromatid without involving the other, only later to involve two chromatids. Although this falls short of telling us that the asymmetrical phase is formed first, it is consistent with it. Incidentally, these data also seem to tell us that it is the chromatid on which initiation occurs which is converted.

The findings from the *b2* locus of *Ascobolus* on this point are more clearly defined. Hamza *et al.* (1981) studied the effects of including a deletion in the middle of the locus. The presence of the deletion had no effect on the segregation of alleles to the left (the high conversion side). Heterozygosity for the deletion reduced aberrant 4:4 segregations to near zero for markers to the right. It also reduced the disparity in the direction of conversion. Both these observations tell us that the heteroduplex is asymmetrical in the presence of heterozygosity for the deletion, while it is often symmetrical when homozygous for deletion or for wild type. Application of the algebra of Paquette and Rossignol (1978) (discussed in Chapter 4) confirms that no symmetrical heteroduplex occurs to the right of the heterology. There is an increase in the frequency of asymmetrical heteroduplex to the right of the heterology, which almost compensates for the loss of symmetrical heteroduplex. This observation will be discussed later. These data are interpreted to mean that (1) the high conversion end of the locus is the site of asymmetrical initiation of the recombination event, (2) the asymmetrical phase is transformed into the symmetrical phase, and (3) the mechanisms of propagation of the asymmetrical and the symmetrical phases are different.

III. MODELS OF HETERODUPLEX FORMATION

A. The Holliday Model

The basic model of recombination by heteroduplex formation and repair is that of Holliday (1964). This model, shown in Fig. 1, is as follows: Homologous chromatids are cut at identical sites (Fig. 1a); the cut chains unwind (Fig. 1b) and may pair with the complementary chains of the homologues (Fig. 1c). The

5. Models of Heteroduplex Formation

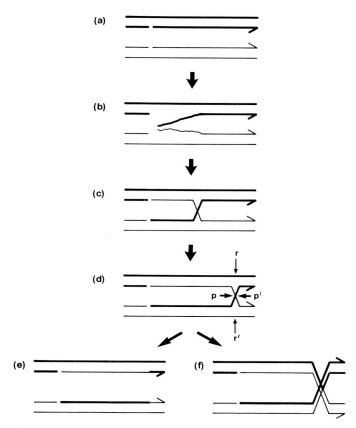

Fig. 1. The Holliday model (Holliday, 1964). See text for commentary.

structure formed by this process may travel in either direction. Travel away from the initiation sites will form a length of symmetrical heteroduplex (Fig. 1d). Eventually the structure will be cut endonucleolytically, either at p and p' or at r and r' (Fig. 1d), to give either a non-crossover (Fig. 1e) or a crossover (Fig. 1f) outcome.

The structure formed by the reciprocal exchange of nucleotide chains is a fundamental concept of recombination theory. It is known as a cross-strand exchange or Holliday structure. It has been seen by electron microscopy *in vitro* (see Potter and Dressler, 1979) and *in vivo* in yeast meiosis (Bell and Byers, 1979). Meselson (1972) has calculated that the structure will travel in either direction by rotary diffusion. Sigal and Alberts (1972) have constructed models of the cross-strand exchange. They found that it can be formed with all bases paired and will be in a state of rapid isomerization such that the crossover and non-crossover configurations are occurring equally commonly.

The heteroduplex formed by this mechanism is necessarily symmetrical. Holliday's model could accommodate the observations implying asymmetrical heteroduplex if correction had occurred at the mismatch on one chromatid, and not on the other, when back-migration of the cross-strand exchange shortened the lengths of heteroduplex.

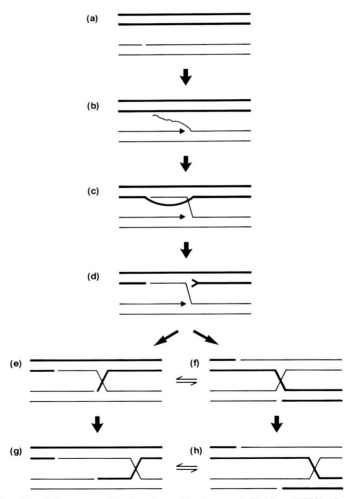

Fig. 2. The Aviemore model, based on Meselson and Radding (1975). See text for commentary.

B. The Aviemore Model

Meselson and Radding (1975) proposed a model to account for the preponderance of asymmetrical events observed in yeast. The Meselson and Radding model is often referred to as the Aviemore model after the site of the meeting where these authors created it. The model is shown in Fig. 2. The process is as follows: One homologous chromatid is nicked (Fig. 2a), allowing entry of a polymerase which, moving in the 5' direction, displaces a length of nucleotide chain (Fig. 2b). The displaced strand invades a homologous chromatid, forming a D-loop (Fig. 2c). This invasion induces a nick, and the chain displaced by invasion is removed by exonuclease action, giving us Fig. 2d. Concerted action of the polymerase and exonuclease causes extension of the length of heteroduplex on one chromatid. This process is driven by the energy of polymerization. Polymerization eventually dies out and the polymerase falls off. This structure may then drift into a cross-strand exchange in the parental configuration (Fig. 2e) in a process which has not been defined, or it may isomerize to give a cross-strand exchange in the crossover configuration (Fig. 2f). In either case, diffusion of the cross-strand exchange away from the origin of the event will propagate lengths of symmetrical heteroduplex (Figs. 2g and h). The cross-strand exchange is subject to Sigal and Alberts's isomerization and endonucleolytic resolution in either isomer to give equality of crossover and non-crossover outcomes, as on the Holliday model in Figs. 1e and f. Mismatches anywhere in the heteroduplex are subject to base-sequence-directed mismatch repair.

Thus, the Meselson and Radding model is a driven formation of asymmetrical heteroduplex which leads into the passively extended Holliday model. If we presume that the initiation is outside the gene at the high conversion end, then we see that the event enters the gene asymmetrically, traveling toward the low conversion end, and is transformed into the symmetric phase, the movement of which toward the low conversion end is by a different mechanism. The crossover, if any, will be at the variable, low conversion end of the heteroduplex lengths.

C. Some Perspective on the Aviemore Model

The process described by Meselson and Radding (1975) is remarkably parallel to the picture of the recombination event in the *b2* locus of *Ascobolus* that emerges from the work of Paquette and Rossignol (1978), Rossignol and Haedens (1980), and Hamza *et al.* (1981), except that the position of the crossover is not defined in *b2*. It was discussed in Chapter 4 that the crossover was placed at the downstream end of the heteroduplex by Savage and Hastings (1981). The observations of the crossover being on either side of a site showing

PMS was explained by these authors by postulating that the cross-strand exchange may migrate back toward the origin and, hence, into the asymmetric heteroduplex. Unlike symmetric heteroduplex, asymmetric heteroduplex would not be canceled out by back-migration and, thus, the crossover formed by resolution of the cross-strand exchange may have heteroduplex on its downstream side.

There are, however, aspects of the recombination event which are not adequately described by the Meselson and Radding process. Modifications and substitutions of mechanisms in various parts of the process seem to be able to accommodate most of the objections. The following issues must be addressed: (1) Evidence has been discussed above which seems to show that the chromatid on which initiation occurs is the recipient, not the donor, in the strand-exchange process. We must change the manner of initiation. (2) When Hamza *et al.* (1981) used a heterology in the *b2* locus to block formation of symmetrical heteroduplex downstream of the heterology, they presumably, on the Meselson and Radding process, blocked the diffusion of the cross-strand exchange after polymerase-driven extension of the asymmetric phase had ended, the polymerase had fallen off the DNA, and the first isomerization had occurred. This mechanism does not allow the frequency of asymmetric heteroduplex beyond the heterology to be increased. Yet a large increase was seen. We need a mechanism of transition to, or extension of, the symmetric phase which would allow the asymmetric phase extension to continue if the transition does not occur. (3) The model predicts an equality of crossover and non-crossover events, since the decision depends on the interaction of Sigal and Alberts's isomerization with cutting by the endonuclease responsible for resolution. It was reported in Chapter 4 that the proportion of crossover to non-crossover events is variable and apparently locus-specific. We need to find another mechanism for making the decision as to crossover or non-crossover outcome. A related point is that the Meselson and Radding mechanism (and all other recombination models) offers no mechanism for crossover position interference. We need a mechanism for controlling the crossover–non-crossover decision. (4) Finally, we must study the evidence that seems to tell us that a crossover is a separate, though correlated, event, not merely one possible outcome of the recombination event described here.

D. The *RecA* Model

Radding (1982) has offered a variant mechanism for the initiation step of the Meselson and Radding model which allows the initiation to occur on the chromatid which will become the recipient. The scheme, shown in Fig. 3, is based on observations *in vitro* of the properties of *RecA* protein from *Escherichia coli*. This protein catalyzes the uptake of single-stranded DNA into homologous duplex. One situation in which it can catalyze this reaction is that shown in Fig. 3a,

5. Models of Heteroduplex Formation 145

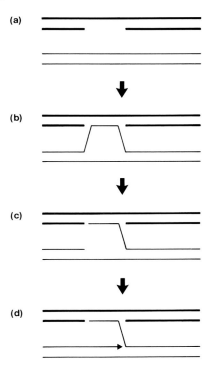

Fig. 3. The *RecA* model version of the early steps of the Aviemore model. Figure 3d corresponds to Fig. 2d and later stages of the two models are the same. Based on Radding (1982). See text for commentary.

where there is a single-strand gap but no free single-stranded end. If initiation occurs as a nick, which is enlarged to a single-strand gap, and this reacts with an intact homologue (Fig. 3b), cutting of the loop shown in Fig. 3b will permit entry of a polymerase and will lead us into the asymmetric phase of the Meselson and Radding process, as shown in Figs. 2d and 3d, except that now the asymmetric heteroduplex will form on the chromatid on which the first nick occurred.

This seems to be a satisfactory solution to the problem of initiation on the recipient chromatid. While Radding's is only one of many possible solutions, it makes it clear that this is not likely to become a serious problem.

E. The Double-Strand Break Repair Model

Szostak *et al.* (1983) proposed that recombination may be initiated by the cutting of both chains of a chromatid, followed by enlargement of the gap.

Invasion of chain ends from the gap into an intact homologue leads into the Meselson and Radding process at either end of the gap. At least the gap-filling part of this process will show the initiating strand to be the recipient in the conversion process. Although arguments given in Chapter 4 make it seem unlikely that double-strand gap filling as a primary event is a major cause of meiotic conversion, even in yeast, analogy with mitotic recombination shows that double-strand gaps in meiosis would, if they occurred, be recombinogenic, and that, therefore, this mode of initiation needs to be borne in mind.

The Szostak *et al.* (1983) model has the initial double-strand gap filled by a mechanism which, like the double-strand break repair model of Resnick (1976), involves the formation of heteroduplex at both ends of the gap. This leads to the formation of a pair of cross-strand exchanges. Now the decision whether the outcome of the event is a crossover or not depends upon whether the resolution of two cross-strand exchanges is the same or different. One need not assume that the behavior of the two structures will be the same. For example, one may be resolved in the same way all the time. It is also possible to consider that the resolution of one structure may influence the resolution of the other. So two cross-strand exchanges give many more possibilities for controlling the outcome than does one cross-strand exchange. One is reminded here of the possibilities which are present in the model of Sobell (1972), in which there is a traveling bubble of symmetrical heteroduplex bounded by two cross-strand exchanges which must migrate together. A recombination event is also flanked by two cross-strand exchanges in the *RecBC* model of Smith *et al.* (1984) described next.

F. The *RecBC* Model

Smith *et al.* (1984) have developed a model which deals with the question of initiation at an earlier step than the *RecA* model. It is based on the study of an *Escherichia coli* nuclease, the *RecBC* enzyme. The early part of the process is shown in Fig. 4. The *RecBC* enzyme enters DNA at an end and travels into the molecule, looping out single nucleotide chains in symmetrical "rabbit's ears." The enzyme cuts one chain when it encounters a specific nucleotide sequence—a chi site. The resulting 3' end, in the presence of *RecA* protein, can invade a homologous molecule (Fig. 4g) to form a D-loop. The 3' end can also act as a primer for polymerization (Fig. 4h), giving extended displacement and a reciprocal invasion. This stage corresponds to the asymmetric phase of the Aviemore model but, like the *RecA* model, has the extensive length of asymmetrical heteroduplex forming on the strand on which initiation occurred. It differs from the *RecA* model in that it uses a free end for invasion, and this free 3' end is the primer for the polymerization responsible for chain displacement. An interesting

5. Models of Heteroduplex Formation

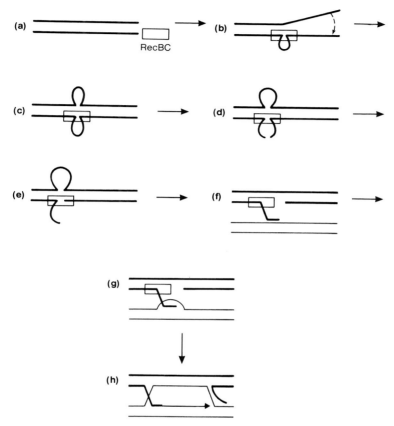

Fig. 4. The *RecBC* model. See text for explanation. Redrawn from Smith *et al.* (1984), with permission.

feature is the short length of symmetrical heteroduplex occurring near the initiation site, possibly evading detection by genetic methods by being outside the gene. A second, downstream, cross-strand exchange now forms in one of the ways proposed by Meselson and Radding (1975) (Fig. 2e). Presumably, this structure could proceed to form symmetric heteroduplex as on the Holliday model, if the D-loop has been cut. Thus, the length of asymmetrical heteroduplex occurs on the chromatid on which initiation occurred and may have symmetrical heteroduplex and cross-strand exchange at both its upstream and its downstream ends.

This scheme appears to have the potential to produce any distribution of heteroduplex which data might require.

IV. CONTROL OF CROSSOVER POSITION BY AN AVIEMORE PROCESS

I shall now try to show that the Aviemore model is capable of explaining a locus-specific inequality of crossovers and non-crossovers and an organism-specific interference between crossovers. The key points in the proposal are: (1) There is a class of recombination event, first described by Stadler and Towe (1971), which stands apart from the Aviemore style of event as seen in the $b2$ locus of *Ascobolus*, in that the events are short, asymmetrical, and non-crossover. (2) These events can be described as incomplete Aviemore events, in which strand displacement has outrun strand uptake. This would prevent the formation of the cross-strand exchange and, hence, give no possibility, by the Aviemore mechanism, of crossover formation. (3) If a factor required for strand uptake is locally limiting, a crossover would reduce the local supply of the factor, reducing the chances of other events nearby becoming complete. (4) A biochemical mechanism for modulating strand uptake has been proposed by Holliday *et al.* (1984). Each of these points will now be expanded upon.

There is a method for determination of the proportions of symmetrical and asymmetrical heteroduplex which applies only to 5:3 and 3:5 segregations in non-crossover events. Figure 5 shows that, in non-crossover events, 5:3 segregations arising from symmetrical heteroduplex involve correction on one chromatid and not on the other. Depending on which heteroduplex is corrected, the tetrad may be either tritype or tetratype, and these two tetrad classes are expected to be equal. If, however, the heteroduplex is formed asymmetrically, 5:3 segregation occurs when it is not corrected, and only tritype tetrads will result. Thus, the proportion of non-crossover 5:3 or 3:5 segregations which arises from symmetrical heteroduplex is twice the proportion of 5:3 or 3:5 tetrads which is tetratype. Using this method, Stadler and Towe (1971) found that non-crossover 5:3 and 3:5 tetrads were almost exclusively asymmetrical in the *w17* locus of *Ascobolus immersus*. Those authors suggested that all symmetrical events were resolved as crossovers and that non-crossover events were asymmetrical events.

The rarity of symmetrical heteroduplex in non-crossover 5:3 tetrads was also seen in *Sordaria brevicollis*, at the *buff* locus, by Sang and Whitehouse (1979). In *Sordaria fimicola*, at the *gray* locus, there is considerable symmetrical heteroduplex in 5:3 segregations (Kitani and Whitehouse, 1974), but aberrant 4:4 tetrads, which are necessarily symmetrical, show more crossing-over than do 5:3 tetrads. This last observation, which has been extended by Nicolas (1982) using data for the *b8* locus in *Ascobolus*, seems to mean that many of the 5:3 tetrads which we see as asymmetrical do not become symmetrical further on in the way which was reported for events in the *b2* locus by Rossignol and Haedens (1980). If all asymmetrical heteroduplex were the first phase of a complete Aviemore event, 5:3 tetrads should not differ from aberrant 4:4 tetrads in the overall outcome of the event. Nicolas calculated the asymmetry of 5:3 tetrads both by

5. Models of Heteroduplex Formation

from $\frac{M\ 1\ N}{m\ +\ n}$

Asymmetrical heteroduplex:

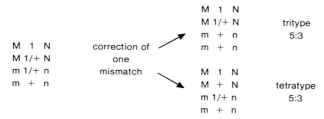

Symmetrical heteroduplex:

Fig. 5. Determination of the proportions of symmetrical and asymmetrical heteroduplex from the proportion of tritype and tetratype 5:3 tetrads. The method is explained in the text.

the tritype:tetratype method and by using the algebra of Paquette and Rossignol (1978). He obtained the same result by the two methods and, in addition, was able to use the algebraic method on the 6:2 tetrads, finding that they too have a low frequency of symmetric heteroduplex correlated with a low frequency of associated crossing-over.

It is possible, in theory, to have a complete Aviemore event without symmetrical heteroduplex, since the cross-strand exchange may diffuse backward into the asymmetric phase, yet still be resolved to form a crossover. This would seem to provide a satisfactory explanation for crossing-over in yeast, where the symmetric phase is not yet identified. However, the low crossover frequency of 5:3 and 6:2 segregations in the data reported above rules out this explanation for the low symmetry in these classes.

We should note in passing that, in both the *gray* locus of *S. fimicola* and the *b8* locus of *Ascobolus*, the amount of crossing-over seen in aberrant 4:4 tetrads is indistinguishable from 50%—the proportion expected from Sigal and Alberts's (1972) isomerization.

The evidence that non-crossover events are shorter than crossover events is of a different type. The data of Murray (1963) on selected prototrophs from *me-2* crosses in *Neurospora*, shown in Table I of Chapter 4, show a steady increase in the proportion of prototrophs which show crossing-over, as one takes two-point crosses further away from the high polarity end of the locus. Equating high polarity with the origin of events, this means that crossover events extend further into *me-2* than do non-crossover events. Hence, assuming that the two types of event originate in the same place, crossover events are, on the average, longer than non-crossover events.

A second example is to be found in the data for *am-1* in *Neurospora* (Fincham, 1974), cited in Chapter 4 as an exception to our ability to explain the properties of intragenic recombination in specific loci by varying the length of heteroduplex and correction tract relative to the distances involved. This locus is immappable, implying long heteroduplex, especially in crossover events, but shows strong polarity by comparison of the numbers of prototrophs in the two non-crossover classes, implying that non-crossover heteroduplex is often shorter than the distance between the alleles.

A further example was reported by Savage and Hastings (1981) in tetrad data showing recombination within the *his1* locus of yeast. Those recombinant tetrads which showed conversion of the allele nearer to the low conversion end showed substantially more recombination of flanking markers (57%) than did the tetrads showing conversion of an allele nearer to the high conversion end of the locus (40%).

Thus, there are several examples to show that some non-crossover events are entirely asymmetrical and other examples to show that non-crossover events tend to be short.

Evidence to support the contention that these events are incomplete, and that they are incomplete because the strand uptake process was limiting, can be found in the data for the *buff* locus of *Sordaria brevicollis*, already cited as an example of asymmetrical non-crossover events (Sang and Whitehouse, 1979, 1983; MacDonald and Whitehouse, 1979; Theivendirarajah and Whitehouse, 1983). In this series of papers, it is shown that the comparison of the two classes of long crossover recombination events is not the same as that described above for the *his1* locus of yeast. In the *buff* locus data, tetrads showing conversion of the left allele, normal 4:4 segregation of the right allele, and a crossover to the right of the locus are greatly in excess of the comparable class showing 4:4 to the left, conversion to the right, and then a crossover (compare the classes R_2p and R_1q in Fig. 8 of Chapter 4). This means that there is a strong excess of apparent restoration over conversion for the marker farthest from the high conversion end.

It is easier to suppose that the marker was never in heteroduplex in these tetrads than to suppose that there is preferential restoration from heteroduplex. Whitehouse and his co-workers conclude that the crossover, separated from the conversion by an unconverted segment, is not part of the conversion event as described in the Aviemore model, but a separate event correlated with conversion in its occurrence. This interpretation will be discussed in Section V. Here, however, we should note that the same evidence can be explained by postulating that the length of heteroduplex was discontinuous within a single Aviemore style

5. Models of Heteroduplex Formation

of event. If the asymmetric phase is propagated by strand displacement, which Meselson and Radding (1975) suggested would be catalyzed by polymerization, and strand uptake into a homologous molecule, which must presumably be catalyzed by a reaction like that of the *RecA* protein from *Escherichia coli*, there would seem to be no reason why displacement and uptake should be coupled. Although no discontinuity in the length of heteroduplex has been detected in the *b2* locus of *Ascobolus* (Hastings *et al.*, 1980), it is apparent that, in the work of Catcheside and Angel (1974) concerning recombination events at an interchange break-point in *Neurospora*, the heteroduplex was not continuous from the origin of the event; either the beginning of strand uptake was delayed until the event had passed the break-point or the heteroduplex was discontinuous across the break-point.

If displacement and uptake are not coupled, and conditions for uptake are limiting, displacement may outrun uptake. In this circumstance, the symmetric phase cannot be formed by isomerization. Nor can it form by a reciprocal invasion, since the uptake cofactor is not available. Hence, there can be no symmetric phase, no cross-strand exchange, and no crossover associated with such an event unless, rarely, a renewed supply of uptake cofactor becomes available later. This would allow reinvasion and a complete event with discontinuous heteroduplex, as seen in the *buff* locus data.

It is a small step from here to suggest that a complete event, involving extensive heteroduplex formation, will deplete the local supply of the limiting uptake cofactor, causing other events initiated nearby to be incomplete and, therefore, non-crossover. Thus the occurrence of a crossover reduces the probability of another crossover occurring nearby. This model readily explains the situation in those organisms which do not show crossover position interference—in those cases the uptake cofactor is not limiting.

Holliday *et al.* (1984) reported some properties of the protein analogous to the *E. coli* RecA protein which they have isolated from *Ustilago maydis*. Like the *RecA* protein, the *U. maydis* protein binds single-stranded DNA, catalyzes its uptake into homologous DNA, and hydrolyzes ATP in the process. It was also found that it is inhibited by ADP. The authors suggest that the ratio of ATP to ADP could provide a modulation of strand uptake which would be sufficiently sensitive to account for position interference. They postulate that linear gradients of ATP and ADP would be established within a length of synaptonemal complex.

V. CONVERSION AND CROSSING-OVER AS SEPARATE EVENTS

While all models of recombination seek to explain crossing-over as an outcome of the recombination event which leads to conversion and postmeiotic segregation, the idea that a crossover is something separate from the conversion event has always been present. Pritchard (1960), in the course of his work on negative interference, suggested that exchanges occurred in clusters. The data for *am-1* in *Neurospora*, which were interpreted above as evidence that non-crossover events are shorter than crossover events, were interpreted by Fincham (1974) in terms of a cluster model in which the crossover is not part of the same event. Similarly, the data for *buff* in *Sordaria brevicollis* were cited earlier as evidence of discontinuity in heteroduplex. However, Sang and Whitehouse (1979) preferred the explanation that the associated crossover was a separate event occurring nearby. This involves a reinitiation of heteroduplex formation (Whitehouse, 1982; Sang and Whitehouse, 1983). These are cases where a crossover seems to be occurring at a physical distance from a short length of heteroduplex, and both styles of interpretation appear to be valid.

There is evidence that, in mitotic recombination in yeast, conversion and the associated crossover occur at separate times. It is clear that, in some mitotic events, the conversion affects a whole chromosome while the crossover concerns chromatids. Esposito (1978) has proposed that the conversion and crossover were set up at the same time, but that the cross-strand exchange was resolved by replication so that it appeared to have occurred in G_2. Recently, Roman and Fabre (1983) have shown that many of these events cannot be explained by a separate time of resolution of the event, but should be thought of as occurring at separate times—a G_1 conversion associated with a G_2 crossover.

New evidence from the *b2* locus in *Ascobolus* seems to leave no room to avoid accepting that there is a crossover which can exist as a separate event, not merely the resolution of a recombination event involving heteroduplex formation. The data are presented, and the case is argued by Rossignol *et al.* (1984), from experiments on the effects on recombination of including large deletions in the *b2* locus. Those authors make the point that previous work selected for the effects of heteroduplex formation, and therefore studied only crossovers which are associated with heteroduplex. Now they have a method for selecting for the crossover and looking for evidence of associated heteroduplex. This is possible because crossovers are formed with high frequency at the left boundary of heterologies in the *b2* locus.

The finding that the symmetric phase of a recombination event does not cross a heterology was described above. This was presented as evidence that the symmetric phase is propagated differently from the asymmetric phase. It is not difficult to imagine that a heterology would block the diffusion of a cross-strand exchange and, therefore, block the extension of symmetric heteroduplex. We were, therefore, not surprised to learn that there was an increased occurrence of reciprocal recombination at the upstream edge of the heterology; we expect to see cross-strand exchanges stop there, and some of them will become crossovers in

5. Models of Heteroduplex Formation

the resolution of an Aviemore event. It is apparent, however, that this is not what is happening: (1) The crossovers found to the left of the heterologies are about 10 times more frequent than can be accounted for by the loss of symmetric events to the right; (2) the crossovers show no association with postmeiotic segregation or conversion of alleles further to the left which must have been involved in Aviemore-style events in the locus; and (3) genotypes which show decreased occurrence of heteroduplex in $b2$ do not show a decrease in these heterology-associated reciprocal events.

What is revealed here appears to be a simple crossover. Since there is no evidence of heteroduplex, there is no evidence that rejoining of recombinant molecules occurred by splicing complementary nucleotide chains together. These crossovers may be forming by a blunt-end breakage and rejoining. It is striking that the newly revealed crossovers show polarity, the same polarity as is shown by heteroduplex events, in that they are found only to the left of a heterology.

There appear to be three points of view which could be taken about these crossovers. First, one might suppose that they do not occur in a normal meiosis, but are induced by the heterology. Rossignol *et al.* (1984) argue against this on the basis of an experiment which included two deletions in the $b2$ locus. The reciprocal events were found to occur only to the left of the upstream deletion. The right-hand deletion was unable to induce crossovers when cut off by a heterology from the left side of the locus. It remains possible, however, that the heterology induced a crossover when exposed to an influence which travels from the left and which can be blocked by heterology farther upstream. The second position is that favored by Rossignol *et al.* (1984), that there are two types of meiotic crossover, those resulting from the resolution of a heteroduplex event and the new type revealed by the heterologies. The evidence they cite for believing that an Aviemore-style event can be resolved as a crossover is the observation that aberrant 4:4 tetrads show 50% recombination of flanking markers. It has been calculated by Hurst *et al.* (1972) and again by Fincham (1983), on the basis of some rather loosely estimated numbers, that the amount of crossing-over which is detected in association with conversion is sufficient to account for the total map length of yeast. On this basis one would not expect simple crossovers not associated with conversion to be very common. The third possible point of view on this evidence is that the crossover normally found in association with heteroduplex events is the same as the crossover revealed by Rossignol *et al.* (1984) by the inclusion of deletions in the cross. That is, an event of heteroduplex formation and conversion does not lead to crossing-over, but a simple crossover is very likely to occur near a heteroduplex event. This hypothesis is in keeping with the evidence of physical and temporal separation of conversion and associated crossing-over in a few systems cited above.

VI. SPECULATION

Rossignol et al. (1984) propose that there is a synaptic intermediate which takes the form of a "duplex-to-duplex pairing step." The intermediate travels in a polarized way (left to right at the $b2$ locus) and its travel is dependent on homology. This intermediate is proposed as a precursor both to heteroduplex formation and to simple crossing-over by cutting and rejoining.

The concept of a precursor to heteroduplex formation was reviewed by Hastings (1975) on the basis of evidence that heterozygosity itself causes heteroduplex to occur. It was suggested then that the precursor to heteroduplex was a traveling transient bubble of symmetrical heteroduplex (see Sobell, 1972). Another possibility is brought to mind by the *RecBC* recombination model of Smith et al. (1984). Nucleotide chains looped out from homologous chromatids by the action of a *RecBC*-like enzyme might interact and be compared. Heterozygosity might lead to heteroduplex formation, perhaps by the induction of a nick. Other recombination events may occur when a nick is encountered, or when a specific sequence is encountered, as proposed for *RecBC*-mediated recombination in bacteriophage λ, when the enzyme encounters a chi sequence (reviewed by Smith et al., 1984). Of course, one does not visualize a structure in DNA traversing a length of synapsed chromosome, but rather a complex of enzymes and other proteins, which constitute a factory for producing recombination. "Recombination nodules," seen in synaptonemal complex, have been interpreted as just such recombination factories (Carpenter, 1975; reviewed elsewhere in this volume).

If the crossover is formed by a traveling recombination nodule in an event separate from the heteroduplex formation activity, one needs to account for the association of crossing-over with heteroduplex and for crossover position interference, since the explanations for these phenomena based on the Aviemore model given above would no longer hold. It was first discussed by Sang and Whitehouse (1979) that the association of heteroduplex with crossing-over could be explained as separate events if the same enzyme aggregate were responsible for both events. Whitehouse (1982) discusses the equation of this enzyme aggregate with recombination nodules. This could be extended by proposing that the enzyme aggregate makes crossovers at random. Then a crossover might be formed with equal probability in any unit of time and is, therefore, most likely to occur where the travel of an aggregate was delayed at a heteroduplex event, or where its travel was blocked at a heterology, as seen in $b2$ in *Ascobolus*.

Crossover position interference would be expected if each aggregate could make only one crossover. I imagine that the aggregate or recombination nodule travels a defined length of synapsed chromosome, perhaps from one to another of the nicks reported in yeast meiotic DNA, which occur at defined positions and have a frequency comparable to the number of crossovers per meiosis (Resnick *et al.*, 1984). If each such length is traversed once, or a low number of times, per meiosis, heteroduplex events may be formed wherever conditions are right, but only one crossover may be dropped per traverse.

ACKNOWLEDGMENTS

I am grateful to the many people who have read and commented on the manuscript for me. I also wish to thank Anne Galloway for assistance with the figures and Dorothy Woslyng for the typing. This work is supported by an operating grant from the Natural Sciences and Engineering Research Council of Canada.

REFERENCES

Bell, L., and Byers, B. (1979). Occurrence of crossed strand-exchange forms in yeast DNA during meiosis. *Proc. Natl. Acad. Sci. U.S.A.* **76**, 3445–3449.
Carpenter, A. T. C. (1975). Electron microscopy of meiosis in *Drosophila melanogaster* females. II. The recombination nodule—A recombination-associated structure at pachytene? *Proc. Natl. Acad. Sci. U.S.A.* **72**, 3186–3189.
Catcheside, D. G., and Angel, T. (1974). A *histidine-3* mutant, in *Neurospora crassa*, due to an interchange. *Aust. J. Biol. Sci.* **27**, 219–229.
Esposito, M. S. (1978). Evidence that spontaneous mitotic recombination occurs at the two-strand stage. *Proc. Natl. Acad. Sci. U.S.A.* **75**, 4436–4440.
Fincham, J. R. S. (1974). Negative interference and the use of flanking markers in fine-structure mapping in fungi. *Heredity* **33**, 116–121.
Fincham, J. R. S. (1983). "Genetics." Jones & Bartlett, Boston, Massachusetts.
Hamza, H., Haedens, V., Mekki-Berrada, A., and Rossignol, J.-L. (1981). Hybrid DNA formation during meiotic recombination. *Proc. Natl. Acad. Sci. U.S.A.* **78**, 7648–7651.
Hastings, P. J. (1975). Some aspects of recombination in eukaryotic organisms. *Annu. Rev. Genet.* **9**, 129–144.
Hastings, P. J., Kalogeropoulos, A., and Rossignol, J.-L. (1980). Restoration to the parental genotype of mismatches formed in recombinant DNA heteroduplex. *Curr. Genet.* **2**, 169–174.
Holliday, R. (1964). A mechanism for gene conversion in fungi. *Genet. Res.* **5**, 282–304.
Holliday, R., Taylor, S. Y., Kniec, E. B., and Holloman, W. K. (1984). Biochemical characterization of *rec1* mutants and the genetic control of recombination in *Ustilago maydis*. *Cold Spring Harbor Symp. Quant. Biol.* **49**, 669–673.
Hurst, D. D., Fogel, S., and Mortimer, R. K. (1972). Conversion associated recombination in yeast. *Proc. Natl. Acad. Sci. U.S.A.* **69**, 101–105.
Kitani, Y., and Whitehouse, H. L. K. (1974). Aberrant ascus genotypes from crosses involving mutants at the *g* locus in *Sordaria fimicola*. *Genet. Res.* **24**, 229–250.
MacDonald, M. V., and Whitehouse, H. L. K. (1979). A *buff* spore colour mutant in *Sordaria brevicollis* showing high-frequency conversion. I. Characteristics of the mutant. *Genet. Res.* **34**, 87–119.
Meselson, M. (1972). Formation of hybrid DNA by rotary diffusion during genetic recombination. *J. Mol. Biol.* **71**, 795–798.

Meselson, M. S., and Radding, C. M. (1975). A general model for genetic recombination. *Proc. Natl. Acad. Sci. U.S.A.* **72,** 358–361.

Murray, N. E. (1963). Polarized recombination and fine structure within the *me-2* gene of *Neurospora crassa*. *Genetics* **48,** 1163–1183.

Nicolas, A. (1982). Variation of crossover association frequencies with various aberrant segregation classes in *Ascobolus*. *Curr. Genet.* **6,** 137–146.

Paquette, N., and Rossignol, J.-L. (1978). Gene conversion spectrum of 15 mutants giving postmeiotic segregation in the *b2* locus of *Ascobolus immersus*. *Mol. Gen. Genet.* **163,** 313–326.

Potter, H., and Dressler, D. (1979). DNA recombination: *In vivo* and *in vitro* studies. *Cold Spring Harbor Symp. Quant. Biol.* **43,** 969–985.

Pritchard, R. H. (1960). Localized negative interference and its bearing on models of gene recombination. *Genet. Res.* **1,** 1–24.

Radding, C. M. (1982). Homologous pairing and strand exchange in genetic recombination. *Annu. Rev. Genet.* **16,** 405–437.

Resnick, M. A. (1976). The repair of double strand breaks in DNA: A model involving recombination. *J. Theor. Biol.* **59,** 97–106.

Resnick, M. A., Chow, T., Nitiss, J., and Game, J. (1984). Changes in chromosomal DNA of yeast during meiosis in repair mutants and the possible role of a deoxyribonuclease. *Cold Spring Harbor Symp. Quant. Biol.* **49,** 639–649.

Roman, H., and Fabre, F. (1983). Gene conversion and associated reciprocal recombination are separable events in vegetative cells of *Saccharomyces cerevisiae*. *Proc. Natl. Acad. Sci. U.S.A.* **80,** 6912–6916.

Rossignol, J.-L., and Haedens, V. (1980). Relationship between asymmetrical and symmetrical hybrid DNA formation during meiotic recombination. *Curr. Genet.* **1,** 185–191.

Rossignol, J.-L., Nicolas, A., Hamza, H., and Langin, T. (1984). Origins of gene conversion and reciprocal exchange in *Ascobolus*. *Cold Spring Harbor Symp. Quant. Biol.* **49,** 13–21.

Sang, H., and Whitehouse, H. L. K. (1979). Genetic recombination at the *buff* spore colour locus in *Sordaria brevicollis*. I. Analysis of flanking marker behaviour in crosses between *buff* mutants and wild type. *Mol. Gen. Genet.* **174,** 327–334.

Sang, H., and Whitehouse, H. L. K. (1983). Genetic recombination at the *buff* spore color locus in *Sordaria brevicollis*. II. Analysis of flanking marker behavior in crosses between *buff* mutants. *Genetics* **103,** 161–178.

Savage, E. A., and Hastings, P. J. (1981). Marker effects and the nature of the recombination event at the *his1* locus of *Saccharomyces cerevisiae*. *Curr. Genet.* **3,** 37–47.

Sigal, N., and Alberts, B. (1972). Genetic recombination: The nature of a crossed strand-exchange between two homologous DNA molecules. *J. Mol. Biol.* **71,** 789–793.

Smith, G. R., Amundsen, S. K., Chaudhury, A. M., Cheng, K. C., Ponticelli, A. S., Roberts, C. M., Schultz, D. W., and Taylor, A. F. (1984). Roles of RecBC enzyme and chi sites in homologous recombination. *Cold Spring Harbor Symp. Quant. Biol.* **49,** 485–495.

Sobell, H. M. (1972). Molecular mechanism for genetic recombination. *Proc. Natl. Acad. Sci. U.S.A.* **68,** 2483–2487.

Stadler, D. R., and Towe, A. M. (1971). Evidence for meiotic recombination in *Ascobolus* involving only one member of a tetrad. *Genetics* **68,** 401–413.

Szostak, J. W., Orr-Weaver, T. L., Rothstein, R. J., and Stahl, F. W. (1983). The double-strand-break repair model of recombination. *Cell* **33,** 25–35.

Theivendirarajah, K., and Whitehouse, H. L. K. (1983). Further evidence that aberrant segregation and crossing over in *Sordaria brevicollis* may be discrete, though associated, events. *Mol. Gen. Genet.* **190,** 432–437.

Whitehouse, H. L. K. (1982). "Genetic Recombination: Understanding the Mechanisms." Wiley, New York.

6

Investigating the Genetic Control of Biochemical Events in Meiotic Recombination

MICHAEL A. RESNICK

Yeast Genetics/Molecular Biology Group
Cellular and Genetic Toxicology Branch
National Institute of Environmental Health Sciences
Research Triangle Park, North Carolina 27709

I. INTRODUCTION

A. Purpose of Review

The reassociation of genetic traits has been one of the most intriguing areas of the biological sciences since the early experiments of Gregor Mendel (1865). While the meiotic stage of development in eukaryotes has been extensively examined, little is known about the actual mechanisms of the associated recombination or the features of meiotic development that are needed for recombination. Meiosis, similar to any set of biological events, should be amenable to genetic investigation. As a collection of biochemical processes, genetic approaches can be used for identifying those genes which are important to recombination, determining the nature and biochemical function of the gene products, and ordering the events in recombination. Since meiosis is a unique developmental stage in the life cycle and since the amount and the pattern of recombination differs from that in mitotic cells (Malone *et al.*, 1980), some of the genetically identified functions might be expected to be unique to meiosis while others might be common to both stages of growth and development.

Generally there have been three approaches to the study of meiosis as it relates to recombination: (1) morphological and cytological investigations, (2) genetic dissection of the recombinational and developmental events in meiosis and inferring the underlying biochemical processes, and (3) characterization of the biochemical processes of meiosis and inferring their relation to development and recombination. The cytological events and structures, particularly the synaptonemal ccomplex, which plays a central role in bringing chromosomes together and presumably mediates recombination, have been reviewed by Dresser in Chapter 8, Westergaard and von Wettstein (1972), von Wettstein et al. (1984), and Lu (1984). Meiotic recombination as analyzed genetically from the meiotic products has been discussed by Fogel et al. (primarily for yeast, 1981), Rossignol (primarily for *Ascobolus*, 1986), Stahl (1979), Whitehouse (1982), and most recently by Orr-Weaver and Szostak (1985) and by Hastings in Chapter 4. An extensive review on the identification of mutants altered in meiosis and meiotic recombination was written by Baker et al. (1976) and this was followed by the review of Esposito and Klapholz (1981), which described the various mutants of yeast. The biochemistry of meiosis with particular reference to DNA metabolism has been examined by Lu (1981, 1984), Stern and Hotta (1977, 1978; Chapter 10, this book), and Pukkila (1977).

An integrated understanding of recombinational events and the associated biochemical processes and their relationship to meiotic development requires a combination of genetic and biochemical analyses. Because of the general lack of systems which would allow a combined approach, there is a relatively small amount of information available on the genetic control of biochemical events in meiotic recombination. This review will, therefore, emphasize approaches that can be taken for such investigations as well as describe recent work which is relevant to this topic. The first half will discuss meiotic recombination, the various factors which may be involved in meiotic recombination, and molecular genetic approaches that can and are being taken for their identification and characterization. The second part of the review addresses work in yeast and results which have utilized genetic and biochemical approaches, particularly in the case of DNA repair mutants.

There are many reasons for a combined genetic and biochemical investigation of meiosis. For example, it is possible to distinguish genetically the controls and processes that are involved specifically in the physiology of meiotic development from those related to meiotic recombination; similarly, the function of recombination in meiotic development can be evaluated. The role that observed meiotic biochemical processes, especially those involving DNA, actually play in meiotic recombination can be evaluated using a mutational analysis and it should be possible to identify and understand successive biochemical events (i.e., biochemical pathways) that take place in recombination. Since more than one type of recombination could occur, biochemical processes can be related to specific recombinational events. It is also possible to assess the roles of mitotically identified repair genes both in the protection of meiotic cells and in normal meiotic recombination.

While there has been substantial progress with vegetative systems, particularly with prokaryotes such as *Escherichia coli,* in terms of understanding the relationship between genetically determined factors in recombination, the corresponding biochemical functions, and the actual process of recombination (see review by Smith, 1983), little is known about the specific molecular events and the associated enzymology of meiotic recombination or their genetic control. The major impediment has been the lack of systems which would enable both the identification of meiotic factors and the consequences at the biochemical level when they are eliminated by mutation. Although meiotic- and recombination-defective mutants have been isolated in various organisms, the organisms typically have not been amenable to biochemical studies during meiosis. Conversely, there are several organisms in which it has been possible to examine meiosis-specific biochemical events or enzymes, however, their importance in recombination has generally not been assessable because the systems have not been genetically well defined. Much of the recent progress has been due to the biochemical and molecular genetic methods available with the yeast *Saccharomyces cerevisiae* which have allowed an integrated approach to studying meiotic recombination. This review will therefore emphasize observations with this organism.

B. Recombination in Relation to Meiosis

Meiosis differs from all other types of cellular development by virtue of the associated DNA metabolic events. While other types of development necessarily involve DNA in an informational context and may lead to changes in utilization of the information, as for the case of methylation of specific bases (reviewed by Razin *et al.,* 1984), meiosis in addition results in a reassociation of informational content and the orderly segregation of a total complement of chromosomes into gametes.

Recombination plays an important role in meiosis in that it is generally required for the segregation of chromosomes during gametogenesis to yield one-half the number of chromosomes per gamete as compared to the parent cells. Presented in Fig. 1 is a schematic diagram of the events of meiosis that relate to homologous chromosome recombination. Each of the boxes represents landmarks in meiosis; described beneath each box are the areas which have been investigated. The major landmarks include signals for switching of cells to the meiotic mode of development, the meiotic round of DNA synthesis (also known as premeiotic synthesis), chromosome processing and association, synapsis maintained by the synaptonemal complex, recombination (discussed more fully in the next section), and subsequent segregation of chromosomes.

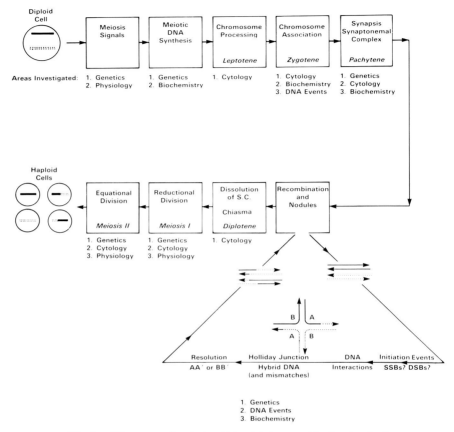

Fig. 1. The events of meiosis which relate to recombination (see text).

Though diagrammed in a linear fashion (Fig. 1), the events of meiosis could occur through independent pathways; for example, the completion of meiotic physiological development need not always require recombination. A commonly cited exception is found for the *Drosophila* male. Recombination does not occur during male gametogenesis while it does in females; even so, there is an orderly segregation of individual chromosomes in males. Recombination per se does not appear to be necessary for meiotic development processes based on genetic evidence from such diverse organisms as *Lilium* (Toledo *et al.*, 1979), yeast (cf. discussion of *spo11* and *rad50* in Sections IV,B and C), as well as *Drosophila* (see summary in Baker *et al.*, 1976).

The selective value of recombination to a species and the rapid reassociation of genetic traits has been discussed extensively (Maynard-Smith, 1978). As suggested by Jinks-Robertson and Petes (1985) and Kohli *et al.* (1984), the high levels of recombinational activity which occur during meiosis might provide a means for the rapid transmittal of information between gene groups or nonhomologous chromosomes (such as common transfer RNAs) through gene conversion (see Section II,B,2).

Recombination during meiosis may also fulfill another less obvious role in the life cycle, that of correcting DNA alterations that arise during the growth of germinal cells (Bernstein *et al.*, 1981; Medvedev, 1981; Holliday, 1984). As suggested by Bernstein (1979), this accounts for why "babies are born young"; the meiotic process may allow for infinite regeneration from organisms with limited life spans (Medvedev, 1981). Holliday has proposed that correction mechanisms could be provided through chromosome interactions (i.e., recombination) which enable corrections to take place in a directed fashion. In particular, an epigenetic change, such as methylation, could signal a recombinational event.

The importance of recombination in DNA repair is well established. Genetic and biochemical evidence has shown that several types of DNA damage, particularly double-strand damage, can only be repaired in mitotic cells through mechanisms associated with recombination (Resnick, 1979). Possibly, meiotic recombination evolved from mitotic repair systems (Bernstein *et al.*, 1981). This latter possibility has been examined indirectly and is discussed later (Section IV,C). Recombinational systems in meiosis might also serve to protect cells during meiosis against potentially harmful intracellular changes in DNA or damage due to external agents.

Regardless of the importance of recombination during meiosis, it is clear that in those organisms which exhibit it normally, recombination is vital for normal chromosomal disjunction and mutants lacking recombination generally or for specific chromosomes exhibit chromosome malsegregation (summarized in Baker *et al.*, 1976; Mason and Resnick, 1985).

C. Types and Proposed Mechanisms of Meiotic Recombination

At the genetic and molecular level there are several types of recombination that can occur meiotically as well as mitotically. The traditional cytological view of recombination is that identified by chiasmata formed between homologues. Reciprocal exchanges have been identified genetically as "crossover" events which lead to a conservative (i.e., no loss of information) reshuffling of information between homologous chromosomes. Gene conversion is the replacement of information in one chromosome by information obtained from the homologous chromosome or a homologous region (see Hastings, Chapters 4 and 5). Because there is a high association of reciprocal genetic exchange with gene conversion (Hurst *et al.*, 1972; also summarized in Whitehouse, 1982), the mechanisms involved in gene conversion (i.e., the formation of heteroduplex DNA and subsequent mismatch repair) have been proposed to be part of the system which can lead to a conservative exchange (no loss of information) of outside markers. This concept is described more fully elsewhere in this volume (Hastings, Chapters 4 and 5) and in other reviews (Whitehouse, 1982; Rossignol *et al.*, 1986). Even though the underlying biochemical events involved in recombination are poorly understood, there is a general consensus (see Fig. 1) in the various models (Holliday, 1964; Meselson and Radding, 1975; Szostak *et al.*, 1983; Hastings, 1984; see Whitehouse, 1982) that recombination requires an initiating event on at least one chromosomal DNA molecule, the formation of interacting DNA molecules with regions of hybrid DNA (from both chromosomes), the development of Holliday junctions (the structure developed at the point of interaction between two recombining DNA molecules), and resolution (the development of the two DNA molecules which have undergone recombination). Depending on which two strands are nicked at the junction of two interacting chromosomes, resolution can lead to reciprocal recombination as measured by outside genetic or molecular markers. Gene conversion occurs when a region of heteroduplex includes mismatched bases or regions and one of the strands is "corrected" using the opposite strand as a template, i.e.,

$$\begin{matrix} ...AAAGAAA... \\ ...TTTATTT... \end{matrix} \rightarrow \begin{matrix} ...AAATAAA... \\ ...TTTATTT... \end{matrix}$$

Based on genetic and molecular evidence, gene conversion can also occur when double-strand breaks or gaps in one chromosome are repaired using information from the homologue (Resnick, 1976; Szostak *et al.*, 1983).

Meiotic recombination can also occur within chromosomes and between nonhomologous chromosomes, however, significant differences have been observed. Using heteroalleles of genes that were present on either the same or homologous chromosomes, Klein (1984) demonstrated that reciprocal exchanges normally associated with interchromosomal gene conversion during meiosis were absent for the case of intrachromosomal gene conversion. An extensive analysis of recombination between duplicated genes (Jackson and Fink, 1985) that enabled a distinction between inter- and intrachromosomal reciprocal exchanges and gene conversions also demonstrated that intrachromosomal reciprocal exchange is rare compared to interchromosomal reciprocal exchange while inter- and intrachromosomal gene conversions are comparable. These observations suggest that the mechanism of meiotic recombination is sensitive to the nature of chromosomal interactions.

Meiotic gene conversion has also been detected between the same genes (tRNAs or protein coding) on nonhomologous chromosomes in both *S. cere-*

visiae and *Schizosaccharomyces pombe* (Kohli *et al.*, 1984; Amstutz *et al.*, 1985; Jinks-Robertson and Petes, 1985). These conversion events in *S. cerevisiae* are frequently associated with reciprocal exchanges (S. Jinks-Robertson and T. D. Petes, personal communication) while in *S. pombe* they are rare.

Other types of recombination which have been identified mitotically and that are relevant here are gene amplification (see Schimke, 1982) and sister chromatid exchange (see reviews in Tice and Hollaender, 1984). Gene amplification has been described in several systems. It was observed in mammalian cells under conditions which select for cells with increased number of resistance genes (Schimke *et al.*, 1978). Fogel *et al.* (1984) have demonstrated that amplification in yeast can occur during meiosis under conditions which do not involve selection for resistance phenotypes; it appears to occur primarily through gene conversion processes. Sister chromatid exchange is also observed in meiotic cells; however, except for the case of exchanges in ribosomal DNA, its frequency is considerably less than recombination between homologous chromosomes (see Section II,B,2). The role and mechanisms for amplification and sister chromatid exchange in normal meiosis are far from understood, although they should be subject to genetic analysis.

II. GENETICALLY IDENTIFIABLE FACTORS AFFECTING MEIOTIC RECOMBINATION

A. General Description

A large number of factors must exist that function in meiotic development and recombination should be genetically identifiable (see Fig. 1). These factors can be divided into those which are meiosis-specific and those which are common to both meiosis and mitosis. This is diagramed in Fig. 2 for meiotic DNA metabolism. The meiotic cycle can be divided generally into two stages which include a complete round of DNA synthesis followed by processing events associated with recombination. Enzymes associated with DNA synthesis might function in both mitotic and meiotic cells (see Section III,B,1) or alternatively there may be unique enzymes for each stage. Many of the recombinational events may be meiosis-specific; however, proteins utilized in mitotic chromosome metabolism such as repair might also function in meiosis. While repair and repair mutants are usually characterized in mitotic cells, repair systems could protect cells undergoing meiosis as well as function in the normal recombinational processes. As shown in Fig. 3, various kinds of DNA damage and/or repair intermediates might resemble some of the DNA configurations which have been proposed to occur during meiotic recombination.

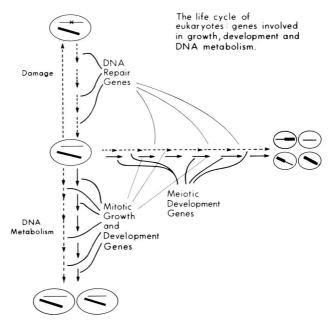

Fig. 2. DNA metabolism in mitotic and meiotic cells. The processing of DNA in the two types of cells is expected to involve several genes, some of which are common and others of which are unique. For example, enzymes involved in replication might be used for DNA synthesis in both cell types while some enzymes involved in meiotic replication could be specific to the meiosis phase of the cell cycle. Genes associated with repair would be expected to function in both parts of the life cycle to protect cells from DNA-damaging agents. They also might function during meiotic and/or mitotic recombination (see Fig. 3).

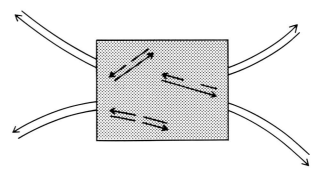

Fig. 3. Intermediates in meiotic recombination and DNA repair. Among the several models that describe recombination, various DNA structures arise which have similarities to either DNA damage or intermediates in repair. The repair genes which are identified in mitotic cells for dealing with DNA damage could function during meiotic recombination by processing such intermediates.

TABLE I
Genetically Identifiable Factors Which Might Be Involved in Meiotic Recombination

I. Factors indirectly affecting recombination
   A. Controls for developmental switch to meiosis
   B. Meiotic development
   C. Meiotic (premeiotic) DNA synthesis
   D. Chromosome processing prior to pairing
   E. Structures required for chromosome interaction
      1. Sites for chromosome interactions and initiation of pairing
      2. The physical components in pairing structures (i.e., synaptonemal complexes)
   F. Termination controls to allow structures devoted to recombination to disperse

II. Chromosomal factors directly affecting recombination
   A. Sequence
      1. Distance between sites (map expansion)
      2. Sequence of sites being recombined (palindromes and/or high recombination sites; gene conversion)
      3. Relation to initiation sites
      4. Nature of site (particularly gene conversion)
      5. Modifications of DNA, such as methylases
      6. DNA damage
   B. Position and chromosomal location
      1. Relation to centromere and telomeres
      2. Heterochromatin versus euchromatin and genetic activity
      3. Sister chromatid exchange
      4. Homologous versus nonhomologous chromosomes

III. Processing factors directly affecting recombination
   A. Recombination proteins
      1. DNA binding proteins: renaturation and denaturation proteins
      2. Topoisomerases
      3. Ligases
      4. DNA polymerases
      5. Mismatch excision enzymes
      6. DNA interaction proteins (i.e., RecA)
      7. Nucleases including endo-, exo-, single-, and double-strand
      8. Resolvase
      9. Initiating enzymes (e.g., nicking or restrictionlike)
      10. Methylases
      11. Proteases or protein processing enzymes
   B. Recombination complex

Based on what is known about the biology and biochemistry of meiosis and the presumed events in recombination, several factors are proposed in Table I and discussed in the following sections for which it should be possible to isolate mutants. This table is not intended as a complete list of all possible factors that might affect or relate to meiotic recombination but rather to illustrate most of the relevant ones which potentially can be identified using current genetic systems and molecular biology approaches (see Section III). Examples of known or possible mutants for these factors are presented below. Table I is divided into indirect and direct factors according to whether or not they are expected to have a direct role in recombination (See Fig. 1). The indirect factors are those that would enable the cells to get to the point of recombination in the meiotic cycle, or facilitate the recombinational process, but would not be expected to be directly involved in the exchange processes. Included in the direct factors are signals for recombination, unique sequences, and enzymes involved in recombination. Approaches to the actual isolation of mutants for these factors, which are presented in Section III, may be complicated by the system being used and the possible pleiotropic effects of the mutants of interest, including effects on viability.

B. Potentially Identifiable Factors

1. Indirect

Among the indirect factors affecting meiotic recombination are those which allow a switch from mitotic activities to meiotic development (see Fig. 1). Several externally supplied signals have been identified which are required for the switch or alternatively the continuation of meiotic development. Signals for shifting to meiosis in lower eukaryotes include reduction in nutrients, exposure to light, and temperature changes. With yeast it has been possible to define genetically metabolic events that take place during mitotic growth which allow entrance into meiosis. For example, using mitotic cell division cycle mutants of yeast, Hirschberg and Simchen (1977) have shown that cells enter into meiosis from the G_1 phase. Similarly there have been extensive studies describing the conditions for entrance into meiosis and several mutants have been identified in both *S. cerevisiae* and the fission yeast *Schizosaccharomyces pombe* which do not enter into meiosis or alternatively are not subject to normal meiosis-inducing or -repressing signals (see Dawes, 1983; Dawes and Calvert, 1984; Calvert and Dawes, 1984; Matsumoto *et al.*, 1983; Nurse, 1985; Iino and Yamamoto, 1985a,b). Included in the latter category are mutants lacking adenylate cyclase (*cyr1*) or *ras2* mutants that appear to modulate this enzyme (Toda *et al.*, 1985). It is interesting to note that mutants of the *RAS* gene, which show considerable homology with mammalian *ras* genes (Defeo-Jones *et al.*, 1983), can be complemented by human ras oncogene sequences (Kataoka *et al.*, 1985), suggesting related mechanisms of developmental control.

Much of meiotic development including the eventual formation of gametes can be separated genetically from the process of recombination. The corresponding factors can be considered as recombination-independent although the eventual expression of recombination generally requires functioning gametes (see Section IV,A). Meiotic mutants have been isolated in various organisms which are

6. Genetics and Biochemistry of Recombination

specifically defective in meiotic developmental processes. While they may have been characterized cytologically, the biochemical defects are generally not known. In the yeast *S. cerevisiae* the SPO series of mutants (described in Section III) was isolated on the basis of altered meiotic sporulation physiology. The *spo12* and *spo13* mutants are deficient in reductional (i.e., meiosis I) but not equational division, yet recombination is normal; two rather than the normal four meiotic spores are detected. They are particularly useful for analyzing recombination in the absence of reductional division.

Events more directly related to recombination should also be subject to genetic and biochemical analysis. The synthesis of DNA is an important step in meiosis. In organisms such as *Coprinus*, DNA synthesis occurs prior to karyogamy. While there are suggestions that recombination in some organisms can be initiated during the meiotic round of DNA synthesis (Grell *et al.*, 1980), it is clear from various genetic studies with yeast that DNA synthesis is required for recombination and the completion of meiosis (summarized in Esposito and Klapholz, 1981). Several genes involved with mitotic DNA synthesis have been identified in yeast which are required for meiotic DNA synthesis; temperature-sensitive mutants of these genes do not exhibit meiotic development (see Section III,A). No mutants have been identified in the various organisms examined which specifically affect meiotic and not mitotic DNA synthesis mechanisms. It is interesting that while two types of DNA polymerases (α and β) have been found in eukaryotic cells, not all organisms have the two polymerases throughout meiosis. The fungi *Corpinus* and yeast have both polymerases during the period of meiotic DNA synthesis (Sakaguchi and Lu, 1982; Resnick *et al.*, 1984b), however, only the β polymerase is also present in *Coprinus* at later stages of meiosis. *Lilium* is also greatly enriched for the β polymerase during meiosis (Sakaguchi *et al.*, 1984). Lu (1981) has suggested that the β polymerase may specifically function during recombination to carry out repair-type synthesis. Polymerase α may also be absent during mouse spermatogenesis, although evidence is conflicting (summarized in Orlando *et al.*, 1984; Sakaguchi *et al.*, 1984). A mutational analysis of polymerases including the use of aphidicolin-resistant mutants (Huberman, 1981) would resolve the importance of α polymerases in meiosis.

Chromosome processing and pairing is an intricate and coordinated series of steps in meiosis. The association of the chromosomes with the nuclear membrane/matrix has been described in several organisms (von Wettstein *et al.*, 1984). Since homologous chromosomes are observed to preferentially associate (Dresser, Chapter 8), it appears that there are interactions or communication between homologous chromosomes long before the formation of synaptonemal complexes. As discussed by Bennett (1983), the interactions could occur in the mitotic cells prior to meiosis. While the nature of these interactions is far from understood (Maguire, 1984; Callow, 1985), the structures involved in attachment should be subject to genetic analysis. Membrane mutants might be candidates for "anchoring" deficiencies since telomeres are associated with nuclear membranes (Westergaard and von Wettstein, 1972; Gillies, 1975). For example, the fatty acid synthetase mutants of yeast described by Henry and Fogel (1971) and Culbertson and Henry (1973) appear to cause extensive mitotic nondisjunction. They also increase meiotic aneuploidy although there appears to be chromosome specificity.

The site of attachment of chromosomes and initiation of pairing appears to be through telomeres. The organization and function of the telomeres at the molecular and cytological level in chromosome segregation and meiotic pairing has been summarized recently (Blackburn and Szostak, 1984; Maguire, 1984). Telomeres are not static structures but are capable of changing size and undergoing recombination (Horowitz and Haber, 1984; Dunn *et al.*, 1984; Walmsley and Petes, 1985). To specifically evaluate the role of telomeres, chromosomes can be generated that have altered telomeres in terms of size or sequence and the organization between centromeres and telomeres can be manipulated (Murray and Szostak, 1983). Recent observations with *Drosophila* by Mason *et al.* (1984) have demonstrated that new types of chromosome ends can be generated as a result of a unique "healing" process of broken chromsomes in *mu-2* mutants. The effect of the new chromosome ends on meiotic recombination and nondisjunction is under investigation (J. M. Mason, personal communication).

The paired chromosomes and associated synaptonemal complexes are one of the landmarks of meiosis and recombination is considered to be associated with these structures (see Dresser,Chapter 8). Mutants have not been identified which are specifically altered in synaptonemal complex. Mutants that do not exhibit synaptonemal complex, such as those described above, may be defective in steps prior to rather than in the actual formation of the synaptonemal complex. Temperature-sensitive mutants have been described in yeast which arrest specifically during the pachytene phase of meiosis including *cdc5* (Schild and Byers, 1978; Simchen *et al.*, 1981) and the naturally occurring *pac1* mutant (Byers and Goetsch, 1982; Davidow and Byers, 1984). Such mutants could be useful for more direct isolation of genes involved in synaptonemal complex production and/or subsequent steps in meiotic prophase development. For example, the *cdc5* mutant used by Simchen *et al.* (1981) was reported to accumulate modified synaptonemal complex at the restrictive temperature.

Associated with paired chromosomes are structures identified by electron microscopy that may be related to recombination and are referred to as the recombination nodule (Carpenter, 1975). While the frequency and nonrandom location of this electron-dense body corresponds to expectations for recombination (Carpenter, 1979), the function in recombination has not been determined. Carpenter (1982), using mutants defective in reciprocal exchange but not gene conversion, concluded that the nodules are involved in gene conversion processes which in wild-type strains can lead to reciprocal exchanges. It should be possible to obtain mutants specific for this structure and therefore directly determine its role. Given the size of this structure, it could accommodate several protein molecules and therefore the nodule may be a complex of molecules (as suggested by Carpenter, 1984) that functions not only in recombination but in other aspects of chromosome metabolism.

At the conclusion of chromosomal interactions and recombination, the dissolution of synaptonemal complexes must occur which would then allow for the subsequent resolution of recombinants and chromosomes. The absence of the dissolution factors or controls would presumably result in chromosome malsegregation or meiotic arrest.

2. *Chromsomal Factors Directly Affecting Recombination*

The DNA sequence plays an important role in several aspects of recombination. It can influence the extent and nature of recombination. The DNA sequence of specific mutant alleles undergoing recombination can directly alter recombination, particularly gene conversion. Specific DNA sequences could be responsible for binding sites for possible recombination complexes (i.e., possibly recombination nodules), the actions of enzymes involved in initiation, the progress of enzymes or complexes along the chromosome, correction of mismatches, and even the resolution of structures. Mutational analysis combined with biochemical characterization enables the detection and determination of the various roles of DNA sequence in recombination.

If there are preferential initiation or resolution sites, the precise sequence in these regions will greatly affect recombination. Such sites have been characterized in prokaryotes. A well-described example is the chi site of the *Escherichia coli* bacteriophage lambda which is a sequence of eight bases (Smith *et al.*, 1981) and which, by its interaction with the RecBC protein, can enhance recombination by as much as 10 times. The interaction requires the nuclease function of the RecBC enzyme (Chaudhury and Smith, 1984).

Except for the mating-type switching locus of yeast (reviewed by Klar *et al.*, 1984b), initiation or resolution sites have not been specifically identified in eukaryotes nor has their biochemical function been determined. The detection of gradients of recombination and regions of high and low recombination activity (cf. Whitehouse, 1982; Fogel *et al.*, 1981; Rossignol *et al.*, 1984) strongly supports the existence of such regions as does the identification of genetic changes that result in "hotspots" of recombination. The hotspot mutational changes such as cog of *Neurospora* (Angel *et al.*, 1970) or M26 of *S. pombe* (Gutz, 1971) are generally locus- or region-specific and based on the directionality of recombination that occurs with such mutants, many of these may relate to enhancers of initiation, although other interpretations are possible. The identification of specific DNA sequences that affect meiotic recombination has not been described. However, it has been shown in yeast that the recombination frequency per nucleotide during meiosis can vary by as much as three- to fourfold (Mortimer and Schild, 1981) and that there are regions between genes that can specifically alter the frequency of recombination between genes by over fivefold (Borts et al., 1984). Sequences within ribosomal genes that are cis-acting have been identified which can enhance mitotic recombination (see Section II,B,2; Keil and Roeder, 1984), but they do not affect meiotic recombination.

The sequence of a mutant allele which is involved in recombination can greatly affect recombination. As shown by Leblon (1972), gene conversion of a specific allele can vary dramatically based on the nature of the allele. Both gene conversion and postmeiotic segregation events were greatly influenced by whether the mutant alleles being examined arose from agents purported to induce frameshift or base-pair substitution events. By varying the mutants the extent of heteroduplex could be altered as well as the site of reciprocal exchange (Rossignol et al., 1984).

In intragenic mapping studies, it has been shown that the extent of recombination is not always directly proportional to the distance between sites. For example, intragenic interference (Fincham and Holliday, 1970) occurs over small distances and this has been interpreted in terms of mechanisms for generating heteroduplexes, such as the length and the likelihood of correction of mismatched regions. Given that repair occurs over a limited region, it is reasonable that recombination should increase considerably as the distance between sites extends beyond the modal length of correction.

While intragenic recombination is generally proportional to physical distance, there can be large disparities which can be related to individual alleles. Unexpected high levels of intragenic recombination can occur for certain pairs of alleles (Moore and Sherman, 1977). While the reason for these disparities is not known, the observations suggest an interaction between the alleles either directly (such as palindromes) or through unique effects on enzymatic processing.

Modifications of DNA might be expected to alter recombination. One type of DNA modification which should be subject to genetic and biochemical analysis is methylation. The extent and pattern of methylation can vary considerably from that found in mitotic cells (Ponzetto-Zimmerman and Wolgemuth, 1984; Jagiello et al., 1982). Just as methylation can affect mutation (Glickman and Radman, 1980; Pukkila et al., 1983), it may affect mechanisms of correction, accessibility to enzymes, and even control of enzymes involved in meiotic recombination (Holliday and Pugh, 1975). Modifications of DNA can also be produced by DNA-damaging agents. Both intragenic and intergenic mitotic recombination are greatly affected by DNA damage (summarized in Resnick, 1979). Meiotic recombination can also be affected, however, it has not been possible except for the case of yeast to specifically relate a particular type of DNA damage to recombination and the capability for the repair of that damage (see Section IV,C).

The position of regions undergoing recombination could markedly affect the nature and extent of recombination. Regions near the centromere generally exhibit low levels of recombination (Mather, 1939; Beadle, 1932; discussed in Szauter, 1984). Recently it was shown in yeast that the centromeric DNA region affects recombination; movement of centromeric DNA from its natural position to another place in the chromosome enhanced recombination in the original region and decreased it in the new region where the centromere was moved (Lambie and Roeder, 1986). By manipulating centromere function and sequence in yeast (summarized in Resnick and Bloom, 1986) it should be possible to assess the components of the centromere and surrounding region that affect recombination. In organisms such as *Drosophila* in which chromosome regions can be defined in terms of euchromatic or heterochromatic regions, the level of recombination is greatly depressed in the latter (Mather, 1936; Sturtevant and Beadle, 1936; discussed in Szauter, 1984). This may relate directly to the genetic activity of these regions or alternatively may relate to the structure and lack of accessibility for recombination. The specific effect of genetic activity on recombination under conditions where a gene can be rendered transcriptionally active or inactive (i.e., turned on or off) during meiosis remains to be determined. The ribosomal DNA regions of both yeast (Petes, 1979) and *Coprinus* (Cassidy et al., 1984), which are highly repeated, exhibit very low levels of homologous chromosome recombination as measured genetically and molecularly; frequent sister chromatid exchange has been observed in the former (Szostak and Wu, 1980; Petes, 1980).

Another important aspect of chromosomal position in recombination relates to the nature of the chromosomal interaction that can lead to recombination. For example, while sister chromatid exchange is genetically neutral, it does provide a means for amplification when there are unequal exchanges. Sister chromatid exchanges (SCEs) have been examined genetically using ring chromosomes (cf. Gatti, 1982; Haber *et al.*, 1984), visually using differential staining techniques (summarized in Tice and Hollaender, 1984), and molecularly using recombinant DNA techniques (Szostak and Wu, 1980; Petes, 1980) or density labeling procedures (Moore and Holliday, 1976). The general conclusion from work in mammals (Allen and Gewaltney, 1985) and in grasshoppers (Tease and Jones, 1979) is that SCEs as measured cytologically occur at a much lower frequency than reciprocal exchanges during meiosis and levels are comparable to those for mitotic cells. Thus, while recombination in meiosis is greatly enhanced, the enhancement is specific to homologous exchanges. The specificity could be mediated through the synaptonemal complex or there could be additional controls. Similar results have been reported for genetically detected SCEs using ring chromosomes in that SCEs occur at a much lower frequency than homologous chromosome exchanges (Morgan, 1933; also see Gatti, 1982; Haber et al., 1984). If homologous recombination is required for normal disjunction, then it might be necessary that exchanges between homologues be higher than between sister chromatids. In the absence of a homologue, considerable intrachromosomal recombination can be observed in yeast, suggesting that homologous recombination suppresses or competes with intrachromosomal or sister chromatid recombinational events (Wagstaff et al., 1985).

The genetic controls and biochemical mechanisms of SCE and its relationship with other meiotic processes remain to be elucidated. One of the more interesting questions is the relationship between DNA synthesis in meiotic cells and SCE. In mitotically growing mammalian cells, SCEs are considered to occur during synthesis (Kato, 1974; see reviews in Tice and Hollaender, 1984). Since homologous chromosome recombination can be separated from DNA synthesis in meiotic cells, the mechanism for SCE may be very different, and this can be determined through a genetic and biochemical analysis. In mitotic cells of yeast it appears that systems responsible for general recombination (i.e., the *RAD52* pathway) are not required for SCEs in the repeat genes of ribosomal DNA (Petes, 1980), although they may play a role in SCE events for single-copy genes (Fassulo and Davis, 1986; Resnick et al., 1986a). While the mechanisms of the genetic control of amplification of DNA in *Drosophila* are not completely understood, the evidence is consistent with SCE processes (Endow et al., 1984).

3. Processing Factors Directly Affecting Recombination

Processing factors include the proteins and enzymes associated with recombination and the structures in which these are organized. Although little is known about the molecular events in meiotic recombination, the combination of genetic evidence, observations with mitotic cells, and results from prokaryotic systems have indicated the types of proteins expected. Some of these have actually been detected and characterized during meiotic development (see chapter 10 by Hotta and Stern; Lu, 1981; Resnick et al., 1984b). It has generally not been possible, except with yeast, to relate meiotic biochemical observations with genetic events during meiosis.

Recombination is generally considered to involve a recognition and initiation event (see Fig. 1). Based on the relatively small number of recombinational events (10 to 100 per genome) among various organisms and the sequence and regional differences, the initiation processes would appear to be coordinated rather than random. However, an excess of initiating events could occur with only a few actually leading to recombination, as has been suggested for *Lilium* (Howell and Stern, 1971). The actual initiation event could be coordinated by an initiation protein that might recognize a particular region or site in DNA. Based on observations from *Lilium*, nicking involves an association of a unique RNA

6. Genetics and Biochemistry of Recombination

species with the DNA sequences that are nicked (Hotta and Stern, 1981). There are several mechanisms that can be proposed for initiation, including restriction enzyme-like cuts, single-strand breaks introduced randomly into localized single-strand regions arising in supercoiled DNA, and nicking of DNA by double-strand endonucleases.

Following or as part of initiation, the recombinational structure could be mediated by a RecA-like protein which can facilitate the interaction and transfer of strands between such combinations as single-strand and double-strand molecules, double-strand linear DNA and nicked molecules (see Radding, 1982), and molecules with double-strand breaks and double-strand molecules (West and Howard-Flanders, 1984). The role of RecA protein in alignment and pairing of DNA molecules during recombination has been discussed by Bianchi *et al.* (1983).

A RecA-like protein has been purified from the fungus *Ustilago maydis* and has been shown to have properties generally similar to those of the RecA protein (Kmiec and Holloman, 1982). In addition it can facilitate the pairing and, in combination with topoisomerase, the linking of molecules. An intermediate is generated in interactions involving single-strand with double-strand molecules that appears to have a Z-DNA configuration based on antibody to Z-DNA (Kmiec and Holloman, 1984). The *rec1* mutants of *Ustilago* are deficient in this protein (Kmiec and Holloman, 1982) and the meiotic products from *rec1* mutants frequently exhibit chromosome breakage associated with reciprocal exchange and aneuploidy although gene conversion is normal (Holliday *et al.*, 1976). Therefore, it appears that this protein is involved in the proper development or regulation of recombinant structures.

The interactions between DNA molecules are expected to be facilitated by DNA binding proteins which could have the effect of either causing local regions of denaturation or enhancing renaturation. Such proteins have been identified in *Lilium* during meiosis (Hotta and Stern, 1971; and see Chapter 8). In yeast single-strand binding proteins (Chang *et al.*, 1978; LaBonne and Dumas, 1983; A. Sugino, unpublished) and ATP-dependent DNA unwinding proteins (A. Sugino *et al.*, 1986) have also been identified in mitotic cells. Many of them, including topoisomerases (Goto and Wang, 1982), are involved in replication. Their role in meiosis remains to be determined. Topoisomerases would be expected to be present to relieve topical constraints during recombination. The topoisomerase I of yeast (DiNardo *et al.*, 1983) is not required for mitotic growth or meiotic development while topoisomerase II is essential in mitotic cells for resolving newly replicated molecules (DiNardo *et al.*, 1984). The role of the topoisomerase II in meiotic replication and recombination remains to be determined.

Among the various models proposed for meiotic recombination, DNA polymerases function to either "drive" heteroduplex formation or correct mismatches. Since mutants have not been isolated which specifically affect DNA polymerization, it has not been possible to assess the role of DNA synthesis in recombination. Included with a DNA polymerization process is a need to ligate the newly synthesized region. While ligase-deficient mutants are available in yeast, it has been difficult to ascertain their specific role in recombination as compared to in normal synthesis because there is one detectable ligase. The absence of functional ligase (at restrictive temperature) in mitotic yeast cells results in a considerable stimulation of recombination, presumably due to the presence of single-strand breaks (Game et al., 1979). Using return to mitotic medium experiments (see Section IV,A) it has been possible to demonstrate a normal level of commitment to meiotic recombination in ligase mutants (Simchen, 1974).

Nucleases would also be expected to be involved during the initiation and formation of heteroduplexes and in the correction of mismatched regions that might arise in heteroduplex regions. Such enzymes have been identified for the repair of DNA damage (summarized in Friedberg, 1985). Increases in nucleases have been observed in *Lilium* (Howell and Stern, 1971) at a time during meiosis when recombination is expected. In yeast, a nuclease appears during normal meiosis at a time corresponding to recombination. Since it is absent in a recombination-defective mutant, this nuclease has been proposed to be directly involved in normal recombination (see Section IV,D,3).

Nuclease α has been identified in *Ustilago maydis* which, based on genetic evidence, may be involved in mismatch repair. The nuclease acts on single- and double-strand supercoiled DNA or in regions of distortion containing pyrimidine dimers or apurinic sites. It does not act on relaxed DNA or DNA with single mismatches, except under conditions of enhanced denaturation (Holloman et al., 1981). In strains that are mutant in both the *NUC1* and the *NUC2* genes, the nuclease is greatly decreased and meiotic intragenic recombination is reduced, although reciprocal exchange appears to be normal (Badman, 1972; discussed in Holliday and Dickson, 1977). While it is not possible to examine *Ustilago* during meiosis, these results are consistent with the nuclease α functioning to correct mismatches during meiosis. Mutants of yeast (*cor* or *pms*; Fogel et al., 1981; Williamson et al., 1985) have been identified which exhibit altered levels of postmeiotic segregation (see Hastings, Chapters 4 and 5); these genetic observations have been proposed as resulting from lowered efficiencies of mismatch repair.

Another unique category of deoxyribonucleases, resolvases, has been proposed for the resolution of recombinational intermediate cruciformlike structures (see Fig. 1). A resolvase activity has been identified with T4 bacteriophage and the protein has been isolated and characterized (Mizuchi et al., 1982). Resolvase-type activity has also been identified in mitotically growing yeast (West and Körner, 1985; Symington and Kolodner 1985), however, the specific action of this enzyme on Holliday-type structures and the role that this plays in mitotic or meiotic recombination remain to be determined.

Two other categories of enzymes, methylases and proteases, could be involved in the control of recombination and the processing of associated enzymes. Methylases could signal preferential direction of repair of mismatched regions as they function in preferential strand repair following exposure of cells to DNA-damaging agents (Glickman and Radman, 1980; Pukkila et al., 1983). Proteases may be important in meiotic processes as evidenced by the *pep* mutants of yeast (Jones, 1977; Zubénko and Jones, 1981). Of particular interest are the *pep4* mutants which abolish meiosis. However, because of the pleiotropic nature of these mutations, the specific role of a protease(s) has not been assessed. While a direct role for proteases in recombination has not been shown, they could function to process enzymes involved in recombination. An example of this is provided by the *uvs3* mutant of *Neurospora crassa,* which has been shown to be meiotically sterile. This mutant has reduced amounts of deoxyribonuclease activity although there is the accumulation of precursor (Chow and Fraser, 1979). The *uvs3* mutant lacks a proteolytic activity which cleaves the precursor protein.

It is reasonable to propose that some of the enzymes and proteins associated with recombination are organized into complexes or structures which carry out recombination in a concerted fashion and which are subject to genetic analyses. Such multiprotein complexes have been identified as being involved in the replication of bacteriophage T4 (Formosa et al., 1983). The recombination nodule originally described in *Drosophila* (Carpenter, 1975; and discussed in Section II,B,1) could be a similar macromolecular structure.

III. METHODS OF ISOLATING AND CHARACTERIZING MUTANTS

There are several approaches which can be used for obtaining mutants which are altered in meiosis and specifically in recombination. Many of these have been discussed extensively in the review by Baker et al. (1976) and are briefly mentioned here. Recent molecular biology techniques have considerably expanded the methods for obtaining mutants in specific functions. Presented in Table II is a list of approaches that are currently being taken or should be possible and of methods that utilize unique features of the genetic systems examined. Where possible, it is useful to isolate temperature-sensitive mutants to enable studies addressing the time during meiosis at which events occur, to order gene functions, and for the identification of the corresponding protein.

TABLE II

Approaches for Isolating Mutants Altered in Meiotic Recombination

1. Identify natural variants
2. Identify mutants in diploids by backcrossing or using homothallic strains
3. Obtain meiotic mutants in haploids which are capable of undergoing meiosis
4. Identify mutants for related gene functions in mitotic cells and determine role in meiotic cells
5. Suppression of mutants that affect meiosis/recombination; the suppressor mutations can be intragenic or extragenic
6. Isolate meiosis-specific transcripts, identify the corresponding genes, and directly alter the genes
7. Isolate meiosis-inducible genes and directly alter the genes
8. Use antibodies from other species to isolate relevant proteins and subsequently to identify the corresponding genes
9. Isolate sequences specifically involved in recombination
10. Introduce genes for enzyme functions which could affect recombination

A. Traditional Approaches

One of the difficulties in isolating recombination-deficient mutants has been that the functional defect is observed in diploid cells. Since most mutations affecting meiosis are expected to be recessive, they must be isolated and characterized in a homozygous condition in diploid organisms. One of the more obvious methods is to utilize natural variants and these have been summarized previously. Among the natural variants that have been particularly useful for studying the relation between meiotic biochemical events and chromosomal interactions is the Black Beauty (Toledo *et al.*, 1979) strain of *Lilium* (see Chapter 8). Chromosomal pairing does not occur in this strain. It has been possible, for example, to study the relation of the renaturing "R" protein and nicking events to chromosomal pairing in this strain (Hotta *et al.*, 1979). Since R protein does not increase nor are nicks observed during meiosis in this strain, it appears that the protein and the nicking events may be important in meiotic chromosomal interactions including recombination.

Other schemes used with diploid organisms include the mutagenizing of diploids, followed by homozygosis through backcrossing to identify mutants associated with meiotic defects or altered repair capabilities (see below). This approach has been used successfully with *Drosophila* to obtain meiosis-defective mutants (Sandler *et al.*, 1968; Baker and Carpenter, 1972); some of these have subsequently been shown to be mutagen-sensitive (Boyd *et al.*, 1976a). Specific biochemical repair defects have been described in diploid cells of some of these mutants including lack of excision repair for *mei-9* (Boyd *et al.*, 1976b) or postreplication repair (Boyd and Setlow, 1976) for *mei-41*. The *mei-41* mutants appear to lack a deoxyribonuclease (Chow *et al.*, 1986; see Section IV,D,4).

6. Genetics and Biochemistry of Recombination

However, the biochemical role of these repair genes in meiosis has not been determined. It is possible that some aspect of the corresponding mitotically defined repair function is required during meiosis. For the case of *mei-9* there is a good correspondence since *mei-9* mutants yield a higher frequency of postmeiotic segregation, consistent with a role for excision repair in recombination; levels of gene conversion are normal, while reciprocal exchange is reduced (Carpenter, 1982).

An alternative approach has been taken with yeast to obtain specifically meiotic mutants by mutagenizing haploid spores of homothallic strains. As the haploid single cells grow, mating-type switches can occur which enable mating of cells and the generation of diploids that are homozygous for all genetic markers. The diploid colonies that arise can be examined physiologically and genetically for meiotic defects. Several *spo* mutants have been isolated with this technique (Esposito and Esposito, 1969; summarized in Esposito and Klapholz, 1981).

Organisms which have life cycles that alternate between haploid and diploid states are particularly useful. As shown in Fig. 2, there are mitotic functions common to both haploid and diploid cells which might be expected to have a role in meiosis. It should be possible to screen for particular categories of mitotic mutants in haploids and, following the appropriate crosses, determine the effects on cells undergoing meiosis. This approach has been used with yeast and several categories of mutants affecting mitotic cell cycle activities (Hartwell, 1974; Simchen, 1978) or responses to DNA-damaging agents have led to the identification of genes involved in meiosis, particular recombination (summarized in Esposito and Klapholz, 1981). Among these are the mitotic *cdc* mutants which affect meiosis and for which a specific enzymatic defect has been described. For example, a *cdc28* gene mutant is arrested at the "start" position of the cell cycle; it is also defective in sporulation (Simchen, 1974; Shilo *et al.*, 1978) due presumably to lack of entry into meiotic development. The *CDC28* is interesting since it is the structural gene for a protein kinase and it shows homology with src oncogenes (Lorincz and Reed, 1984).

As discussed earlier, cell division cycle mutations have been especially useful for examining meiotic DNA metabolism. Most of the CDC mutants which affect DNA synthesis are also deficient in meiotic replication (Simchen, 1974; see Esposito and Klapholz, 1981, for a summary of the various *cdc* mutants and effects on meiosis). There may be DNA metabolic activities specifically associated with meiosis or meiotic recombination, and the *mei* (Roth, 1973, 1976) and *spo11* mutants (Klapholz *et al.*, 1985) might correspond to such genes. The products of some CDC genes involved with mitotic replication have been identified. The *CDC9* gene is responsible for DNA ligase (Johnston and Nasmyth, 1978), and haploidization does not occur in mutants at the restrictive temperature (Simchen, 1974). The *CDC21* gene corresponds to the structural gene for thymidylate synthetase (Bisson and Thorner, 1976); meiotic DNA synthesis does not occur in *cdc21* mutants. Similarly, meiosis is prevented in *cdc8* mutants (Zamb and Roth, 1977); this gene codes for thymidylate kinase (dTMP → dTTP; Scalafani and Fangman, 1984).

Several mutants have been isolated in haploid cells which exhibit increased sensitivity to a variety of agents including UV, ionizing radiation, and alkylating chemicals. Among the mutants are those which affect meiosis or meiotic recombination (Baker *et al.*, 1976). While it has been possible to examine their mitotic biochemical defects and in some cases to infer their function in meiosis (see Fig. 2), the role of the corresponding genes in meiosis has not been generally established. Many of the yeast mutants (Baker *et al.*, 1976; Haynes and Kunz, 1981; Game, 1983) which are sensitive to ionizing radiation exhibit defective meiotic recombination; included among these are mutants which lack double-strand break (DSB) repair (see Section IV,D,1). Similar mutants have been isolated in haploid *U. maydis* in that they are defective as homozygous diploids in mitotic recombination, meiosis, and the repair of radiation-induced DSBs (Leaper *et al.*, 1980). While this may indicate that DSBs have a role in meiotic recombination (Resnick, 1976; Szostak *et al.*, 1983), the mitotically observed defects should be treated judiciously in the interpretation of meiotic results (see Section IV,D).

Additional approaches for isolating meiotic mutants in haploids include the direct isolation of mutants that are defective in mitotic recombination using haploid strains that are disomic for a particular chromosome. Several mutants have been identified (see M. S. Esposito *et al.*, 1984) in which changes in DNA binding proteins have been observed; these mutants appear to alter meiotic recombination as well. One of the useful features of yeast is that meiosis can also be induced in haploids that are of the appropriate genetic construction. It should, therefore, be possible to identify directly recessive mutants that are defective in meiotic processes. This approach has been utilized by Fogel and Roth (1974), who developed strains that were disomic for the chromosome containing the mating-type alleles (to allow sporulation); several mutants were isolated that did not exhibit recombination between genes present on the disomic chromosome. An alternative approach is to use the *mar* mutation (Klar *et al.*, 1979) in combination with *spo13* (dispenses with the meiotic reductional division; Klapholz and Esposito, 1980a,b; see Section IV,B) in haploids. The *mar* haploids express complementary copies of the mating-type alleles on the same chromosome which enables cells to undergo meiosis. The combination of *spo13 mar* in a haploid allows meiosis and the production of two viable haploid spores. In addition to being useful for identifying genes involved in meiosis, the strains can be developed to monitor intrachromosomal recombination (Wagstaff *et al.*, 1985) or even homologous recombination if the strains are also disomic.

Having identified mutant genes it should be possible to use the mutants to isolate further mutants in the same genes or as intragenic or intergenic suppressors. With a mutation present in a heterozygous condition in a diploid, further mutations for the gene in question can be preferentially induced although recombination may result in homozygosity of the mutation. Intragenic suppressors can potentially provide information about the gene domain and interactions within the protein. As discussed by Botstein and Maurer (1982), intergenic suppressors provide the opportunity to determine interactions with other proteins or to identify alternative pathways. Complementation of a temperature-sensitive mutation by another temperature-sensitive mutation could arise by an interaction between different regions of interacting proteins. This approach has been used successfully to identify genes and interacting protein products associated with the morphogenesis of bacteriophage p22 (Jarvik and Botstein, 1973) in *Salmonella typhimurium*. Intergenic suppression can also be used to identify new pathways. The isolation of such suppressor mutations has been used extensively in *Escherichia coli* to characterize several facets of alternative recombinational pathways from the originally identified recA/recBC pathways (Clark, 1980). Such mutations have enabled the identification of specific nucleases involved in recombination. The suppression of mutations by complementation is clearly an important step in obtaining cloned genes for a particular trait and has enabled the isolation in yeast of several genes involved in meiosis and recombination. Selection schemes requiring the production of viable spores that exhibit recombination (i.e., prototrophy) and proper chromosome disjunction are possible. Such an approach has been used for the cloning of the *SPO11* gene (Giroux, 1984; DiDomenico *et al.*, 1984), which is generally required for meiotic recombination (Section IV,B).

B. Molecular Genetic Methods

Recent molecular genetic methods have involved the isolation of meiosis-specific RNA transcripts. Since much of the development that occurs during meiosis is specific to that stage of the life cycle, it is expected that certain genes would only be active during that time. It has been shown that the meiosis-specific *SPO11, -12,* and *-13* genes are induced during meiosis (DiDomenico *et al.*, 1984; Wang *et al.*, 1984). Enrichment for specific proteins (see below) and RNA transcripts has been demonstrated during meiosis in yeast (Kaback and Feldberg, 1985; Percival-Smith and Segall, 1984; Weir-Thompson and Dawes, 1984; Kurtz and Lindquist, 1984), *Coprinus* (Yashar and Pukkila, 1985), *Lilium* (Porter *et al.*, 1983), and *Tetrahymena thermophilia* (Martindale and Bruns, 1983). Of particular interest is a category of poly(A) RNA which is transcribed from a late replicating (during zygotene) region of DNA in both *Lilium* and mouse and which may have unique function in the subsequent events of meiosis (Hotta *et al.*, 1985). It should therefore be possible to identify, using cDNA libraries from mitotic cells and cells that are in various stages of meiosis, genes that are preferentially transcribed and function in the latter. Using this approach, several genes that appear to have meiosis-specific functions have been identified in yeast (Kaback and Feldberg, 1985; Percival-Smith and Segall, 1984). However, they are not essential since deletion mutations in these genes do not prevent meiosis or meiotic recombination. It is possible that they are, in fact, associated with developmental or recombinational functions in meiosis but that there is redundancy in the genes that can affect these functions.

As expected from the above observation with meiotic-specific mRNAs, it should be possible to identify meiotic proteins and subsequently determine the function of these proteins. Proteins have been examined directly from cells using 2-D gels (Petersen *et al.*, 1979; Kraig and Haber, 1980) or after *in vitro* translation of mRNA from meiotic cells (Weir-Thompson and Dawes, 1984). These approaches have led to the identification of several meiotic proteins, however, very few have been isolated and characterized further (summarized in Dawes, 1983). Meiotic-specific proteins might be expected to accumulate and these could be analyzed further. Recent molecular biology techniques enable the identification of a gene corresponding to a particular protein if the protein can be purified in sufficient quantities. Antibody can be raised against the protein and a gene expression library can be probed with the antibody. An alternative but more laborious procedure involves sequencing a portion of the protein, developing corresponding redundancy DNA sequences, and probing the appropriate DNA library.

Another method for identifying proteins involved with mitotic and meiotic functions also involves antibody detection. Since protein sequences are likely to be conserved between species, it may be possible to use antibodies raised against proteins of one species to identify antigenically similar proteins in another species. Having identified the protein in the organism of interest it should then be possible as mentioned above to isolate the gene in the appropriate expression library. This approach has been taken in the isolation of a nuclease from a yeast that is antigenically similar to a nuclease from *Neurospora;* the yeast enzyme appears to be involved in meiosis and possibly recombination (Chow and Resnick, 1983; see Section IV,D,4). Antibodies have been raised against mouse synaptonemal complexes (see Dresser, Chapter 8); using these antibodies, cross-reacting material has been identified in extracts of meiotically developing yeast (Giroux *et al.*, 1986). It has also been possible to identify protein in yeast that cross-reacts with antibodies raised against RecA protein of *E. coli* (Angul *et al.*, 1984). However, caution must be exercised in the use of antibodies for isolating related proteins between species since a protein identified with RecA antibody has been shown to be ribonucleotide reductase (S. Ellidge and R. Davis, personal communication).

Another related approach for looking at meiotic-specific functions is to isolate meiosis-inducible genes. This has been successfully ex;lored for the identification of genes (referred to as S.O.S.) that are inducible by DNA-damaging agents in *E. coli* and *S. typhimurium* (Walker, 1984). When inducible genes are cloned into a plasmid containing the *lacZ* gene, the inducing agent can lead to the production of readily detectable sugar metabolites. A similar method was used recently with yeast to isolate genes that are inducible by UV and alkylating agents (Ruby and Szostak, 1985). Adaptations of this method might reveal genes that are specifically induced during meiosis although as noted earlier the activities of nonspecific genes may also be enhanced (as was found by C. N. Giroux, C. Martinez-Arias, and M. Casadeban, personal communication).

Sequence can play an important role in recombination. By appropriate manipulation of cloned regions, "mutants" of the region can be developed and the effects on recombination examined. In yeast it is possible by the technique of specific gene transplacement (Rothstein, 1983) to insert a linear sequence into any part of the genome provided there are regions of homology with the ends of the sequence. Questions can then be asked about the effect of chromosomal location on recombination (see Section II,B,2). Similarly, regions of high recombination can be examined when they are placed in new environments. As mentioned earlier, experiments of this nature have been conducted with the centromere region of yeast. It is also possible to clone random regions of the genome and study the specific effects on recombination. This has been done by Keil and Roeder (1984), who cloned yeast DNA into a plasmid containing two copies of the *HIS4* gene each having a different mutation. They found that a fragment of ribosomal gene strongly stimulated mitotic but not meiotic recombination.

As mentioned earlier, the informational content of DNA can be changed by altering the pattern of methylation. One approach to evaluating the role of methylation is to block it genetically or alternatively to use azacytidine, which inhibits methylation. Another approach has been taken by Hoekstra and Malone (1985) by introducing DNA adenine methylase of *E. coli* into yeast. This "mutant" transformant is able to methylate GATC sequences whereas the parent does not. While variable (i.e., locus-specific) effects in mitotic recombination were observed, no effect on meiotic reciprocal exchange was detected. Even so, this approach emphasizes the possibility of using trans-species proteins for analyzing meiotic events.

Many genetic systems are limited in that it is not possible to relate specific mutational changes to biochemical defects. The *in vitro* assay system described by Symington *et al.* (1983, 1984b) may provide a general means of investigation. They developed methods for producing yeast extracts which were capable of catalyzing recombination reactions between plasmids. Recombination was detected by transforming $recA^-$ *E. coli* strains with the plasmid DNA. Using this system transformants were increased by 1000-fold when the plasmids were first exposed to extracts from wild-type yeast. While extracts of recombination-deficient yeast (Esposito *et al.*, 1984) including *rad52* yielded a factor of two to three times less than the Rad$^+$ strains, no extracts were completely deficient in recombination ability. Extracts of wild-type cells taken at a time corresponding to cellular recombination in meiosis resulted in a three- to fourfold increase over mitotic controls. Based on these observations, it is possible that other genetic systems could be analyzed in a similar fashion either directly or by complementation between species. Such systems could also be amenable to investigating sequence-specific effects.

IV. MOLECULAR AND GENETIC STUDIES OF MEIOTIC RECOMBINATION IN YEAST

The yeast *S. cerevisiae* is ideal for studying the genetic control of biochemical events in meiotic recombination and many examples of its use have already been cited. Included among the reasons are that various types of recombination can be examined, it is well characterized genetically, recombinant DNA/molecular biology studies can be performed readily (Struhl, 1983), it can be maintained as a haploid or a diploid and with the appropriate shift in medium it undergoes meiosis (see Dawes, 1983, for genetic controls), mitotic and meiotic mutants can be obtained relatively easily, it is easy to manipulate, and, as noted above, it is amenable to biochemical studies.

While it may be possible to obtain mutants that affect the products of meiosis both physiologically and genetically, the specific role of a particular gene in meiosis and recombination can only be assessed by correlating changes in the biochemical events during meiosis with changes in recombination. Because many meiotic mutants might be expected to yield either no gametes or inviable gametes, it may be necessary to relate a biochemical defect to recombinational processes taking place during meiosis. Such studies are possible in yeast for the following reasons: cells can be examined throughout meiosis for biochemical and molecular changes, it is possible to follow recombinational events at the DNA level, and the progress of genetic recombination can be monitored during meiosis in return to growth experiments (Sherman and Roman, 1963; also referred to as return to mitotic medium, RMM; Resnick *et al.*, 1986b).

A. Meiotic Events and RMM Experiments

RMM experiments involve plating cells to a medium that is diagnostic for recombination at various times during meiosis rather than examining the accumulated events in cells (spores) at the completion of meiosis. For example, recombination between heteroalleles of genes involved in amino acid metabolism can give rise to a wild-type copy of the gene through gene conversion or reciprocal exchange processes. Prior to recombination the cells will not grow on medium lacking the particular amino acid, while after recombination (or after a commitment to recombination) the cells will grow on the medium. Sherman and Roman (1963) and subsequently Simchen et al. (1972) and Esposito and Esposito (1974) were able to show that recombination could be detected in the cells long before spore formation. In fact, at sufficiently early times high levels of commitment to recombination could be detected even though the cells were still diploid. Cells that had undergone this process were referred to as meiototic (Esposito and Esposito, 1978) and it was possible under these conditions to order temporally recombination in various genes.

The RMM type of experiment provides the opportunity to relate specific biochemical events and mutational defects to recombination. This has been done with strains derived from SK-1 (Kane and Roth, 1974) that exhibit both an efficient and synchronous meiosis. As shown in Fig. 4, over 95% of the cells can give rise to asci within 12 hr following a shift from mitotic growth to meiotic medium (Resnick et al., 1984b). The increase in the number of cells that have become mature asci (nearly all contain three or four spores) occurs in a syn-

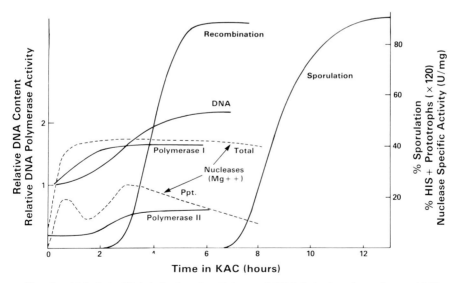

Fig. 4. Meiosis in SK-1-derived strains. Cultures of SK-1-derived strains undergo a highly efficient and synchronized meiosis in terms of DNA synthesis, recombination, sporulation, and the appearance of enzymes that are presumably associated with these processes, including polymerase I, polymerase II, and deoxyribonucleases. Recombination was measured during meiosis using the RMM procedures described in the text. (Adapted from Resnick et al., 1984a.)

chronous fashion with over 80% of the cells forming asci in less than 3 hr after the first appearance. The time for sporulation can be reduced if the mitotic cells are synchronized prior to incubation in the meiotic medium (M. A. Resnick, unpublished results). Using the high-efficiency meiosis strains, it has been possible to relate recombination to DNA synthesis, DNA polymerases, and nuclease activity. As shown in Fig. 4, there is a large increase in Mg^{2+}-dependent nuclease activity shortly after cells are introduced into meiotic medium and this remains high throughout much of meiosis. About an hour later DNA polymerases I and II increase by a factor of two to three and remain at a nearly constant level. Following this, total cellular DNA begins to increase to a level which is approximately twice the mitotic level; this is the meiotic round of DNA synthesis. Commitment to meiotic gene conversion can be detected shortly after the beginning of DNA synthesis. The recombinants which occur appear to be completed because incubation in a rich growth medium following incubation in meiotic medium but prior to plating to diagnostic medium does not alter the recombination frequency (see Section IV,D,3 and Fig. 5).

Using less synchronous strains, it has been possible to demonstrate that the appearance of recombinants in RMM experiments corresponds with the detection of recombined molecules. Borts *et al.* (1984, 1986) have examined recombination in the mating-type region of chromosome III. The region is detected by a specific probe following restriction of DNA. This region contains an insert in one of the homologous chromosomes as well as restriction site differences. Recombination within the region will give rise to a new band that can be detected at a level of about 1%. Recombinant molecules are observed at a time corresponding approximately to genetic recombination for other markers in the genome. While these experiments measure recombination, they can only detect reciprocal exchange.

B. Mutants and Biochemical/Genetic Effects

The RMM type of experiment has been used to characterize the defects in meiotic mutants, to examine the effects of DNA damage or role of repair systems in protecting cells during meiosis, and also to determine the role of repair genes in the progression of meiosis. Several meiotic mutants have been examined for effects on recombination and these have been summarized by Esposito and Klapholz (1981). Of particular interest are the *spo11, -12,* and *-13* mutants, *pac1,* which arrests in meiotic prophase, and *cdc5,* a mitotic cell division cycle mutant that also arrests at the pachytene stage of meiosis, as well as the *cdc* replication mutants (Sections II,B,1 and III,A). The *spo11* mutant is deficient in meiotic recombination based on RMM experiments and observations with spores (Klapholz *et al.,* 1985; Wagstaff *et al.,* 1985), although it does exhibit premeiotic DNA synthesis (Giroux, 1984; Klapholz et al., 1985). The naturally occurring *spo12* and *spo13* mutations eliminate the meiosis I division leaving only an equational meiosis II division (Klapholz and Esposito, 1980a,b). As might be expected of such mutants, events prior to the meiosis I division can occur normally, particularly the meiotic round of recombination. The two (rather than four) meiotic spores which arise are diploid but the chromosomes which are present have had the opportunity to undergo meiotic recombination. The *spo11* is clearly meiotic recombination-deficient (Klapholz et al., 1985) since the diploid spores arising from meiosis in *spo11 spo13* diploids lack any recombination. The *spo11* mutants are meiotic-specific in that they are not radiation-sensitive (C. N. Giroux, unpublished), do not markedly affect mitotic recombination (Bruschi and Esposito, 1983; Klapholz et al., 1985), and a meiosis-specific transcript of the *SPO11* gene is detected (DiDomenico et al., 1984). Meiotic recombination is considerably enhanced in the *pac1* mutant at its restrictive temperature (Davidow and Byers, 1984). However, the nature of recombination is different from normal recombination in that the region of heteroduplex formation as determined genetically appears to become substantially longer than that for wild type and also more postmeiotic segregation is observed for *pac1*, indicative of a long-lived heteroduplex. Since there is no increase in the frequency of reciprocal exchange, it appears that recombinant structures which are formed in the *pac* mutant have a longer time to interact. These results can be interpreted in terms of longer regions of branch migration (Davidow and Byers, 1984).

Another mutant has been identified recently that specifically affects meiosis without altering meiotic recombination. The mutation was created as part of a search for functions related to centromere stability. A 2200-base-pair transcript was identified that extended to within a few hundred base pairs of the centromere of chromosome XI. A disruption of the corresponding gene (by introducing a large segment of DNA) did not affect mitotic cells but it did result in the abolishment of meiotic asci; a single nucleus was observed in cells that had been incubated in meiotic medium (Yeh et al., 1986). Using the RMM experimental approach, it was shown that the cells were capable of a normal round of meiotic recombination (E. Yeh, unpublished); therefore, the gene appears to function at a stage later than the completion of recombination and is required for development to anaphase I.

C. Repair Mutants and Meiosis

It has been well established that exposure to DNA-damaging agents can alter the genetic outcome of meiosis. Using the RMM approach it is possible with yeast to examine the genetic and biochemical consequences of DNA-damaging agents during meiosis in Rad$^+$ (Machida and Nakai, 1980; Kelly et al., 1983a,b;

TABLE III

Mitotic and Meiotic Phenotypes of *RAD52* Epistasis Group Mutants

	rad50	rad51	rad52	rad54	rad55	rad57
Mitotic						
γ-sensitivity[a,b]	−	− −	− −	− −	±	−
DSB repair[c]	−	−	−	−		±
Mating-type switching[a]	+*	−	−	−		+
Spontaneous mitotic recombination	+ → + +[d]	±[e]	− → ±[d,f,g]			
Induced mitotic recombination	− → ±[h,i]	−[i]	−[g,i,j]	− → ±[h,i]	±[h,i]	− → ±[h,i]
SS-nuclease/Ab sensitive[k]	+	±	−	+		+
Meiotic						
Viability of spores	−[f]	−[a,e]	−[f]	±[a]	−	−[f]
Viability of radX spo13 spores	+[l]	−[a]	−[l]		−[m]	−[a]
DNA synthesis	+[f]	+[e]	+[f]			+[f]
Recombination:						
RMM	−[n]	±[e]	±[n]		+[m]	±[o]
radX spo13	−	NA	NA			NA
Single-strand breaks	None[n]		Yes[n,o]			Yes[o]

* + = wild-type levels; NA = not applicable since spores are dead.
[a] Summarized in Game (1983).
[b] Game and Mortimer (1974).
[c] See text.
[d] Malone (1983).
[e] Morrison and Hastings (1979).
[f] Game *et al.* (1980).
[g] Prakash *et al.* (1980).
[h] Saeki *et al.* (1980).
[i] Summary by Haynes and Kunz (1981) of Strike (1978).
[j] Resnick (1975).
[k] Chow and Resnick (1983, 1986).
[l] Malone and Esposito (1981).
[m] S. Lovett (personal communication).

Salts *et al.*, 1976) and various mitotically defined repair-deficient mutants (Resnick *et al.*, 1983a,b). The RMM method is particularly useful since viability of gametes is often low after treatment with various agents. More significantly there has been the opportunity to evaluate the role that repair genes play in normal meiosis and to use repair mutants to examine the events that take place in meiosis.

	rad50	rad51	rad52	rad54	rad55	rad57
Other						
Mitotic aneuploidy	+[p]	High[e]	High[q]		High[r]	High[r]
Spontaneous mitotic mutation	+	++[s]	++[t]	++[u]	++[u]	++[u]
Spontaneous unequal sister chromatid recombination[v]	+		−	−		

[n] Resnick et al. (1981, 1984a, 1986b).
[o] J. Nitiss and M. A. Resnick (unpublished).
[p] D. Schild and R. K. Mortimer (personal communication).
[q] Mortimer et al. (1981).
[r] Discussed in Game (1983).
[s] Hastings et al. (1976).
[t] von Borstel et al. (1971).
[u] S.-K. Quah and R. C. van Borstel, discussed in Haynes and Kunz (1981).
[v] Resnick et al. (1986a).

1. Categories of Repair Mutants in Relation to Biochemical Defects

Over 50 loci have been identified in *S. cerevisiae* that affect sensitivity to a variety of DNA-damaging agents (see reviews by Haynes and Kunz, 1981; Game, 1983). Among these are three sets of genes that have been assigned to common epistasis groups or "pathways" based on their interactions with diagnostic mutants in these pathways: *RAD1*, *RAD52*, and *RAD6*. The *RAD1* epistasis group is involved with excision repair processes, particularly the removal of UV-induced pyrimidine dimers. One of the genes, *RAD3*, has been found to be essential for mitotic growth since deletion mutants are inviable (Higgins et al., 1983; Naumovski and Friedberg, 1983). The *RAD6* group of genes is identified by the absence of damage-induced mutagenesis in these mutants. The *RAD52* group has been identified by sensitivity to ionizing radiation; a large proportion of the mutants for genes in this group lack the ability to repair radiation-induced double-strand breaks (Section IV,C and Table III).

Several genes in the *RAD1* group (Dowling et al., 1985; DiCaprio and Hastings, 1976) have been examined for their effects on meiosis in terms of recombination, postmeiotic segregation (which would be diagnostic of an involvement in heteroduplex repair), and viability. Other than a small reduction in viability among some double-mutant combinations, there is no apparent involvement of these genes in meiosis or meiotic recombination. These results are consistent with molecular investigations of the chromosomal DNA during meiosis. Using highly synchronous strains, it was shown that DNA synthesis patterns and the size distribution of DNA during meiosis were comparable in Rad$^+$ and *rad1* strains; the recombination was also comparable (Resnick *et al.*, 1983a,b).

The *rad6* mutation has a marked effect on meiosis including the abolishment of sporulation. Based on results from RMM experiments, gene conversion and/or reciprocal exchange does not occur although there is meiotic DNA synthesis (Game *et al.*, 1980). Other mutants (including *rad5, rad9,* and *rad18;* Dowling *et al.*, 1985; DiCaprio and Hastings, 1976) in the *RAD6* epistasis group which have varying effects on mutagenesis do not affect meiosis. Electron microscopic investigations of the *rad6* mutants have shown an absence of chromosome pairing and the absence of a proper metaphase spindle (Kundu and Moens, 1982; Moens, 1982). The *RAD6* gene has been cloned and sequenced (Reynolds *et al.*, 1985); the gene has been shown to be nonessential in mitotic growth, although it is essential in meiosis (Game *et al.*, 1980; Malone and Esposito, 1981). The carboxy terminus sequence corresponds to a long stretch of acidic amino acids, particularly aspartate. The corresponding protein would be expected to have unique properties in terms of chromosomal interactions and may be similar to the nucleosome high-mobility proteins (HMG) which are involved in nucleosome assembly (see Johns, 1982). For example, it might mediate the action of proteins involved in processing DNA damage or even recombination. Since the meiotic round of DNA replication occurs in *rad6* mutants, it is possible that the *RAD6* protein functions to allow interactions with DNA-metabolizing proteins or complexes involved in signaling recombination.

2. Protection and Genetic/Biochemical Effects

The DNA metabolic and genetic consequences of ultraviolet light (UV), which induces pyrimidine dimers as the major type of DNA damage have been extensively examined in repair proficient and defective cells (Resnick *et al.*, 1983a). The same *RAD1* system which was shown to be involved in the repair of pyrimidine dimers in mitotic cells was also involved in the repair of damage in meiotic cells. Cells were irradiated at the beginning or at various stages of meiosis and allowed to continue in meiosis. In the absence of pyrmidine dimer excision, cells were highly sensitive to UV. Low UV doses to the *rad1* mutants early in meiosis reduced the extent of the subsequent meiotic gene conversion and appeared to abolish reciprocal exchange; the molecular exchanges expected in chromosomal DNA during normal meiosis also were not detected. These results suggested that while the excision repair genes were not important in normal meiosis, they were essential for the protection of cells undergoing meiosis. They also indicated that small amounts of DNA damage could have a large effect on meiosis and the damage could exert its effect over a considerable distance; in addition, unrepaired DNA damage could dissociate gene conversion from reciprocal recombinational events in meiotic cells.

D. The *RAD52* Pathway in Meiosis

The mutants of genes in the *RAD52* group were originally identified by increased sensitivity to ionizing radiation (summarized in Game and Mortimer, 1974). The genes in this epistasis group which have been studied extensively include *RAD50, -51, -52, -54, -55,* and *-57*. Various mutants in this epistasis group are important in several meiotic and mitotic functions including mutagenesis (see Table III). Many of the mutants show altered mitotic and meiotic recombination as well as high levels of mitotic aneuploidy. The aneuploidy events may be related to unresolved or improperly resolved recombinational events (Mortimer *et al.*, 1981; discussed in Mason and Resnick, 1985). Similarly, the absence of mating-type switching which involves DSBs (Strathern *et al.*, 1982; Klar *et al.*, 1984a) indicates the general involvement of these genes in some types of recombination. The meiotic deficiencies are not related to entry into meiosis but are due to an absence or an alteration in recombination.

1. General Description: Mitotic and Meiotic Defects

Among the mutants which have been examined, *rad50* (J. Nitiss, unpublished), *rad51* (Jachymcyzk *et al.*, 1981), *rad52* (Resnick and Martin, 1976; Ho, 1975), *rad54* (Budd and Mortimer, 1982), and *rad57* (J. Nitiss, unpublished), there is the general observation that mitotic cells lack the ability to repair DNA double-strand breaks induced in chromosomal DNA and are deficient in ionizing radiation-induced mitotic recombination. Several observations led to the conclusion that DSBs in chromsomal DNA were repaired through a process of recombination between homologous chromosomes or sister chromatids and resulted in a model for DNA double-strand break repair which was also proposed for initiating events in normal meiotic recombination (Resnick, 1976; Szostak *et al.*, 1983). The sensitive component of a Rad$^+$ haploid cell population was explained as being due to a population of cells (G_1) that were incapable of recombinational repair (Resnick, 1979). Subsequent experiments strongly supported this view, including the observation that DSBs in duplicated chromosomes are effectively repaired (i.e., G_2; Brunborg *et al.*, 1980). In addition, it was found that mating-type switching appeared to be initiated by a DSB (Strathern *et al.*, 1982) and that switching in *rad52* mutants did not occur; attempts to switch led to lethality (Malone and Esposito, 1980). Using a much more defined system, it was shown that DSBs in plasmid DNA were highly recombinogenic and that the recombinational process required the *RAD52* gene product (Orr-Weaver *et al.*, 1981). These results led to the highly innovative method of gene transplacement (Rothstein, 1983).

Because of the importance that the *rad* mutants had on mitotic recombination, investigations were initiated (Game *et al.*, 1980) to determine the role that these genes might play in meiosis and meiotic recombination. The *rad50*, *rad52* (also see Prakash *et al.*, 1980), *rad51* (Morrison and Hastings, 1979), and *rad57* mutants all led to a reduction in asci and the spores in these asci were inviable. These genes and *RAD6* (see Section IV,C) were concluded to be essential for normal meiosis but not for mitotic growth. Since meiotic DNA synthesis was observed, these genes are not required for the switch to meiotic development. Based on results from RMM experiments it was concluded that the mutants lacked meiotic recombination. Consistent with this was the observation that recombined molecules are not detected in *rad50* (Borts et al., 1984) and much reduced in *rad52* strains (Borts *et al.*, 1986). It is interesting to note that not all genes which are required for the repair of DSBs in mitotic cells are essential in meiosis. The *rad54* mutants are radiation-sensitive, lack DSB repair, and exhibit complete meiosis (Game and Mortimer, 1974).

2. Molecular Changes and Distribution

The *rad50*, *rad52*, and *rad57* mutations have been particularly useful for examining some of the biochemical events in meiosis and relationships between recombinational events at the molecular level and eventual genetic consequences. Since these mutants were unable to repair DSBs in mitotic cells, it was anticipated that such breaks might occur during normal meiosis and that the meiotic defect might be due to a processing problem. Sucrose gradient procedures were developed for analyzing full-size chromosomal DNA during meiosis in strains that exhibited high levels of meiotic synchrony (see Fig. 4). Neither double-strand nor single-strand breaks (SSBs) were detected in Rad$^+$ strains (Resnick *et al.*, 1981; J. Nitiss and M. A. Resnick, unpublished data). While some sort of break would be expected to occur in association with recombination, the inability to detect them could be due to rapid resolution when they occur. Using techniques discussed below, alternative methods of analysis may enable their detection. Single-strand breaks have been identified in *Lilium* (Howell and Stern, 1971; Hotta and Stern, 1974); however, because of their excessive numbers in relation to the number of exchanges that occur and the inability to relate them to genetic recombination, their function in recombination is not clear. Single-strand breaks have also been reported in yeast (Jacobson *et al.*, 1975; Kassir and Simchen, 1980), however, their relationship to recombination is not established, partly because of technical difficulties (for example, it was not possible to determine the number average molecular weight of chro-

6. Genetics and Biochemistry of Recombination

mosomal DNA). There have been observations recently of a small number of DSBs during meiosis (Høgset and Øyen, 1984), but these may be due to technical problems in isolating and characterizing chromosomal DNA from late meiotic cells (J. Nitiss, unpublished).

In *rad52* mutants single-strand interruptions have been detected after the meiotic round of DNA synthesis has begun (Resnick *et al.*, 1981); they appear in newly synthesized as well as parental DNA in comparable numbers (J. Nitiss and M. A. Resnick, unpublished data). No interruptions are detected in *rad50* mutants (Resnick *et al.*, 1984a). The interruptions which were detected with alkaline sucrose gradient methodologies were subsequently shown to be breaks and not alkaline-sensitive sites or gaps (Resnick *et al.*, 1984a). Because the number of events was so small relative to the length of DNA (see below), it was necessary to isolate the DNA without the artifactual introduction of breaks in the DNA in order to characterize the interruptions. To do this the DNA was sedimented through a neutral sucrose gradient containing a restriction enzyme (blunt-end) layer. The resulting DNA which was isolated could then be handled without further breakage (either SSBs or DSBs). The presence of nicks or gaps was indicated by the ability of the isolated DNA to act as primer in a nick translation DNA synthesis assay. Since DNA synthesis took place with polymerase I from *E. coli* but not with the Klenow fragment of this enzyme or with T4 polymerase (which lack nick translation activity), it was concluded that the interruptions were due to breaks and not gaps in the DNA. Furthermore, some of the breaks could be ligated.

The total number of breaks observed in *rad52* mutants (Resnick *et al.*, 1981), approximately 200 to 400 per cell, is comparable to the number of recombinational events expected in a wild-type strain during meiosis (assuming 5000 centimorgans and two chromosomes involved in each exchange event). These breaks are proposed to be directly involved in recombination based on their frequency and the time of appearance relative to recombinational events in wild-type cells (Resnick *et al.*, 1984a,b); results with *rad50* mutants and additional genetic observations (discussed below) lend further support to this proposal. It is interesting that the recombinationally silent ribosomal DNA region does not exhibit the SSBs during meiosis (Høgset and Øyen, 1984).

In the initial report of SSBs in a *rad52* mutant (Resnick *et al.*, 1981), it was not possible for technical reasons (Resnick *et al.*, 1984b) to analyze their distribution. These problems have been partly overcome by probing for specific chromosomes from a sucrose gradient (Resnick *et al.*, 1984a). Unbroken single strands of DNA from a specific chromosome should sediment to a particular position in an alkaline sucrose gradient. Random breaks would result in random-size pieces of DNA while unique break sites would yield DNA in specific fractions. Based on a mathematical analysis of number and sizes of pieces of DNA from the fraction of a sucrose gradient that hybridize with a *SUP4–CDC8* probe, it was concluded that the breaks which are present in this region of chromosome X of a *rad52* mutant are not randomly distributed. Probing of other chromosomes in recent experiments has yielded similar results (J. Nitiss, unpublished). These results indicate that if these breaks are directly related to recombination, the recombinational sites may have regional or site specificity; this is in agreement with genetic evidence for polarity of recombination in yeast (discussed in Fogel *et al.*, 1981).

3. RMM Experiments with rad52, rad50, and rad57 Mutants

While the *RAD50* and *RAD52* genes may be required in meiotic recombination, it is not possible to discern from the above experiments whether they are required for the initiation of recombination or are involved in intermediate steps. To more fully demonstrate the role of *RAD50* and *RAD52* in meiotic recombination it was necessary to analyze the mutants under conditions where the absence of recombination would not lead to inviability of the meiotic spores. Strains were developed which were mutant for one or both of these genes as well as the *SPO13* gene (Malone and Esposito, 1981; Malone, 1983; see Section IV,B). Meiotic recombination is not necessary for proper disjunction in the double mutants because the only chromosome segregation that occurs is equational. Malone and Esposito (1981) showed that the *spo13 rad50* diploids, unlike S_1 *rad50*, gave rise to viable spores. Genetic analysis of the diploid spores revealed that meiotic recombination had not occurred although recombination did occur in Rad$^+$ *spo13* mutants. These results demonstrate that the *RAD50* gene is required for recombination, but not for meiotic physiological development, and recombination is required for the production of viable haploid meiotic products; recombination presumably assures proper reductional disjunction. Similar experiments were done with *rad52 spo13* double mutants; it was concluded that *rad52* mutants were phenotypically different from *rad50* mutants because the *rad52 spo13* spores which arose were inviable. Since the triple mutant *rad50 rad52 spo13* was viable, it was proposed that in a genetic sense the *RAD50* function precedes *RAD52* in meiotic recombination. The observation that SSBs are not detected in *rad50* mutants during meiosis is consistent with a blockage at an earlier step (Resnick *et al.*, 1984a). In addition, the *rad50* mutants lack synaptonemal complex and paired chromosomal structures during meiosis while they are present in *rad52* (B. Byers, personal communication). In some way, blockage at a later step in meiosis as occurs in *rad52* mutants is deleterious.

Since the *rad50* and *rad52* mutants could be distinguished in terms of genetic and molecular consequences and order of function, the effects of the mutants on recombination in RMM experiments were reexamined in strains which exhibited greater meiotic synchrony (see Fig. 4), therefore transient events could be more

6. Genetics and Biochemistry of Recombination

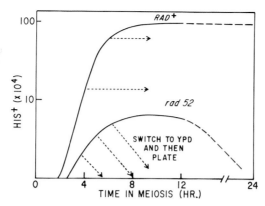

Fig. 5. The induction of recombination in a *rad52* mutant by meiosis. The *rad52* mutant was originally identified as being recombination-defective during meiosis in RMM types of experiments (Game *et al.*, 1980). Using more synchronous strains, it has been possible to demonstrate that genetic recombination can be detected during meiosis at a time when recombination occurs in a wild-type strain. The maximum level in the *rad52* strain is about 10 to 20 times lower than that in the Rad⁺ strain. Unlike the situation in Rad⁺ strains, the recombinants require incubation in mitotic medium under nongrowth conditions for expression. This is shown by the experiment in which cells are transferred to rich growth medium (YPD) before plating to the mitotic restrictive medium on which only recombinants will grow. Only recombinants will grow on the latter. As shown in this figure, a short period of incubation in YPD results in the loss of the recombinants. These results suggest that there is a potential for recombination in the *rad52* mutants during meiosis but the actual development of the "complete" recombinants requires additional functions. (Summarized from experiments of Resnick *et al.*, 1986.)

readily detected (Resnick *et al.*, 1986b). No increased recombination over the mitotic controls was observed with the *rad50* mutants. However, meiotic specific recombination could be detected with the *rad52* mutants under these conditions; the number of recombinants increased during meiosis at about the time of meiotic recombination in Rad⁺ strains and then decreased. The meiotic frequency increased as much as 20 times over the mitotic levels (Fig. 5), however, the maximum was at most 5–10% of the Rad⁺ level. While it was surprising to observe that meiosis could induce recombination in *rad52* mutants, previously considered to be recombinationless, it was not surprising that the *rad52* would differ from *rad50* mutants. The results were interpreted in terms of the SSBs described above. Possibly the SSBs correspond to initiating or intermediate events in recombination and the plating of cells to the diagnostic mitotic medium in the RMM experiments allows for the completion of the recombinational process. To test this, cells were incubated for a brief period in rich groth medium (YPD) prior to plating on diagnostic medium. As shown in Fig. 5, this resulted in the complete loss of recombinants. Since there was no reduction in Rad⁺ strains, it would appear that recombinational events are completed in the meiotic medium for wild-type cells but not *rad52* cells.

Based on methods used for studying the *rad52* defect, recent observations with *rad57* mutants suggest that the *RAD57* gene functions later in meiosis than *RAD52*. The *rad57-1* mutant is cold-sensitive for the repair of radiation-induced DSBs. During meiosis cells die at both the "cold" temperature (23°C) and the temperature permissive for DSB repair in mitotic cells (34°C). It had previously been shown that recombination in RMM experiments could be detected at both temperatures (Game *et al.*, 1980). Using more synchronous strains, wild type levels were detected at the two temperatures. At 23°C the recombinants were unstable when challenged with YPD while at 34°C they were stable (Nitiss, 1986; Resnick *et al.*, 1986a). Single-strand interruptions were observed and the frequencies at both temperatures were comparable to those with *rad52* mutants (Nitiss, 1986). Given the number of single strand interruptions, the high levels of recombination and the development of stable recombinants at the high temperature, it appears that the *RAD57* gene functions later in the recombinational process than *RAD52*. The *RAD50*, *RAD52*, and *RAD57* genes which were originally identified in mitotic cells as being required for the repair of ionizing radiation damage, specifically DSBs, now appear to define different steps, respectively, in the process of meiotic recombination.

It has therefore been possible to relate the functions of specific genes involved in meiotic recombination to the temporal events of recombination and molecular changes in DNA. As part of a complete understanding of the role that these genes play in recombination, it is important to determine the products of these genes and their action during meiosis. Considerable progress has been made in terms of cloning several of the genes in the *RAD52* epistasis group (Calderon *et al.*, 1983; Kupiec and Simchen, 1984). In addition the *RAD52* gene has been sequenced (Adzuma *et al.*, 1984). While the corresponding proteins or functions have not been identified, a nuclease has been characterized that is controlled by the *RAD52* gene.

4. Nucleases and RAD52 Control

Several deoxyribonucleases have been identified in yeast; however, their genetic control and relevance to cellular activities have not been established. Because of the large amounts of material needed for enzyme analysis, most studies have focused on mitotic cells. Using the SK-1-derived strains described above (see Fig. 4), alkaline (pH 8) single-strand deoxyribonuclease activity has been followed throughout meiosis and related to the timing of recombination. Under the premeiotic growth conditions employed, the level of activity prior to switching to meiotic medium is low (Resnick *et al.*, 1984b). Within 1 hr following the switch to meiotic medium, there is a 5- to 10-fold increase in activity which

6. Genetics and Biochemistry of Recombination

remains high throughout the remainder of meiosis (Fig. 4). This increase occurs shortly before an increase in DNA polymerase and also before the appearance of recombinants as measured in RMM experiments. The increase in activity is meiosis-specific since strains which do not undergo meiosis because of homozygosity for mating type do not exhibit the increase. The nuclease activities can be divided into two categories, Mg^{2+}-dependent and -independent, and both groups increase during meiosis.

Within the category of Mg^{2+}-dependent nuclease activities in both mitotically growing and meiotically developing wild-type cells is an activity that is absent in *rad52* mutants. This activity was identified utilizing a method described in Section III,B involving the cross species identification of antigenically similar enzymes. A single-strand deoxyribonuclease had been isolated and characterized from *Neurospora* which was greatly increased in *uvs3* mutants (Chow and Fraser, 1983). The interesting feature of the *uvs3* mutant of *Neurospora* was that it was radiation-sensitive and it did not undergo meiosis. Antibody raised against the nuclease (Fraser *et al.*, 1986) was utilized to test for antigenically similar enzyme activity in yeast extracts. Approximately 40 to 50% of the Mg^{2+}-dependent single-strand nuclease activity in logarithmically growing yeast cells was precipitated by the *Neurospora* antibody (Chow and Resnick, 1983). A comparable percentage of activity was precipitable from meiotically developing cells within 1 hr after introducing cells into meiotic medium, however, the amount of activity which could be precipitated decreased at later times (Resnick *et al.*, 1984b).

Using this antibody it has been possible to purify the corresponding nuclease from logarithmically growing yeast (the enzyme activity is absent from stationary phase cells). The purified protein has a molecular weight of 72,000 (Resnick *et al.*, 1984a; Chow and Resnick, 1986). The enzyme exhibits both single-strand and double-strand nuclease activity, however, it is about five times more active toward the former. It also exhibits single-strand DNA endonuclease activity and ribonuclease activity. These properties, which are similar to those of the corresponding *Neurospora* nuclease, are presumably important in DNA repair and/or recombination.

Based on the observations with *Neurospora*, several radiation-sensitive mutants of yeast have been examined for the absence of the *Neurospora*-like nuclease. Only the *rad52* mutants have been shown to completely lack activity in mitotic and also meiotic cells. It has been concluded from these results that the *rad52* defect may arise from an absence of deoxyribonuclease (Chow and Resnick, 1983, 1986). Depending on the model of recombination invoked, this nuclease could function to produce strand breaks in DNA, lead to the formation of double-strand breaks, or even modify breaks through single-strand or double-strand exonuclease activity (see Resnick *et al.*, 1984a). It is important to note that since the protein is significantly larger than predicted from the reading frame of the *RAD52* gene (Adzuma et al., 1984), it is probably not the product of this gene. If this is so, then it would indicate that the *RAD52* gene may serve a control or regulatory function within the cell and the nuclease may be only one of the proteins influenced by this gene. More information about the nuclease will come with the identification of the corresponding nuclease gene (T. Y.-K. Chow and M. A. Resnick, unpublished) and the analysis of derived mutants. It will also be interesting to determine whether a similar nuclease exists in other organisms. Recent evidence (Chow et al., 1986) suggests that indeed an antigenically similar activity is present in *Drosophila* and that mutants of one of the meiotic loci, *mei-41*, lack this activity.

The *RAD52* gene has been shown to play an important role in meiotic recombination as well as in mitotic repair. Using a combined genetic, molecular, and enzymological approach with mutants of this gene it has been possible to gain insight into the process of meiotic recombination. As discussed above, intermediate events in recombination (i.e., compare *rad52* with *rad50* mutants) have been identified and under the appropriate conditions (i.e., RMM) they can lead to recombination. The *RAD50* gene product may provide a signal or enable chromosome interactions to occur so that the subsequent recombination events can occur. The *RAD52* gene product could play an important role in either directly processing recombinational intermediates or regulating the proteins involved. The *RAD52* gene product appears to function later in the recombination, possibly near its completion. While it is possible to suggest other reasons, the results with the *rad52* mutants are consistent with the idea that a deficiency in a particular nuclease under the control of the *RAD52* gene prevents the subsequent processing of the intermediates. The events which involve the accumulation of single-strand breaks are of particular interest since the breaks are not randomly distributed, which suggests (as noted above) unique sites or regions involved with recombination.

V. CONCLUSIONS

A complete understanding of the relationship between the recombinational events which occur during meiosis as determined in the meiotic products and the underlying biochemical processes requires a genetic dissection of the relevant biochemical pathways. Among the various organisms which have been examined there are large numbers of mutants that affect meiosis, particularly recombination. Because of the inaccessibility of most systems to a combined genetic and biochemical investigation or the lack of opportunity to relate biochemical to genetic events, progress has generally been slow in understanding the underlying biochemical mechanisms of genetic recombination. Recent investigations with the *RAD52* epistasis group of yeast represent the only integrated approach.

As discussed within this review, various methods are now available for identifying genes involved in meiosis and meiotic recombination, determining the products, and subsequently evaluating their role in meiosis. There are only a small number of systems, particularly yeast, which will allow a detailed evaluation of the mechanisms of *in vivo* genetic recombination. It may, however, be possible to express genes from one organism in other more genetically accessible systems; those involved with synaptonemal complexes would be good candidates. An approach which may have general applicability involves the detection of recombination *in vitro* and possible complementation between systems. Thus, while there have been limits in many systems on the study of meiosis and meiotic recombination, current techniques should enable considerable progress in the near future.

It is clear that there are several meiosis-specific genes and these may include genes involved in recombination. This category is very different from those that also function mitotically. In an evolutionary sense it is reasonable to expect that mitotically active genes could be included in the evolution of meiotic systems. Common functions such as DNA synthesis might be expected to utilize previously existing genes. Unique functions would need to be evolved and these are exemplified by the various genes that are meiosis-specific and are involved in meiotic control mechanisms for development and recombination. The genes identified originally as being involved in mitotic repair and damage-induced recombination are interesting in that some of them are mitotically dispensable (i.e., not required for growth) but meiotically essential. They presumably serve two functions in meiosis, the repair of DNA damage and participation in meiotic recombinational processes. It is not clear whether they also function to correct DNA alterations which might occur during normal mitotic growth (see Section I,B).

The investigations into meiotic gene functions can therefore be seen as a means for understanding not only meiotic processes but also the genetic and biochemical interrelationships between various stages of the life cycle, developmental events, and mechanisms of control. Recent advances with molecular biology techniques will considerably aid in our understanding of the various aspects of meiosis.

ACKNOWLEDGMENTS

I greatly appreciate the important suggestions and critical reviews by Drs. Mason, Chaudhury, Nitiss, Giroux, and Chow, and the opportunity to cite unpublished results by several colleagues.

This chapter was written by the author in his private capacity. No official support or endorsement by the National Institute of Environmental Health Sciences is intended or should be inferred.

REFERENCES

Adzuma, K., Ogawa, T., and Ogawa, H. (1984). Primary structure of the *RAD52* gene in *Saccharomyces cerevisiae*. *Mol. Cell. Biol.* **4,** 2735–2744.

Allen, J. W., and Gewaltney, C. W. (1985). Sister chromatid exchanges in mammalian meiotic chromosomes. *In* "Sister Chromatid Exchanges" (R. R. Tice and A. Hollaender, eds.), pp. 629–645. Plenum, New York.

Amstutz, H., Munz, P., Heyer, W.-D., Leupold, U., and Kohli, J. (1985). Concerted evolution of tRNA genes: Intergenic conversion among three unlinked serine tRNA genes in *S. pombe*. *Cell* **40,** 879–886.

Angel, T., Austin, B., and Catcheside, D. G. (1970). Regulation of recombination of the his-3 locus in *Neurospora crassa*. *Aust. J. Biol. Sci.* **23,** 1229–1240.

Angulo, J., Schwencke, J., Moreau, P., Moustacchi, E., and Devoret, R. (1985). A yeast protein analogous to *Escherichia coli* RecA protein whose cellular level is enhanced after UV irradiation. *MGG, Mol. Gen. Genet.* **201,** 20–24.

Badman, R. (1972). Deoxyribonuclease-deficient mutants of *Ustilago maydis* with altered recombination frequencies. *Genet. Res.* **20,** 213–229.

Baker, B., and Carpenter, A. T. C. (1972). Genetic analysis of sex-chromosomal mutants in Drosophila melanogaster. *Genetics* **71,** 255–286.

Baker, B. S., Carpenter, A. T. C., Esposito, M. S., Esposito, R. E., and Sandler, L. (1976). The genetic control of meiosis. *Annu. Rev. Genet.* **10,** 53–134.

Beadle, G. (1932). A possible influence of the spindle fiber on crossing over in *Drosophila*. *Proc. Natl. Acad. Sci. U.S.A.* **18,** 160–165.

Bennett, M. D. (1983). The spatial distribution of chromosomes. *In* "Kew Chromosome Conference III" (P. E. Brandham and M. D. Bennett, eds.), pp. 71–79. George Allen and Unwin, London.

Bernstein, C. (1979). Why are babies young? Meiosis may prevent aging of the germ line. *Perspect. Biol. Med.* **22**(4), 539–544.

Bernstein, H., Byers, G. S., and Michod, R. E. (1981). Evolution of sexual reproduction: Importance of DNA repair, complementation, and variation. *Am. Nat.* **117,** 537–549.

Bianchi, M., Das Gupta, C., and Redding, C. M. (1983). Synapsis and the formation of paranemic joints by *E. coli* recA protein. *Cell* **34,** 931–939.

Bisson, L., and Thorner, J. (1976). Thymidine 5′-monophosphate-requiring mutants of *Saccharomyces cerevisiae* are deficient in thymidylate synthetase. *J. Bacteriol.* **132,** 44–50.

Blackburn, E. H., and Szostak, J. W. (1984). The molecular structure of centromeres and telomeres. *Annu. Rev. Biochem.* **53,** 164–194.

Borts, R. H., Lichten, M., Hearn, M., Davidow, L. S., and Haber, J. E. (1984). Physical monitoring of meiotic recombination in Saccharomyces cerevisiae. *Cold Spring Harbor Symp. Quant. Biol.* **49,** 67–76.

Borts, R. H., Lichten, M., and Haber, J. E. (1986). Analysis of meiosis-defective mutations in yeast by physical monitoring of meiotic recombination. *Genetics* **113,** 551–567.

Botstein, D., and Maurer, R. (1982). Genetic approaches to the analysis of microbial development. *Annu. Rev. Genet.* **16,** 61–83.

Boyd, J. B., and Setlow, R. B. (1976). Characterization of postreplication repair in mutagen-sensitive strains of *Drosophila melanogaster*. *Genetics* **84,** 507–526.

Boyd, J. B., Golino, M. D., Nguyen, T. D., and Green, M. M. (1976a). Isolation and characterization of X-linked mutants of *Drosophila melanogaster* which are sensitive to mutagens. *Genetics* **84,** 485–506.

Boyd, J. B., Golin, M. D., and Setlow, R. B. (1976b). The *mei-9*[a] mutant of *Drosophila melanogaster* increases mutagen sensitivity and decreases excision repair. *Genetics* **84**, 527–544.

Brunborg, G., Resnick, M. A., and Williamson, D. H. (1980). Cell Cycle specific repair of DNA double-strand breaks in *Saccharomyces cerevisiae*. *Radiat. Res.* **82**, 547–558.

Bruschi, C. V., and Esposito, M. S. (1983). Enhancement of spontaneous mitotic recombination by the meiotic mutant *spo11-1* in *Saccharomyces cerevisiae*. *Proc. Natl. Acad. Sci. U.S.A.* **80**, 7566–7570.

Budd, M., and Mortimer, R. K. (1982). Repair of double-strand breaks in a temperature conditioned radiation-sensitive mutant of *Saccharomyces cerevisiae*. *Mutat. Res.* **103**, 19–26.

Byers, B., and Goetsch, L. (1982). Reversible pachytene arrest of *Saccharomyces cerevisiae* at elevated temperature. *Mol. Gen. Genet.* **187**, 47–53.

Calderon, I. L., Contopoulou, C. R., and Mortimer, R. K. (1983). Isolation and characterization of yeast DNA repair genes. II. Isolation of plasmids that complement the mutations *rad50-1*, *rad51-1*, *rad54-3*, and *rad55-3*. *Curr. Genet.* **7**, 93–100.

Callow, R. S. (1985). Comments on Bennett's model of somatic chromosome disposition. *Heredity* **54**, 171–177.

Calvert, G. R., and Dawes, I. W. (1984). Initiation of sporulation in *Saccharomyces cerevisiae*. Mutations preventing initiation. *J. Gen. Microbiol.* **130**, 615–624.

Carpenter, A. T. C. (1975). Electron microscopy of meiosis in *Drosophila melanogaster* females. II. The recombination nodule—A recombination-associated structure at pachytene. *Proc. Natl. Acad. Sci. U.S.A.* **72**, 3186–3189.

Carpenter, A. T. C. (1979). Synaptonemal complex and recombination nodules in wild type *Drosophila melanogaster* females. *Genetics* **92**, 511–541.

Carpenter, A. T. C. (1982). Mismatch repair, gene conversion and crossing-over in two recombination defective mutants in *Drosophila melanogaster*. *Proc. Natl. Acad. Sci. U.S.A.* **79**, 5961–5965.

Carpenter, A. T. C. (1984). Recombination nodules and the mechanism of crossing-over in *Drosophila*. *In* "Controlling Events in Meiosis" (C. W. Evans and H. G. Dickinson, eds.), pp. 233–243. Company of Biologists, Ltd., Cambridge, England.

Cassidy, J. R., Moore, D., Lu, B. C., and Pukkila, P. J. (1984). Unusual organization and lack of recombination in the robosomal RNA genes of *Coprinus cinereus*. *Curr. Genet.* **8**, 607–613.

Chang, L. M. S., Lurie, K., and Plevani, P. (1978). A stimulatory factor for yeast DNA polymerase. *Cold Spring Harbor Symp. Quant. Biol.* **43**, 587–595.

Chaudhury, A. M., and Smith, G. R. (1984). A new class of *Escherichia coli recBC* mutants: Implications for the role of RecBC enzyme in homologous recombination. *Proc. Natl. Acad. Sci. U.S.A.* **81**, 7850–7854.

Chow, T. Y.-K., and Fraser, M. J. (1979). The major intracellular alkaline deoxyribonuclease activities expressed in wild-type and *Rec*-like mutants of *Neurospora crassa*. *Can. J. Biochem.* **57**, 899–901.

Chow, T. Y.-K., and Fraser, M. J. (1983). Purification and properties of ssDNA-binding endoexonuclease in *Neurospora crassa*. *J. Biol. Chem.* **258**, 12010–12018.

Chow, T., and Resnick, M. A. (1983). The identification of a deoxyribonuclease controlled by the *RAD52* gene of *Saccharomyces cerevisiae*. *In* "Cellular Responses to DNA Damage" (E. Friedberg and B. Bridges, eds.), pp. 447–455. Alan R. Liss, Inc., New York.

Chow, T. Y.-K., and Resnick, M. A. (1986). The *RAD52* gene in yeast controls the synthesis of a single-strand deoxyribonuclease. Submitted for publication.

Chow, T. Y.-K., Yamamoto, A. H., Mason, J. M., and Resnick, M. A. (1986). Repair and recombination defective mutants of *Drosophila* lack a DNase which is related to nucleases from fungi. *J. Cell. Biochem.* Suppl. **10B**, 211.

Clark, A. J. (1980). A view of the RecBC and RecF pathway of *E. coli* recombination. *ICN-UCLA Symp. Mol. Cell. Biol.* **19**, 891–899.

Culbertson, M. R., and Henry, S. A. (1973). Genetic analysis of hybrid strains trisomic for the chromosome containing a fatty acid synthetase gene complex (*fas1*) in yeast. *Genetics* **75**, 441–458.

Davidow, L. S., and Byers, B. (1984). Enhanced gene conversion and postmeiotic segregation in pachytene-arrested *Saccharomyces cerevisiae*. *Genetics* **106**, 165–183.

Dawes, I. W. (1983). Genetic control and gene expression during meiosis and sporulation in *Saccharomyces cerevisiae*. *In* "Yeast Genetics: Fundamental and Applied Aspects" (J. F. T. Spencer, D. M. Spencer, and A. R. W. Smith, eds.), pp. 29–64. Springer-Verlag, Berlin and New York.

Dawes, I. W., and Calvert, G. R. (1984). Initiation of sporulation in *Saccharomyces cerevisiae*. Mutations causing derepressed sporulation and G1 arrest in the cell division cycle. *J. Gen. Microbiol.* **130**, 605–613.

Defeo-Jones, D., Scolnick, E., Koller, R., and Dhar, R. (1983). ras-related gene sequences identified and isolation from *Saccharomyces cerevisiae*. *Nature (London)* **306**, 707–709.

DiCaprio, L., and Hastings, P. J. (1976). Postmeiotic segregation in strains of *Saccharomyces cerevisiae* unable to excise pyrimidine dimer. *Mutat. Res.* **37**, 137–140.

DiDomenico, B., Kowalisyn, J., Frackman, S., Jensen, L., Easton-Esposito, R. E., and Elder, R. (1984). The *SPO11* gene encodes a developmentally regulated sporulation specific transcript. *Conf. Yeast Genet. Mol. Biol., 12th, 1984*, p. 267.

DiNardo, S., Thrash, C., Voelkel, K. A., and Sternglanz, R. (1983). Identification of yeast DNA topoisomerase mutants. *In* "Mechanisms of DNA Replication and Recombination" (N. Cozzarelli, ed.), pp. 29–42. Alan R. Liss, Inc., New York.

DiNardo, S., Voelkel, K., and Sternglanz, R. E. (1984). DNA topoisomerase II mutant of *Saccharomyces cerevisiae:* Topoisomerase II is required for segregation of daughter molecules at the termination of DNA replication. *Proc. Natl. Acad. Sci. U.S.A.* **81**, 2616–2620.

Dowling, E. L., Maloney, D. H., and Fogel, S. (1985). Meiotic recombination and sporulation in repair-deficient strains of yeast. *Genetis* **109**, 283–302.

Dunn, B., Szauter, P., Pardue, M. L., and Szostak, J. W. (1984). Transfer of yeast teleomeres to linear plasmids by recombination. *Cell* **39**, 191–201.

Elder, R., Frackman, S., Jensen, L., and Easton-Esposito, R. (1984). Transcription in the region which encodes the *SPO12* gene. *Int. Conf. Yeast Genet. Mol. Biol., 12th, 1984*, p. 268.

Endow, S. A., Komma, D. J., and Atwood, K. C. (1984). Ring chromosomes and rDNA magnification in *Drosophila*. *Genetics* **108**, 969–983.

Esposito, M. S., and Esposito, R. E. (1969). The genetic control of sporulation in *Saccharomyces*. I. The isolation of temperature-sensitive sporulation-deficient mutants. *Genetics* **61**, 79–89.

Esposito, M. S., and Esposito, R. E. (1978). Aspects of the genetic control of meiosis and ascospore development inferred from the study of *spo* (sporulation deficient) mutants of *Saccharomyces cerevisiae*. *Biol. Cell.* **33**, 93–102.

Esposito, M. S., Hosada, J., Golin, J., Moise, H., Bjornshad, K., and Maleas, D. T. (1984). Recombination in *S. cerevisiae:* REC-gene mutants and DNA-binding proteins. *Cold Spring Harbor Symp. Quant. Biol.* **49**, 41–48.

Esposito, M. S., Maleas, D. T., Bjornstad, K. A., and Holbrook, L. L. (1986). The *REC46* gene of *Sacchromyces cerevisiae* controls mitotic chromosomal stability, recombination, and sporulation: Cell-type and life cycle stage-specific expression of the *rec46-1* mutation. *Curr. Genet.* **10**, 425–433.

Esposito, R. E., and Esposito, M. S. (1974). Genetic recombination and commitment to meiosis in *Saccharomyces*. *Proc. Natl. Acad. Sci. U.S.A.* **71**, 3172–3176.

Esposito, R. E., and Klapholz, S. (1981). Meiosis and ascospore development. *In* "The Molecular Biology of Yeast and *Saccharomyces cerevisiae:* Life Cycle and Inheritance" (J. N. Strathern, E. W. Jones, and J. R. Broach, eds.), pp. 211–288. Cold Spring Harbor Lab., Cold Spring Harbor, New York.

Fasullo, M., and Davis, R. (1986). Yeast recombination substrates designed to study non-allelic recombination *in vivo.* Submitted for publication.

Fincham, J. R. S., and Holliday, R. (1970). An explanation of fine structure map expansion in terms of excision repair. *Mol. Gen. Genet.* **109,** 309–322.

Fogel, S., and Roth, R. (1974). Mutations affecting meiotic gene conversion in yeast. *Mol. Gen. Genet.* **130,** 189–201.

Fogel, S., Mortimer, R. K., and Lusnak, K. (1981). Mechanisms of meiotic gene conversion or "wanderings on a foreign strand." *In* "Molecular Biology of the Yeast *Saccharomyces cerevisiae:* Life Cycle and Inheritance" (J. N. Strathern, E. W. Jones, and J. R. Broach, eds.), pp. 289–339. Cold Spring Harbor Lab., Cold Spring Harbor, New York.

Fogel, S., Welch, J. W., and Louis, E. J. (1984). Meiotic gene conversion mediates gene amplification in yeast. *Cold Spring Harbor Symp. Quant. Biol.* **49,** 55–65.

Formosa, T., Burke, R. L., and Alberts, B. (1983). Affinity purification of T4 bacteriophage proteins essential for DNA replication and genetic recombination. *Proc. Natl. Acad. Sci. U.S.A.* **80,** 2442–2446.

Frackman, S., Jensen, L., Easton-Esposito, R., and Elder, R. (1984). Transcription of the *SPO*13 gene during sporulation. *Int. Conf. Yeast Genet. Mol. Biol., 12th, 1984;* p. 287.

Fraser, M. J., Chow, T. Y.-K., Cohen, H., and Koa, H. (1986). An immunochemical study of *Neurospora* nucleases. *Biochem. Cell. Biol.* **64,** 106–116.

Friedberg, E. C. (1985). "DNA Repair." Freeman, New York.

Game, J. C. (1983). Radiation sensitive mutants and repair in yeast. *In* "Yeast Genetics: Fundamental and Applied Aspects" (J. F. T. Spencer, D. M. Spencer, and A. R. W. Smith, eds.), pp. 109–137. Springer-Verlag, Berlin and New York.

Game, J. C., and Mortimer, R. K. (1974). A genetic study of X-ray sensitive mutants in yeast. *Mutat. Res.* **24,** 281–292.

Game, J. C., Johnston, L. H., and von Borstel, R. C. (1979). Enhanced mitotic recombination in a ligase-defective mutant of the yeast *Saccharomyces cerevisiae. Proc. Natl. Acad. Sci. U.S.A.* **76,** 4589–4592.

Game, J. C., Zamb, T. J., Braun, R. J., Resnick, M., and Roth, R. M. (1980). The role of radiation (*rad*) genes in meiotic recombination in yeast. *Genetics* **94,** 51–68.

Gatti, M. (1982). Sister chromatid exchanges in *Drosophila. In* "Sister Chromatid Exchanges" (S. Wolff, ed.), pp. 267–297. Wiley, New York.

Gillies, C. B. (1975). Synaptonemal complex and chromosome structure. *Annu. Rev. Genet.* **9,** 91–109.

Giroux, C. N. (1984). Analysis of *SPO11,* a gene required for the early events of meiosis in yeast. *Int. Conf. Yeast Genet. Mol. Biol., 12th, 1984,* p. 266.

Giroux, C. N., Tiano, H. F., Dresser, M. E., and Moses, M. (1986). Molecular cloning and analysis of genes required for meiotic recombination and DNA metabolism in yeast. *J. Cell. Biochem.* Suppl. **10B,** 215.

Glickman, B. W., and Radman, M. (1980). *Escherichia coli* mutator strains deficient in methylation directed mismatch correction. *Proc. Natl. Acad. Sci. U.S.A.* **77,** 1063–1067.

Goto, T., and Wang, J. C. (1982). Yeast DNA topoisomerase II. An ATP-dependent type II topoisomerase that catalyzes the catenation, decatenation, unknotting, and relaxation of double-stranded DNA rings. *J. Biol. Chem.* **257,** 5866–5872.

Grell, R. F., Oakberg, E. F., and Genoroso, E. E. (1980). Synaptonemal complexes at premeiotic interphase in the mouse spermatocyte. *Proc. Natl. Acad. Sci. U.S.A.* **77,** 6720–6723.

Gutz, H. (1971). Site specific induction of gene conversion in *Schizosaccharomyces pombe*. *Genetics* **69**, 317–337.

Haber, J. E., Thorburn, P. C., and Rogers, D. (1984). Meiotic and mitotic behavior of dicentric chromosomes in *Saccharomyces cerevisiae*. *Genetics* **106**, 185–205.

Hartwell, L. (1974). *Saccharomyces cerevisiae* cell cycle. *Bacteriol. Rev.* **38**, 164–198.

Hastings, P. J. (1984). Measurement of restoration and conversion: Its meaning for the mismatch repair hypothesis of conversion. *Cold Spring Harbor Symp. Quant. Biol.* **49**, 49–53.

Hastings, P. J., Quah, S.-K., and von Borstel, R. C. (1976). Spontaneous mutation by mutagenic repair of spontaneous lesions in DNA. *Nature (London)* **264**, 719–722.

Haynes, R. M., and Kunz, B. A. (1981). DNA repair and mutagenesis in yeast. In "The Molecular Biology of the Yeast *Saccharomyces cererisiae:* Life Cycle and Inheritance" (J. N. Strathern, E. W. Jones, and J. R. Broach, eds.), pp. 371–414. Cold Spring Harbor Lab., Cold Spring Harbor, New York.

Henry, S. A., and Fogel, S. (1971). Saturated fatty acid mutants in yeast. *Mol. Gen. Genet.* **113**, 1–19.

Higgins, D. R., Prakash, S., Reynolds, P., Polakowska, R., Weber, S., and Prakash, L. (1983). Isolation and characterization of the *RAD3* gene of *Saccharomyces cerevisiae* and inviability of *rad3* deletion mutants. *Proc. Natl. Acad. Sci. U.S.A.* **80**, 5680–5684.

Hirschberg, J., and Simchen, G. (1977). Commitment to mitotic cell cycle in yeast in relation to meiosis. *Exp. Cell Res.* **105**, 245.

Ho, K. S. Y. (1975). Induction of double-strand breaks by X-rays in a radiosensitive strain of yeast. *Mutat. Res.* **20**, 45–51.

Hoekstra, M. F., and Malone, R. E. (1985). Expression of the *E. coli dam* methylase in *Saccharomyces cerevisiae:* The effect of *in vivo* adenine methylation on genetic recombination and mutation. *Mol. Cell. Biol.* **5**, 610–618.

Høgset, A., and Øyen, T. (1984). Correlation between suppressed meiotic recombination and the lack of DNA strand breaks in the rDNA genes of *Saccharomyces cerevisiae*. *Nucleic Acids Res.* **12**, 7199–7213.

Holliday, R. (1964). A mechanism for gene conversion in fungi. *Genet. Res.* **5**, 282–304.

Holliday, R. (1984). Biological significance of meiosis. In "Controlling Events in Meiosis" (C. W. Evans and H. G. Dickinson, eds.), pp. 381—394. Company of Biologists, Ltd., Cambridge, England.

Holliday, R., and Dickson, J. M. (1977). The detection of post-meiotic segregation without tetrad analysis in *Ustilago maydis*. *Mol. Gen. Genet.* **153**, 331–335.

Holliday, R., and Pugh, J. E. (1975). DNA modification mechanisms and gene activity during development. *Science* **187**, 226–232.

Holliday, R., Halliwell, R. E., Evans, M. W., and Rowell, V. (1976). Genetic characterization of *rec-1*, a mutant of *Ustilago maydis* defective in repair and recombination. *Genet. Res.* **27**, 413–453.

Holloman, W. K., Rowe, T. C., and Rusche, J. R. (1981). Studies on nuclease α from *Ustilago maydis*. *J. Biol. Chem.* **256**, 5987–6094.

Horowitz, H., and Haber, J. E. (1984). Subtelomeric regions of yeast chromosomes contain a 36-base pair tandemly repeated sequence. *Nucleic Acids Res.* **12**, 7105–7122.

Hotta, Y., and Stern, H. (1971). A DNA-binding protein in meiotic cells of *Lilium*. *Dev. Biol.* **26**, 87–99.

Hotta, Y., and Stern, H. (1974). DNA scission and repair during pachytene in *Lilium*. *Chromosoma* **46**, 279–296.

Hotta, Y., and Stern, H. (1981). Small nuclear RNA molecules that regulate nuclease accessibility in specific chromatin regions of meiotic cells. *Cell* **27**, 309–319.

Hotta, Y., Bennett, M. D., Toledo, L. A., and Stern, H. (1979). Regulation of R-protein and

6. Genetics and Biochemistry of Recombination

endonuclease activities in meiocytes by homologous chromosome pairing. *Chromosoma* **72,** 191–201.

Hotta, Y., Tabata, S., Stubbs, L., and Stern, H. (1985). Meiosis-specific transcripts of a DNA component replicated during chromosome pairing: Homology across the phylogenetic spectrum. *Cell* **40,** 785–793.

Howell, S. H., and Stern, H. (1971). The appearance of DNA breakage and repair activities in the synchronous meiotic cycle of *Lilium*. *J. Mol. Biol.* **55,** 357–378.

Huberman, J. A. (1981). New views of the biochemistry of eukaryotic DNA replication revealed by aphidicolin, an unusual inhibitor of DNA polymerase α. *Cell* **23,** 647–648.

Hurst, D. D., Fogel, S., and Mortimer, R. K. (1972). Conversion associated recombination in yeast. *Proc. Natl. Acad. Sci. U.S.A.* **69,** 101–105.

Iino, Y., and Yamamoto, M. (1985a). Negative control for the initiation of meiosis in *Schizosaccharomyces pombe*. *Proc. Natl. Acad. Sci. U.S.A.* **82,** 2447–2451.

Iino, Y., and Yamamoto, M. (1985b). Mutants of *Schizosaccharomyces pombe* which sporulate in the haploid state. *Mol. Gen. Genet.* **198,** 416–421.

Jachymczyk, W. J., von Borstel, R. C., Mowat, M., and Hastings, P. J. (1981). Repair of interstrand cross-links in DNA of *Saccharomyces cerevisiae* requires two systems of repair: The RAD3 system and the RAD51 system. *Mol. Gen. Genet.* **182,** 196–205.

Jackson, J., and Fink, G. R. (1985). Meiotic recombination between duplicated genetic elements in *Saccharomyces cerevisiae*. *Genetics* **109,** 303–332.

Jacobson, G., Pinon, R., Esposito, R. E., and Esposito, M. S. (1975). Single strand scissions of chromosomal DNA during commitment to recombination at meiosis. *Proc. Natl. Acad. Sci. U.S.A.* **72,** 1887–1891.

Jagiello, G., Tantravahi, U., Fang, J. S., and Erlanger, B. F. (1982). DNA methylation patterns of human pachytene spermatocytes. *Exp. Cell Res.* **141,** 253–259.

Jarvik, J., and Botstein, D. (1973). Conditional lethal mutations that suppress genetic defects in morphogenesis by altering structural proteins. *Proc. Natl. Acad. Sci. U.S.A.* **72,** 2738–27424.

Jinks-Robertson, S., and Petes, T. D. (1985). High frequency meiotic gene conversion between repeated genes on nonhomologous chromosomes in yeast. *Proc. Natl. Acad. Sci. U.S.A.* **82,** 3350–3354.

Johns, E. W., ed. (1982). "The HMG Chromosomal Proteins," pp. 69–87. Academic Press, New York.

Johnston, L. H., and Nasmyth, K. (1978). *Saccharomyces cerevisiae* cell cycle mutant *cdc9* is defective in DNA ligase. *Nature (London)* **274,** 891–893.

Jones, E. W. (1977). Proteinase mutants of *Saccharomyces cerevisiae*. *Genetics* **85,** 23–33.

Kaback, D. B., and Feldberg, L. R. (1985). *Saccharomyces cerevisiae* exhibits a sporulation-specific temporal pattern of transcript accumulation. *Mol. Cell. Biol.* **5,** 751–761.

Kane, S. M., and Roth, R. (1974). Carbohydrate metabolism during ascospore development in yeast. *J. Bacteriol.* **118,** 8–14.

Kassir, Y., and Simchen, G. (1980). Single-strand scissions of nuclear yeast DNA occur without meiotic recombination. *Curr. Genet.* **2,** 79–80.

Kataoka, T., Powers, S., Cameron, S., Fasano, O., Goldfarb, M., Broach, J., and Wigler, M. (1985). Functional homology of mammalian and yeast *RAS* genes. *Cell* **40,** 19–26.

Kato, H. (1974). Possible role of DNA synthesis in formation of sister chromatid exchanges. *Nature (London)* **252,** 739–741.

Keil, R. L., and Roeder, G. S. (1984). *Cis*-acting, recombination-stimulating activity in a fragment of the ribosomal DNA of *S. cerevisiae*. *Cell* **39,** 377–386.

Kelly, S. L., Merrill, C., and Parry, J. M. (1983a). Cyclic variation in sensitivity to X-irradiation during meiosis in *Saccharomyces cerevisiae*. *Mol. Gen. Genet.* **91,** 314–318.

Kelly, S. L., Tippins, R. S., Parry, J. M., and Waters, R. (1983b). *N*-Methyl-*N'*-nitro-*N*-nitrosoguanidine induced genetic change during the meiotic cell cycle of *Saccharomyces cerevisiae:* An absence of S-phase specificity. *Carcinogenesis (London)* **4,** 851–856.

Klapholz, S., and Esposito, R. E. (1980a). Isolation of *spo12-1* and *spo13-1* from a natural variant of yeast that undergoes a single meiotic division. *Genetics* **96,** 567–568.

Klapholz, S., and Esposito, R. E. (1980b). Recombination and chromosome segregation during the single division meiosis in *spo12-1* and *spo13-1* diploids. *Genetics* **96,** 589–611.

Klapholz, S., Waddell, C. S., and Esposito, R. E. (1985). The role of the *SPO11* gene in meiotic recombination in yeast. *Genetics* **110,** 187–216.

Klar, A. J. S., Fogel, S., and KacCleod, M. (1979). *MAR1*, a regulator of the HMa and HMα loci in *Saccharomyces cerevisiae*. *Genetics* **92,** 37–50.

Klar, A. J. S., Strathern, J. N., and Abraham, J. A. (1984a). The involvement of double-strand chromosomal breaks for mating type switching in *Saccharomyces cerevisiae*. *Cold Spring Harbor Symp. Quant. Biol.* **49,** 77–88.

Klar, A. J. S., Strathern, J. N., and Hicks, J. (1984b). Developmental pathways in yeast. *In* "Microbial Development" (R. Losick and L. Shapiro, eds.), pp. 151–195. Cold Spring Harbor Lab., Cold Spring Harbor, New York.

Klein, H. L. (1984). Lack of association between intrachromosomal gene conversion and reciprocal exchange. *Nature (London)* **310,** 748–753.

Kmiec, E., and Holloman, W. K. (1982). Homologous pairing of DNA molecules promoted by a protein from *Ustilago*. *Cell* **29,** 367–374.

Kmiec, E. B., and Holloman, W. K. (1984). Synapsis promoted by *Ustilago* rec1 protein. *Cell* **36,** 593–598.

Kohli, J., Munz, P., Aebi, R., Amstutz, H., Gysler, C., Heyer, W.-D., Lehmann, L., Schuchert, P., Szankasi, P., Thuriaux, P., Leupold, U., Bell, J., Gamulin, V., Hottinger, H., Pearson, D., and Söll, D. (1984). Interallelic and intergenic conversion in three serine tRNA genes of *Schizosaccharomyces pombe*. *Cold Spring Harbor Symp. Quant. Biol.* **49,** 31–40.

Kraig, E., and Haber, J. E. (1980). Messenger ribonucleic acid and protein metabolism during sporulation of *Saccharomyces cerevisiae*. *J. Bacteriol.* **144,** 1098–1112.

Kundu, S. C., and Moens, P. B. (1982). The ultrastructural meiotic phenotype of the radiation sensitive mutant *rad6-1* in yeast. *Chromosoma* **87,** 125–132.

Kupiec, M., and Simchen, G. (1984). Cloning and mapping of the RAD50 gene of *Saccharomyces cerevisiae*. *Mol. Gen. Genet.* **193,** 525–531.

Kurtz, S., and Lindquist, S. (1984). Changing patterns of gene expression during sporulation in yeast. *Proc. Natl. Acad. Sci. U.S.A.* **81,** 7323–7327.

LaBonne, S. G., and Dumas, L. B. (1983). Isolation of a yeast single-strand deoxyribonucleic acid binding protein that specifically stimulates yeast DNA polymerase I. *Biochemistry* **22,** 3214–3219.

Lambie, E. J., and Roeder, G. S. (1986). Repression of meiotic crossing over by a centromere (*CEN3*). Submitted for publication.

Leaper, S., Resnick, M. A., and Holliday, R. (1980). Repair of double-strand breaks and lethal damage in DNA of *Ustilago maydis*. *Genet. Res.* **35,** 291–307.

Leblon, G. (1972). Mechanism of gene conversion in *Ascobolus immersus*. I. Existence of a correlation between the origin of mutants induced by different mutagens and their conversion spectrum. *Mol. Gen. Genet.* **115,** 36–48.

Lorincz, A., and Reed, S. I. (1984). Primary structure homology between the product of yeast cell division control gene *CDC28* and vertebrate oncogenes. *Nature (London)* **307,** 183–185.

Lu, B. C. (1981). Replication of deoxyribonucleic acid and crossing over in *Coprinus*. *In* "Basidium and Basidiocarp: Evolution, Cytology, Function, and Development" (K. Wells and E. K. Well,s eds.), pp. 93–112. Springer-Verlag, Berlin and New York.

6. Genetics and Biochemistry of Recombination

Lu, B. C. (1984). The cellular program for the formation and dissolution of the synaptonemal complex in *Coprinus*. *J. Cell Sci.* **67**, 25–43.
Machida, I., and Nakai, S. (1980). Differential effect of UV irradiation in induction of intragenic and intergenic recombination during commitment to meiosis on *S. cerevisiae*. *Mutat. Res.* **73**, 69–79.
Maguire, M. B. (1984). The mechanism of meiotic homologue pairing. *J. Theor. Biol.* **106**, 605–615.
Malone, R. E. (1983). Multiple mutant analysis of recombination in yeast. *Mol. Gen. Genet.* **189**, 405–412.
Malone, R. E., and Esposito, R. E. (1980). The *RAD52* gene is required for homothallic interconversion of mating types and spontaneous mitotic recombination in yeast. *Proc. Natl. Acad. Sci. U.S.A.* **77**, 503–507.
Malone, R. E., and Esposito, R. E. (1981). Recombinationless meiosis in *Saccharomyces cerevisiae*. *Mol. Cell. Biol.* **1**, 891–901.
Malone, R. E., Golin, J. E., and Esposito, M. S. (1980). Mitotic versus meiotic recombination in *Saccharomyces cerevisiae*. *Curr. Genet.* **1**, 241–248.
Martindale, D. W., and Bruns, P. J. (1983). Cloning of abundant mRNA species present during conjugation of *Tetrahymena thermophila*: Identification of mRNA species present exclusively during meiosis. *Mol. Cell. Biol.* **3**, 1857–1865.
Mason, J. M., and Resnick, M. A. (1985). Mechanisms and detection of chromosome malsegregation using *Drosophila* and the yeast *Saccharomyces cerevisiae*. In "Aneuploidy: Etiology and Mechanisms" (V. Dellarco and A. Hollaender, eds.), pp. 433–444. Plenum Press, New York.
Mason, J. M., Strobel, E., and Green, M. M. (1984). *mu-2*: Mutator gene in *Drosophila* that potentiates the induction of terminal deficiencies. *Proc. Natl. Acad. Sci. U.S.A.* **81**, 6090–6094.
Mather, K. (1936). The determination of position in crossing-over. I. *Drosophila melanogaster*. *J. Genet.* **33**, 207–235.
Mather, K. (1939). Crossing over and heterochromatin in the X chromosome of *Drosophila melanogaster*. *Genetics* **24**, 413–435.
Matsumoto, K., Uno, I., and Ishikawa, T. (1983). Initiation of meiosis in yeast mutants defective in adenylate cyclase and cyclic AMP dependent protein kinase. *Cell* **21**, 417–423.
Maynard-Smith, J. (1978). "The Evolution of Sex." Cambridge Univ. Press, London and New York.
Medvedev, Z. A. (1981). On the immortality of the germ line: Genetic and biochemical mechanisms. A review. *Mech. Ageing Dev.* **17**, 331–359.
Mendel, G. J. (1865). Versuche uber Pflanzen-Hybriden. *Verh. Naturforsch. Ver. Brunn* **4**, 3–47.
Meselson, M., and Radding, C. (1975). A general model for genetic recombination. *Proc. Natl. Acad. Sci. U.S.A.* **72**, 358–361.
Mizuchi, K., Kemper, B., Hays, J., and Weisberg, R. (1982). T4 endonuclease VII cleaves Holliday structures. *Cell* **29**, 357–365.
Moens, P. B. (1982). Mutants of yeast meiosis (*Saccharomyces cerevisiae*). *Can. J. Genet. Cytol.* **24**, 243–256.
Moore, C. W., and Sherman, F. (1977). Role of DNA sequences in genetic recombination in the iso-1-cytochrome *c* gene of yeast. II. Discrepancies between physical and genetic distances determined by five mapping procedures. *Genetics* **79**, 397–418.
Moore, P., and Holliday, R. (1976). Evidence for the formation of hybrid DNA during mitotic recombination in Chinese hamster cells. *Cell* **8**, 573–579.
Morgan, L. V. (1933). A closed X chromosome in *Drosophila melanogaster*. *Genetics* **18**, 250–283.
Morrison, D. P., and Hastings, P. J. (1979). Characterization of the mutator mutation *mut5-1*. *Mol. Gen. Genet.* **175**, 57–65.

Mortimer, R. K., and Schild, D. (1981). Genetic mapping in *Saccharomyces cerevisiae*. In "The Molecular Biology of the Yeast *Sadcharomyces cerevisiae:* Life Cycle and Inheritance" (J. N. Strathern, E. W. Jones, and J. R. Broach, eds.), pp. 11–26. Cold Spring Harbor Lab., Cold Spring Harbor, New York.

Mortimer, R. K., Contopoulou, R., and Schild, D. (1981). Mitotic chromosome loss in a radiation sensitive strain of the yeast *Saccharomyces cerevisiae. Proc. Natl. Acad. Sci. U.S.A.* **78,** 5778–5782.

Murray, A. W., and Szostak, J. W. (1983). Construction of artificial chromosomes in yeast. *Nature (London)* **305,** 189–193.

Naumovski, L., and Friedberg, E. C. (1983). A DNA repair gene required for the incision of damaged DNA is essential for viability in *Saccharomyces cerevisiae. Proc. Natl. Acad. Sci. U.S.A.* **80,** 4818–4821.

Nitiss, J. (1986). Strand breaks and recombination during meiosis in radiation sensitive strains of *Saccharomyces cerevisiae*. Ph.D. thesis, Illinois Institute of Technology, Chicago, Illinois.

Nurse, P. (1985). Mutants of the fission yeast *Schizosaccharomyces pombe* which alter the shift between cell proliferation and sporulation. *Mol. Gen. Genet.* **198,** 497–502.

Orlando, P., Grippo, P., and Geremia, R. (1984). DNA repair synthesis-related enzymes during spermatogenesis in the mouse. *Exp. Cell Res.* **153,** 499–505.

Orr-Weaver, T. L., and Szostak, J. W. (1985). Fungal recombination. *Microbiol. Rev.* **49,** 35–58.

Orr-Weaver, T. L., Szostak, J. W., and Rothstein, R. J. (1981). Yeast transformation: A model system for the study of recombination. *Proc. Natl. Acad. Sci. U.S.A.* **78,** 6358–6362.

Percival-Smith, A., and Segall, J. (1984). Isolation of DNA sequences preferentially expressed during sporulation in *Saccharomyces cerevisiae. Mol. Cell. Biol.* **4,** 142–150.

Petersen, J. G., Kielland-Brandt, M. C., and Nilsson-Tillgren, T. (1979). Protein patterns of yeast during sporulation. *Carlsberg Res. Commun.* **44,** 149.

Petes, T. D. (1979). Meiotic mapping of yeast ribosomal deoxyribonucleic acid in chromosome XII. *J. Bacteriol.* **138,** 185–192.

Petes, T. D. (1980). Unequal meiotic recombination within tandem assays of yeast ribosomal genes. *Cell* **19,** 765–774.

Ponzetto-Zimmerman, C., and Wolgemuth, D. J. (1984). Methylation of satellite sequences in mouse spermatogenic and somatic DNAs. *Nucleic Acids Res.* **12,** 2807–2822.

Porter, E. K., Parry, D., and Dickinson, H. G. (1983). Changes in poly(a)$^+$ RNA during male meiosis in *Lilium. J. Cell Sci.* **62,** 177–186.

Prakash, S., Prakash, L., Burke, W., and Monteleone, B. A. (1980). Effects of the *RAD52* gene on recombination in *Saccharomyces cerevisiae. Genetics* **94,** 31–50.

Pukkila, P. J. (1977). Biochemical analysis of genetic recombination in eukaryotes. *Heredity* **39,** 193–217.

Pukkila, P. J., Peterson, J., Herman, G., Modrich, P., and Meselson, M. (1983). Effects of high levels of DNA adenine methylation on methyl-directed mismatch repair in *Escherichia coli. Genetics* **104,** 571–582.

Radding, C. M. (1982). Homologous pairing and strand exchange in genetic recombination. *Annu. Rev. Genet.* **16,** 405–437.

Razin, A., Cedar, H., and Riggs, A. D., eds. (1984). "DNA Methylation: Biochemistry and Biological Significance." Springer-Verlag, Berlin and New York.

Resnick, M. A. (1975). The repair of double-strand breaks in chromosomal DNA of yeast. In "Molecular Mechanisms for Repair of DNA" (P. Hanawalt and R. B. Setlow, eds.), Part B, pp. 549–556. Plenum, New York.

Resnick, M. A. (1976). The repair of double-strand breaks in DNA: A model involving recombination. *J. Theor. Biol.* **59,** 97–106.

Resnick, M. A. (1979). The induction of molecular and genetic recombination in eukaryotic cells. *Adv. Radiat. Biol.* **8,** 175–217.

Resnick, M. A., and Bloom, K. (1986). Lessons learned from yeast: A molecular and genetic analysis of centromere function. In "Aneuploidy—Incidence and Etiology" (A. Sandberg and B. V. Vig, eds.). Alan R. Liss, New York. (In press.)

Resnick, M. A., and Martin, P. (1976). The repair of double-stranded breaks in the nuclear DNA of *Saccharomyces cerevisiae* and its genetic control. *Mol. Gen. Genet.* **143,** 119–129.

Resnick, M. A., Kasimos, J. N., Game, J. C., Braun, R. J., and Roth, R. M. (1981). Changes in DNA during meiosis in a repair-deficient mutant (*rad52*) of yeast. *Science* **212,** 543–545.

Resnick, M. A., Stasiewicz, S., and Game, J. (1983a). Meiotic DNA metabolism in normal and excision deficient yeast following UV exposure. *Genetics* **104,** 583–601.

Resnick, M. A., Game, J. C., and Stasiewicz, S. (1983b). The genetic effects of UV on meiosis in yeast during meiosis in the presence and absence of excision repair. *Genetics* **104,** 603–618.

Resnick, M. A., Chow, T., Nitiss, J., and Game, J. (1984a). Changes in the chromosomal DNA of yeast during meiosis in repair mutants, and the possible role of a deoxyribonuclease. *Cold Spring Harbor Symp. Quant. Biol.* **49,** 639–649.

Resnick, M. A., Sugino, A., Nitiss, J., and Chow, T. (1984b). DNA polymerases, deoxyribonucleases and recombination during meiosis in yeast. *J. Mol. Cell. Biol.* **4,** 2811–2817.

Resnick, M. A., Chaudhury, A., and Nitiss, J. (1986a). Genetic and molecular analyses of recombination using mutants altered in DNA repair and sister chromatid recombination. In "Current Communications in Molecular Biology" Banbury Center Cold Spring Harbor Laboratory (in press).

Resnick, M. A., Nitiss, J., Edwards, C., Game, J. C., and Malone, R. (1986b). Meiosis can induce recombination in *rad52* mutants of *Saccharomyces cerevisiae* yeast. *Genetics* **113,** 531–550.

Reynolds, P., Weber, S., and Prakash, L. (1985). *RAD6* gene of *Saccharomyces cerevisiae* encodes a protein containing a tract of 13 consecutive aspartates. *Proc. Natl. Acad. Sci. U.S.A.* **82,** 168–172.

Rossignol, J.-L., Nicolas, A., Hamza, H., and Languin, T. (1984). Origins of gene conversion and reciprocal exchange in *Ascobolus. Cold Spring Harbor Symp. Quant. Biol.* **49,** 13–21.

Rossignol, J.-L., Nicolas, A., Hamza, H., and Kaloeropoulos, A. (1986). Recombination and gene conversion in *Ascobolus*. In "The Recombination of Genetic Material" (B. Low, ed.). Academic Press, New York. (In press.)

Roth, R. (1973). Chromosome replication during meiosis: Identification of gene functions required for premeiotic DNA synthesis. *Proc. Natl. Acad. Sci. U.S.A.* **70,** 3087–3091.

Roth, R. (1976). Temperature-sensitive yeast mutants defective in meiotic recombination and replication. *Genetics* **83,** 675–686.

Rothstein, R. (1983). One step gene disruption in yeast. In "Methods in Enzymology" (R. Wu, L. Grossman, and K. Moldave, eds.), Vol. 101, pp. 202–211. Academic Press, New York.

Ruby, S. W., and Szostak, J. W. (1985). Specific yeast genes are expressed in response to DNA-damaging agents. *Mol. Cell. Biol.* **5,** 75–84.

Saeki, T., Machida, I., and Nakai, S. (1980). Genetic control of diploids recovery after γ-irradiation in the yeast *Saccharomyces cerevisiae*. *Mutat. Res.* **73,** 251–265.

Sakaguchi, K., and Lu, B. C. (1982). Meiosis in *Coprinus:* Characterization and activities of two forms of DNA polymerase during meiotic stages. *Mol. Cell. Biol.* **2,** 752–757.

Sakaguchi, K., Hotta, Y., and Stern, H. (1980). Chromatin-associated DNA polymerase activity in meiosis cells of lily and mouse: Its stimulation by helix-destabilizing protein. *Cell Struct. Funct.* **5,** 323–334.

Salts, Y., Simchen, G., and Pinon, R. (1976). DNA degradation and reduced recombination following UV irradiation during meiosis in yeast (*Saccharomyces*). *Mol. Gen. Genet.* **145,** 55–59.

Sandler, L., Lindsley, D. L., Nicoletti, B., and Trippa, G. (1968). Mutants affecting meiosis in natural populations of *Drosophila melanogaster*. *Genetics* **60,** 525–558.

Scalafani, R. A., and Fangman, W. L. (1984). Yeast gene *CDC8* encodes thymidilate kinase and is

complemented by herpes thymidine kinase gene TK. *Proc. Natl. Acad. Sci. U.S.A.* **81**, 5821–5825.
Schild, D., and Byers, B. (1978). Meiotic effects of DNA-defective cell division cycle mutations of *Saccharomyces cerevisiae*. *Chromosoma* **70**, 109–130.
Schimke, R. T., ed. (1982). "Gene Amplification." Cold Spring Harbor Lab., Cold Spring Harbor, New York.
Schimke, R. T., Kaufman, R. J., Alt, F. W., and Kellems, R. F. (1978). Gene amplification and drug resistance in cultured mammalian cells. *Science* **202**, 1051–1055.
Sherman, F., and Roman, H. (1963). Evidence for two types of allelic recombination in yeast. *Genetics* **48**, 225–261.
Shilo, V., Simchen, G., and Shilo, B. (1978). Initiation of meiosis in cell-cycle initiation mutants of *Saccharomyces cerevisiae*. *Exp. Cell Res.* **112**, 241–248.
Simchen, G. (1974). Are mitotic functions required in meiosis? *Genetics* **76**, 745–753.
Simchen, G. (1978). Cell cycle mutants. *Annu. Rev. Genet.* **12**, 161–191.
Simchen, G., Pinon, R., and Salts, Y. (1972). Sporulation in *Saccharomyces cerevisiae:* Premeiotic DNA synthesis, readiness and commitment. *Exp. Cell Res.* **75**, 207–218.
Simchen, G., Kassir, Y., Horesh-Cabill, O., and Friedmann, A. (1981). Elevated recombination and pairing structures during meiotic arrest in yeast of the nuclear division mutant *cdc5*. *Mol. Gen. Genet.* **184**, 46–51.
Smith, G. R. (1983). General recombination. *In* "Lambda II" (R. W. Hendrix, F. W. Stahl, and R. Weisberg, eds.), pp. 175–209. Cold Spring Harbor Lab., Cold Spring Harbor, New York.
Smith, G. R., Kunes, S. J., Schultz, D. W., Taylor, A., and Triman, K. L. (1981). Structure of Chi hotspots of generalized recombinaation. *Cell* **24**, 429–436.
Stahl, F. W. (1979). "Genetic Recombination. Thinking about It in Phage and Fungi." Freeman, San Francisco, California.
Stern, H., and Hotta, Y. (1977). Biochemistry of meiosis. *Phil. Trans. R. Soc. Lond. B.* **277**, 277–293.
Stern, H., and Hotta, Y. (1978). Regulatory mechanisms in meiotic crossing-over. *Ann. Rev. Plant Physiol.* **29**, 415–436.
Strathern, J. N., Klar, A. J. S., Hicks, J. B., Abraham, J. A., Ivy, J. M., Nasmyth, K. A., and McGill, C. (1982). Homothallic switching of yeast mating type cassettes is initiated by a double-stranded cut in the MAT locus. *Cell* **31**, 183–192.
Strike, T. L. (1978). Characteristics of mutants sensitive to x-rays. Ph.D. Thesis, University of California, Davis.
Struhl, K. (1983). The new yeast genetics. *Nature (London)* **305**, 391–397.
Sturtevant, A. H., and Beadle, G. W. (1936). The relations of inversions in the X chromosome of *Drosophila melanogaster* to crossing over and nondisjunction. *Genetics* **21**, 554–604.
Sugino, A., Ryu, B. H., Sugino, T., Naumovski, L., and Friedberg, E. (1986). A new DNA-dependent ATPase which stimulates yeast DNA polymerase I and has DNA-unwinding activity. *J. Biol. Chem 261,* (in press).
Symington, L. S., Fogarty, L., and Kolodner, R. (1983). Genetic recombination of homologous plasmids catalyzed by cell free extracts of *Saccharomyces cerevisiae*. *Cell* **35**, 805–813.
Symington, L., and Kolodner, R. (1985). Partial purification of an enzyme from *Saccharomyces cerevisiae* that cleaves Holliday junctions. *Proc. Natl. Acad. Sci. U.S.A.* **82**, 7247–7251.
Symington, L. S., Morrison, P. T., and Kolodner, R. (1984). Genetic recombination catalyzed by cell-free extracts of *Saccharomyces cerevisiae*. *Cold Spring Harbor Symp. Quant. Biol.* **49**, 805–814.
Szauter, P. (1984). An analysis of regional constraints on exchange in *Drosophila melanogaster* using recombination-defective meiotic mutants. *Genetics* **106**, 45–71.

Szostak, J. W., and Wu, R. (1980). Unequal crossing-over in the ribosomal DNA of *Saccharomyces cerevisiae*. *Nature (London)* **284**, 426–430.

Szostak, J. W., Orr-Weaver, T. L., Rothstein, R. J., and Stahl, F. W. (1983). The double-strand break repair model for recombination. *Cell* **33**, 25–35.

Tease, C., and Jones, G. H. (1979). Analysis of exchanges in differentially stained meiotic chromosomes of *Locusta migratoria* after BrdU-substitution and FPG staining. II. Sister chromatid exchanges. *Chromosoma* **73**, 75–84.

Tice, R. R., and Hollaender, A., eds. (1984). "Sister Chromatid Exchanges." Plenum, New York.

Toda, T., Uno, I., Ishikawa, T., Powers, S., Kataoka, T., Broek, D., Cameron, S., Broach, J., Matsumoto, K., and Wigler, M. (1985). In yeast, RAS proteins are controlling elements of adenylate cyclase. *Cell* **40**, 27–36.

Toledo, L. A., Bennett, M. D., and Stern, H. (1979). Cytological investigations of the effect of colchicine on meiosis in *Lilium* hybrid cv. "Black Beauty" microsporocytes. *Chromosoma* **72**, 157–173.

von Borstel, R. C., Cain, K. T., and Steinberg, C. M. (1971). Inheritance of spontaneous mutability in yeast. *Genetics* **69**, 17–27.

von Wettstein, D., Rasmussen, R. W., and Holm, P. B. (1984). The synaptonemal complex in genetic recombination. *Adv. Genet.* **18**, 331–414.

Wagstaff, J. E., Klapholz, S., Weddell, C. S., Jensen, L., and Esposito, R. E. (1985). Meiotic change with and between chromosomes requires a common rec function in *Saccharomyces cerevisiae. Mol. Cell. Biol.* **5**, 3532–3544.

Walker, G. C. (1984). Mutagenesis and inducible responses to deoxyribonucleic acid damage in *Escherichia coli. Microbiol. Rev.* **48**, 60–93.

Walmsley, R. M., and Petes, T. D. (1985). Genetic control of chromosome length in yeast. *Proc. Natl. Acad. Sci. U.S.A.* **82**, 506–510.

Wang, H.-T., Frackman, S., Jensen, L., Esposito, R. E., and Elder, R. (1984). Transcription of the *SPO13* gene during sporulation. *Int. Conf. Yeast Genet. Mol. Biol., 12th, 1984*, p. 268.

Weir-Thompson, E. M., and Dawes, I. W. (1984). Developmental changes in translatable RNA species associated with meiosis and spore formation in *Saccharomyces cerevisiae. Mol. Cell. Biol.* **4**(4), 695–702.

West, S. C., and Howard-Flanders, P. (1984). Duplex–duplex interaction catalyzed by recA protein allow strand exchanges to pass double-strand breaks in DNA. *Cell* **37**, 683–691.

West, S. C., and Körner, A. (1985). Cleavage of cruciform DNA structures by an activity from *Saccharomyces cerevisiae. Proc. Natl. Acad. Sci. U.S.A.* **82**, 6445–6449.

Westergaard, M., and von Wettstein, D. (1972). The synaptonemal complex. *Annu. Rev. Genet.* **6**, 71–110.

Whitehouse, H. K. (1982). "Genetic Recombination: Understanding the Mechanisms." Wiley, New York.

Williamson, M. S., Game, J. C., and Fogel, S. (1985). Meiotic gene conversion mutants in *Saccharomyces cerevisiae*. I. Isolation and characterization of *pms1-1* and *pms1-2*. *Genetics* **110**, 609–646.

Yashar, B. M., and Pukkila, P. J. (1985). Changes in polyadenylated RNA sequences associated with fruiting body morphogenesis in *Coprinus cinereus. Trans. Br. Mycol. Soc.* **84**, 215–226.

Yeh, E., Carbon, J., and Bloom, K. (1986). Tightly centromere-linked gene (*SPO15*) essential for meiosis in the yeast *Saccharomyces cerevisiae. Mol. Cell. Biol.* **6**, 158–167.

Zamb, T. J., and Roth, R. (1977). Role of mitotic replication genes in chromosome duplication during meiosis. *Proc. Natl. Acad. Sci. U.S.A.* **74**, 3951–3955.

Zubénko, G. S., and Jones, E. W. (1981). Protein degradation, meiosis and sporulation in proteinase-deficient mutants of *Saccharomyces cerevisiae. Genetics* **97**, 45–64.

NOTES ADDED IN PROOF

RecA-like proteins have been purified from mouse and *lilium* meiotic and mitotic cells. The activity is considerably enhanced in meiotic *lilium* cells [Y. Hotta, S. Tabata, R. Bouchard, R. Piñon, and H. Stern (1985). *Chromosoma* **93,** 140–151]. A strand-exchange protein has recently been purified from yeast that increases 10- to 20-fold during meiosis. The increase is meiosis specific, does not occur in strains that are homozygous for mating type, and requires a functional *RAD50* gene (A. Sugino, J. Nitiss, and M. Resnick, submitted for publication).

III
Chromosomes

7

Chiasmata

G. H. JONES

Department of Genetics
University of Birmingham
Birmingham B15 2TT, England

I. HISTORICAL AND DESCRIPTIVE

The earliest studies of meiosis, which were conducted during the last two decades of the nineteenth century, resulted in a confusing mass of observations and theories from which the present-day understanding of meiosis only gradually emerged (Hughes, 1959; Whitehouse, 1965). Prominent landmarks in this progress were the discoveries that homologous chromosomes pair side-by-side (parasynapsis) rather than end-to-end, and that homologues are usually conjoined into bivalents from diplotene to metaphase I by chiasmata.

Chiasmata were first fully described and named by Janssens (1909) from his studies of meiosis in urodele amphibians, although Rückert (1892) had earlier described such structures in the oocytes of sharks. Janssens noted that homologous chromosomes in the bivalents of *Batrachoseps attenuatus* came together at points or nodes which he called chiasmata, and where they did so two chromatids could be seen to cross one another. He defined a chiasma in the following terms: "the word chiasma may be taken in the general sense of two threads which cross one another in a cross of St. Andrew" (translation by Darlington, 1932). Darlington (1932) noted additionally that the same exchange of partners which appears as a St. Andrew's cross will appear as an open Greek cross from another point of view (Fig. 1), that is, the same structure can assume two alternative topological forms. The Greek cross is a bilaterally symmetrical figure and without the advantage of differential chromatid staining (see Section III) it is plainly impossible to identify the recombinant and parental chromatids. However, it has usually been assumed that chiasmata displaying the St. Andrew's cross configu-

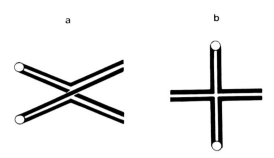

Fig. 1. The alternative topological forms which chiasma can assume: the St. Andrew's cross (a) and the open Greek cross (b).

ration have the two recombinant chromatids crossing one another, the two outer chromatids being nonrecombinant. On the other hand, Darlington (1937) as part of his torsion hypothesis of chiasma formation proposed the reverse to be the case, that the outer chromatids were recombinant and the crossing chromatids parental. With the advent of molecular labeling (see Section III), which can distinguish parental from recombinant chromatids, both these suppositions have been shown to be erroneous since both types of configuration appear regularly in differentially stained bivalents of *Locusta migratoria* following the BrdU/FPG procedure. Evidently the chiasma is a complex three-dimensional figure which can assume either of the two St. Andrew's crosses or the Greek cross configuration depending on the plane in which it is viewed.

Another structural feature of chiasmata is the characteristic dissociation of chromatids in their immediate vicinity, in contrast to the close cohesion of sister chromatids elsewhere in bivalents. This is especially noticeable at early to mid-diplotene when the chromosomes are in a relatively extended state (Fig. 2) and contributes to the fact that a chiasma cannot be accurately located to a precise point, but instead occupies a region of a bivalent.

Although chiasmata fall well within the resolving power of light microscopy and have a long history of investigation by this technique, the application of electron microscopy to the study of meiosis has inevitably led to investigations of the ultrastructural characteristics of chiasmata. Synaptonemal complexes are generally shed from most regions of bivalents during early diplotene, but some mammalian species, including the mouse, apparently retain the single axes of the lateral elements in the disjoined homologous chromosomes. This singular advantage enabled Solari (1970, 1980a) to identify and characterize chiasmata by electron microscopy in diplotene bivalents of male mice. He found that the axes were attached at each end to the nuclear membrane and the axes of homologous chromosomes were widely separated for most of their lengths except for short stretches where they closely approached each other at convergence regions.

7. Chiasmata

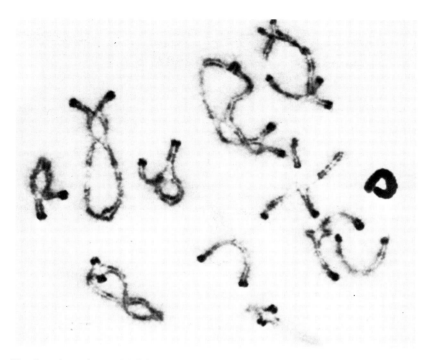

Fig. 2. An early to mid-diplotene nucleus of *Schistocerca gregaria* showing 11 autosomal bivalents and a univalent X chromosome. Numerous chiasmata are present, some of which show the characteristic dissociation of sister chromatids in their immediate vicinity.

During early diplotene these convergence regions each contained a short retained stretch of modified synaptonemal complex (SC). This "first phase" of the convergence regions of axes is followed by a "second phase" in which the remnants of the central space of the SC pieces are lost and replaced by bridges of chromatin fibers. From these observations Solari inferred that the convergence points must correspond to chiasmata. More or less coincidentally, retained SC fragments were observed at diplotene in the fungus *Neotiella* and inferred to represent the positions of chiasmata (Westergaard and von Wettstein, 1970), and subsequently similar observations have been made in a variety of species, including maize (Gillies, 1975), the fungi *Sordaria* (Zickler, 1977) and *Neurospora* (Gillies, 1979), *Ascaris* (Kundu and Bogdanov, 1979), the crane fly, *Pales* (Fuge, 1979), the silkworm, *Bombyx* (Holm and Rasmussen, 1980), the fungus *Coprinus* (Holm et al., 1981), and man (Holm and Rasmussen, 1983). Most of these studies have concluded, with varying degrees of confidence, that these retained SCs, or some of them, represent chiasma sites, but because chromosomal axes are not present in diplotene bivalents of any of these species, this conclusion is to some extent conjectural. Three-dimensional reconstruction analyses of entire bivalents have in some cases established correspondences between the total numbers of retained SCs and the numbers of chiasmata or crossovers observed by conventional means (e.g., Zickler, 1977; Gillies, 1979; Kundu and Bogdanov, 1979; Holm and Rasmussen, 1980, 1983). In other studies the numbers of retained SCs considerably exceeded the best estimates of numbers of chiasmata (Sotello *et al.*, 1973; Gillies, 1975; Fuge, 1979).

The association of retained SCs with chiasmata might indicate, as tentatively suggested by Solari (1970), that they are essential and integral components of chiasmata, or that they are somehow necessary for the transition from a molecular recombination event to a cytological chiasma. On the other hand, they may simply result from localized delayed SC shedding at chiasmata due to the entrapment of the SC in the crossing chromatids. In *Bombyx, Coprinus,* and humans, all lacking continuous axes along their diplotene bivalents, the transformation of retained SCs into stable "chiasmata" has been described in some detail. This process involves the elimination of SC fragments and their replacement by chromatin bridges. In *Bombyx* and *Coprinus,* recombination nodule (RN) derivatives, including complex structures termed "circular components," appear to be involved in this transformation. However, the indirectness of the association between these ultrastructural features and classical chiasmata makes their status and role difficult to evaluate.

II. THE CHIASMATYPE THEORY

The origin of the chiasmatype theory is directly attributable to Janssens (1905, 1909, 1924), who proposed that chiasmata resulted from exchanges between homologous non-sister chromatids of bivalents. Janssens also proposed, with strong support from the *Drosophila* school of T. H. Morgan (e.g., Morgan and Cattell, 1912), that chiasmata were the cytological expression of genetical crossover events. However, the "classical" interpretation of chiasmata, which derived from the cytological research of Wenrich (1916) and Robertson (1916), argued that chiasmata represented the convergences of different modes of opening out or separation of chromatids, on the one side between homologous chromatids and on the other between sister chromatids, and were therefore not related to crossing-over.

The modern, and simplified, version of Janssens' theory was first clearly stated by Darlington (1932), based largely on his own observations (Darlington, 1930) and those of Belling (1927, 1928, 1931). As Darlington stated in 1932, "the assumption that *sister* chromatids are associated on both sides of any chiasma and that therefore crossing-over takes place between two chromatids of different chromosomes at the chiasma, before it is formed, is the only one which stands the test of observation." More recently, the evidence supporting the chiasmatype interpretation of chiasmata has been ably summarized by Whitehouse (1965). This evidence includes the occurrence of diagnostic chiasmate configurations in structurally and numerically altered chromosome complements such as unequal chromosome pairs, inversion heterozygotes and triploids, certain critical interlocked bivalent configurations, and corresponding chiasma and crossover frequencies in genetically well mapped organisms (see Whitehouse, 1965, for details).

The chiasmatype theory confers on chiasmata certain advantages as a means of studying recombination provided that chiasmata are conserved at their original locations, that is, they do not terminalize (see Section IV). Chiasmata provide a very direct and rapid approach to studying recombination. In suitable organisms they can give detailed and accurate information on the total amount of recombination and on its distribution through the genome. They also permit the analysis of recombination in a range of species, unlimited by the availability of genetic markers, and in sterile hybrids which cannot be analyzed genetically. On the other hand, chiasmata are relatively imprecise indicators of recombination positions. At the molecular level crossovers are localized events limited to a few hundred or thousand base pairs, while chiasmata are fairly gross cytological structures. Another drawback to chiasmata is that they equate only to reciprocal crossover exchanges; nonreciprocal recombination events are undetectable cytologically.

III. THE MECHANISM OF CHIASMA FORMATION

While the evidence summarized in the preceding section convincingly disproved the classical theory and led to the widespread acceptance of the chiasmatype theory, it had no bearing on the actual mechanism of exchange leading to chiasma formation. One popular mechanism of genetic exchange at chiasmata which was first proposed by Belling (1928) and later adopted in a modified form by geneticists to account for nonreciprocal recombination (e.g., Lederberg, 1955) proposed that exchanges occurred during conservative chromosome replication by the switching of copying from one chromosome template to the other homologue. This became known as the copy-choice hypothesis, but its popularity was short-lived because of the many glaring anomalies between its basic assumptions and the known or emerging facts of chromosome replication and recombination, most notably the semiconservative mechanism of chromosome replication.

Positive evidence for the alternative breakage—rejoining mechanism of chiasma formation was lacking until the advent of molecular labeling methods for differentiating meiotic chromatids along their entire lengths (reviewed by Jones and Tease, 1980). Taylor (1965) pioneered the application of tritiated-thymidine labeling to meiotic chromosomes and was the first to establish that meiotic chromosomes underwent semiconservative replication like the chromosomes of somatic cells. He found that chromosomes labeled with tritiated thymidine in the penultimate S phase before meiosis showed label segregation and entered meiosis with one labeled and one unlabeled chromatid per chromosome, and furthermore he predicted that half the chiasmata formed would, on average, generate a reciprocal exchange of labeled and unlabeled segments if breakage and rejoining occurred at chiasmata. The relatively low resolution of tritium autoradiography prevented the direct analysis of chromatid labeling patterns in diplotene bivalents containing chiasmata and consequently label exchanges indicative of breakage–rejoining events could not be directly related to chiasmata. Despite this limitation, several studies established clear, albeit indirect, correlations between chiasmata scored at diplotene/diakinesis and label exchanges observed at the more condensed first anaphase and second metaphase stages (Taylor, 1965; Peacock, 1968, 1970; Church and Wimber, 1969; Craig-Cameron and Jones, 1970a). In spermatocytes of the grasshopper *Stethophyma grossum* a clear correlation was established between proximally localized chiasmata and similarly located label exchanges (Jones, 1971).

The analysis of meiotic exchange events was taken a significant step further by the development of BrdU labeling combined with FPG staining to differentially *stain* the sister chromatids of chromosomes (Tease, 1978). The application of this technique to meiotic chromosomes enabled a precise analysis of meiotic exchange events involving breakage and rejoining and their relationship to chiasmata could be determined directly by observations on diplotene/diakinesis

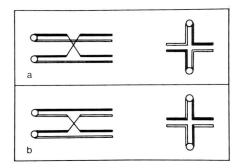

Fig. 3. The origins and appearances of hidden chiasmata (a) and visible chiasmata (b) in differently stained bivalents after BrdU incorporation and FPG staining.

Fig. 4. A single monochiasmate bivalent of *Locusta migratoria* showing differential staining of sister chromatids and a "visible" exchange between dark and light chromatids which coincides exactly with the chiasma (compare with Fig. 3b).

bivalents (reviewed by Jones and Tease, 1980). Following BrdU labeling and FPG staining the two chromatids of each homologous chromosome at meiosis are differentially stained, one dark, the other light, and this differentiation can be clearly discerned at all stages of meiosis from diplotene onward. Provided that chiasma formation involves exchange between two randomly selected non-sister chromatids, it is expected that on average half the chiasmata observed at diplotene/diakinesis should have a visible exchange associated with them as a result of breakage and rejoining of light- and dark-staining chromatids. The remaining chiasmata, resulting from exchanges between similarly staining chromatids (light/light or dark/dark), should not contain visible exchanges; these are termed "hidden" chiasmata (see Fig. 3). These expectations were fully realized in an extensive study of several hundred chiasmata in *Locusta migratoria* (Tease and Jones, 1978; Jones and Tease, 1980; see Fig. 4), and similar findings have been reported from studies on female mice (Polani *et al.*, 1981) and male hamsters (Allen, 1982). However, in male mice (Kanda and Kato, 1980) and in some individual hamsters (Allen, 1982), an unexplained excess of hidden chiasmata was observed. This may be due to an inherent scoring bias favoring hidden chiasmata; alternatively, under certain conditions non-sister chromatid exchanges may preferentially involve chromatids of the same labeling status. The overall conclusion from these studies is, however, clear, that chiasma formation occurs by a breakage–rejoining mechanism, the precise molecular details of which are yet to be fully determined.

IV. DO CHIASMATA TERMINALIZE?

The chiasma terminalization hypothesis was first advanced by Darlington (1929) with the objective of reconciling parasynapsis, or the side-by-side pairing of chromosomes during zygotene/pachytene, with the observation that in many species homologous chromosomes are associated end-to-end at metaphase I. Darlington proposed that chiasmata could move from their sites of origin toward the bivalent ends and that this process led to a progressively more distal distribution of chiasmata and, in extreme cases, a reduction in total numbers of chiasmata as multiple chiasmata merged together at bivalent ends. Despite many modifications and qualifications of the hypothesis (Darlington, 1932, 1937) it is still popularly held that some degree of terminalization occurs in all species, and terminalization is regularly presented as a normal and invariable feature of meiosis.

However, an objective assessment of the available evidence raises some serious doubts as to the extent of terminalization and indeed questions whether it occurs at all. Much of the earlier evidence supporting the terminalization hypothesis consisted of comparisons of chiasma frequencies and distributions at successive stages of meiosis, purporting to show progressive movement and in some cases loss of chiasmata between diplotene and the later stages of diakinesis and metaphase I. This evidence is inherently weak because of its indirect and inferential nature and the considerable difficulties of interpretation involved. The principal criticisms of this evidence are summarized below; a more detailed critique of the terminalization hypothesis will appear separately (Tease and Jones, 1986).

1. Chiasma number at diplotene may be overestimated due to the inclusion of relational twists which are frequently impossible to distinguish from chiasmata. This is especially true in plant species, where clear resolution of the chromatid structure of diplotene bivalents is generally not possible.

2. Because stage comparisons are inevitably made between nonidentical cells or groups of cells, chiasma frequency differences between different stages may be confounded with developmentally or environmentally determined differences (see Section VI).

3. Stage-related variations in bivalent length reflecting progressive condensation can be confounded with variations in length associated with chiasma frequency variation (e.g., Fox, 1973). Hence, apparent reductions in chiasma frequency with advancing stages may in fact reflect a direct causal influence of bivalent length on chiasma frequency.

In contrast to these indirect tests of the terminalization hypothesis, which are at best unconvincing, direct experimental tests for its occurrence have almost without exception proved negative. Tests based on differential labeling with tritiated thymidine combined with autoradiography (e.g., Jones, 1971, 1977) and the exploitation of C-band heteromorphisms (Loidl, 1979; Santos and Giraldez, 1978; Orellana and Giraldez, 1981; Seuanez et al., 1983) have consistently failed to detect terminalization. The more exact and rigorous test based on differential chromatid staining after BrdU incorporation (reviewed by Jones and Tease, 1980) convincingly showed that in a range of organisms, including locusts, mice, and hamsters, chiasmata retain their associations with exchange points, that is, they do not terminalize (see Fig. 4).

Comparisons of chiasma distributions from air-dried nuclei at diakinesis/metaphase I with recombination nodule (RN)/bar distributions from three-dimensional reconstructions of pachytene nuclei have suggested that in human males chiasmata show a more distal distribution than do RNs/bars, and a degree of chiasma terminalization has been advanced as an explanation for this difference (Holm and Rasmussen, 1983). On the other hand, observations on the frequency and distribution of bars on microspread SCs of human spermatocytes showed a reasonably good agreement with the chiasma data (Solari, 1980b). The changes which have been reported in the frequency, distribution, and morphology of RNs/bars during pachytene (Holm and Rasmussen, 1983) add considerably to the difficulties of interpreting these ultrastructural features and assessing their relationships to classical chiasmata. Unfortunately, direct experimental tests of the terminalization hypothesis have so far proved impossible to apply to human spermatocytes.

There have been, in addition, repeated assertions that chiasma terminalization is a regular feature of meiosis in organisms having holocentric chromosomes (John and Lewis, 1965; White, 1973; John, 1976). Since the groups of organisms involved show many atypical features of chromosome organization and meiotic behavior, including diffuse centromeric activity and inverted meiotic sequences (i.e., equational first division and reductional second division), it is very difficult to relate these observations to the general question of chiasma terminalization in conventional meiotic systems, and they are probably best considered as special cases. Whether a phenomenon which is directly analogous to chiasma terminalization occurs in these groups is a point which requires separate detailed and careful consideration.

V. THE NATURE OF "TERMINAL" BIVALENT ASSOCIATIONS

If, as suggested in the previous section, chiasmata do not terminalize but are conserved in position from the time of their first appearance at early diplotene until late metaphase I, the apparently "terminal" end-to-end associations of chromosomes, which are a common feature of metaphase I in some species, require some explanation. One explanation is that these associations are distally localized chiasmata. Chiasma localization was recognized as an important and widespread phenomenon by Darlington (1932) and Mather (1938). Henderson (1969a) also argued that "many cases of presumed terminalisation are likely to be examples of distal localisation." Chiasmata involve an exchange of chromatid segments and consequently they cannot, by definition, originate as truly terminal associations (see John, 1976). The localization hypothesis assumes that in some cases chiasmata occur in such extremely subterminal locations, involving the exchange of very small segments, that they are superficially indistinguishable from end-to-end associations.

This view has received support from a recent study of differentially stained, terminally associated bivalents of *Locusta migratoria* (Jones and Tease, 1984). These bivalents showed precisely the patterns of light and dark chromatid associations expected as a result of subterminal chiasma formation. In addition, a proportion of these bivalents showed minute exchanges of differentially stained chromatid segments, due to marginally subterminal visible exchanges, while in other bivalents the exchanged segments were too small to be resolvable. It is therefore clear that, even at late diplotene/diakinesis, a genuinely chiasmatic association may appear as a terminal end-to-end conjunction of chromosomes, without lateral projections, in conventionally stained preparations.

In other situations, for example, in pollen mother cells of rye (*Secale* spp.), many chiasmata form in substantially subterminal locations and show distinct lateral projections at late diplotene/diakinesis, but frequently appear as terminal, end-to-end associations at metaphase I, an observation which seems to support the terminalization hypothesis. However, detailed examinations of C-banded bivalents of rye have shown that while chiasmata frequently form immediately adjacent and proximal to terminal C-bands, giving prominent C-banding lateral projections at late diplotene/diakinesis, the C-bands are simply absorbed into the width of the metaphase I bivalents as a consequence of bivalent condensation. In this situation, there is no terminalization in the strict sense and the phenomenon has been termed pseudoterminalization (Jones, 1978). The strictly linear appearance of bivalent associations in several other situations has encouraged the view that these may be achiasmatic associations of chromosome ends. Mammalian sex bivalents have been variously interpreted, with several strong claims for achiasmy in species such as the mouse, where the X and Y chromosomes have a simple, apparently end-to-end association (reviewed by Solari, 1974). However, the occurrence of a short SC associated with sex vesicle in male mice suggested that the association was in fact chiasmate (Solari, 1974), although SC formation does not invariably lead to chiasma formation in other situations. However, the chiasmate nature of this association has been confirmed beyond reasonable doubt by recent cytogenetic studies of male mice carrying the Sxr sex-reversing factor (Evans et al., 1982). Nevertheless, genuinely achiasmatic XY associations appear to be the rule in marsupial mammals (Sharp, 1982) and in at least one species of eutherian mammal (Solari and Ashley, 1977), since the axial elements of the sex chromosomes associate end-to-end without SCs. Associations of terminal heterochromatic blocks have also been interpreted as achiasmate end-to-end contacts (e.g., John and King, 1982), on the grounds that no visible chiasmata could be discerned at the conjunctions, that chiasma formation has never been convincingly demonstrated within heterochromatic blocks, and that associations between heterochromatin are not confined to meiosis. Nevertheless, the occurrence of chiasmata in short terminal euchromatic segments cannot be convincingly excluded in these cases.

VI. CHIASMA VARIATION: AN OVERVIEW

Chiasmata are extremely variable features of meiosis, in terms of their numbers (frequencies), their distributions between cells and between bivalents, and their positional distributions along bivalents (see Section VII). The causes of chiasma variation are themselves variable and include genetic determinants (reviewed by Rees, 1961; Baker et al., 1976) and environmental causes, both external and internal to the organisms. The different types and causes of chiasma variation can be expressed at a variety of different levels, viz., between species, between individuals, between cells within individuals, between bivalents, and also between sexes.

Chiasma variation between species is very pronounced and widespread. To a large extent, variation in chiasma *frequency* between species reflects diversity of karyotype organization, particularly variation in chromosome number. However, even quite closely related species with identical chromosome numbers and very similar karyotypes may show significant differences in chiasma frequency, for example, truxaline grasshoppers of the genera *Chorthippus, Omocestus,* and *Myrmeleotettix* (Hewitt, 1964), *Lolium* species (Rees and Jones, 1967), *Triticum* species (Hillel et al., 1973), and *Diabrotica* species (Ennis, 1972). Striking differences in chiasma *distribution* are also found between some closely related species, for instance, *Allium fistulosum* has mostly proximal chiasmata while most chiasmata in *Allium cepa* are located in interstitial and distal regions of bivalents. These interspecific differences are usually presumed to have a genetic basis.

Chiasma frequency and distribution also vary quite considerably between individual plants or animals within species (e.g., Laurie and Jones, 1981; Whitehouse et al., 1981), but genetic and environmental causes can be difficult to separate, particularly in the case of individuals from outbreeding natural populations. However, interindividual differences within inbred lines maintained by strict self-pollination are presumed to be largely environmental in origin (Rees, 1955) and differences between sister inbred lines are presumed to have a genetic basis. Pronounced and sometimes dramatic variations in chiasma frequency and distribution can be experimentally induced by environmental changes, notably by temperature treatments (e.g., Elliot, 1958; Henderson, 1962; King and Hayman, 1978) but also by a wide range of other environmental variables (reviewed by Westerman, 1967). In some cases, evidence for developmental or environmental effects on chiasma formation have been obtained by sampling large numbers of individuals at different ages or in different environments. In this way, chiasma frequency has been shown to decline markedly with age in male grasshoppers (Maudlin, 1972) and in female mice (Speed, 1977) and marked seasonal effects on chiasma frequency and distribution have been demonstrated in the Australian lizard *Phyllodactylus marmoratus* (King and Hayman, 1978).

The numbers and distribution patterns of chiasmata formed in particular cells can vary quite considerably within the same individual, and as these cells have identical genotypes it follows that any chiasma variation they show is strictly developmental (environmental) in origin. The extent of intraindividual variation which is evident at a given time or stage of development varies considerably between different inbred lines of rye and also between inbred lines and their hybrids, reflecting variation in the degree of control which is exercised over chiasma formation (Rees and Thompson, 1956). Significant differences in chiasma frequency have also been demonstrated between testis follicles of the grasshopper *Chorthippus brunneus* (Maudlin, 1972) and between capitula (inflorescences) of *Crepis capillaris* (Tease, 1977) within the same individuals. Systematic differences in chiasma frequency have been demonstrated between different anther segments in *Secale cereale* and in four tulip species, which were clearly related to the developmental ages of pollen mother cells (Rees and Naylor, 1960; Couzin and Fox, 1974).

An important and often very striking category of chiasma variation is that which occurs between sexes (reviewed by Callan and Perry, 1977). Male and female meiocytes may differ in chiasma frequency and/or chiasma distribution, and the differences may be slight or substantial. Generally male meiocytes show the lower chiasma frequencies and also the greater restrictions on chiasma distribution, but there are some exceptions to this rule (Callan and Perry, 1977). The greatest restriction on chiasma formation is found in those species which display achiasmate meiosis (see Table 4 in John and Lewis, 1965). Generally this is restricted to one sex (the heterogametic sex), but in a few species of enchytraeid worms both male and female meiocytes are achiasmate (Christensen, 1961).

VII. INDICATIONS OF CHIASMA CONTROL

In the absence of any form of control, chiasma distribution is expected to be entirely random, a situation which with very rare exceptions is never met in reality. Constraints on chiasma distribution resulting in nonrandomness are therefore outward expressions of controls, and the identification of these constraints is a useful preliminary to discussing models and mechanisms of control.

In the absence of control, the distribution of chiasma number per cell should approximate a Poisson distribution, since the numbers and sizes of chromosomes are identical in different cells. In reality cell chiasma frequencies show much narrower (in statistical terms, overdispersed) distributions (e.g., Jones, 1967), which are indicative of control. It is also known that the extent of cell-to-cell variation in chiasma frequency, and hence the degree of control, shows genetically determined variation (Rees and Thompson, 1956).

The distribution of chiasmata between *similar-sized* bivalents is also generally much narrower (overdispersed) than the Poisson distribution predicts on the basis of random uncontrolled events (Haldane, 1931; Jones, 1967). In such situations, bivalents have similar numbers of chiasmata with a shortage or absence of bivalents with high chiasma numbers, and no univalents (i.e., chromosome pairs lacking chiasmata). The dearth of bivalents with high chiasma numbers is generally attributed to the action of positive chiasma interference (Haldane, 1931; see Section VIII) which, because of its influence on chiasma distribution along bivalents, also effectively limits the total numbers of chiasmata occurring in bivalents. The absence of univalents points to the existence of a further constraint, namely, the invariable formation of at least one chiasma per bivalent, that is, an obligate chiasma. In the majority of species this is essential for regular bivalent orientation and chromosome disjunction.

Most species possessing asymmetrical karyotypes, that is, with chromosomes of widely different lengths, show distributions of chiasmata between bivalents which are generally more or less proportional to chromosome lengths with the important proviso, noted above, that all bivalents—irrespective of length—have at least one chiasma (Mather, 1937). The dependence of *mean* chiasma frequencies of different bivalents on the lengths of the chromosomes concerned is not itself indicative of control, since just this relationship would be expected in the absence of control, with a random distribution. However, a control mechanism of a particular kind is indicated by the *regular* formation of certain numbers of chiasmata in bivalents of different lengths. This relationship has had a particularly notable influence on the development of models and mechanisms for chiasma control (e.g., Mather, 1936a, 1937, 1938; Henderson, 1963; Fox, 1973).

Chiasma distribution between bivalents may also be influenced by interbivalent effects, that is, nonindependence of the chiasma frequencies of different bivalents occupying the same nuclei. Various statistical approaches have been used to study this question including interclass and intraclass correlations and contingency analyses, applied to normal and abnormal individuals. However, these studies have yielded very variable and inconsistent results (reviewed by Sybenga, 1975). While some normal diploid species appear to show significant negative correlations between certain bivalents or groups of bivalents, suggesting interbivalent competition (e.g., Mather, 1936b; Harte, 1956; Basak and Jain, 1963), many other species examined have shown no significant interbivalent effects at all (e.g., Hewitt, and John, 1965; Hulten, 1974) or inconsistent and variable effects including significant positive and negative intraclass correlations in different plants (Rowlands, 1958). However, in abnormal situations where one or more chromosome pairs show altered chiasma frequencies for structural or genetic reasons, associated changes in the chiasma frequencies of the other bivalents frequently occur. Inversion heterozygotes of *Drosophila melanogaster* commonly show this effect (Schultz and Redfield, 1951). Heterozygotes for certain interchanges in a variety of plant and animal species show increased chiasma frequencies of the interchange multiples and, in some cases, of the noninterchanged chromosomes too (Hewitt, 1967; Arana *et al.*, 1980), while in other cases the chiasma frequencies of the noninterchanged chromosomes are unaffected (Ross and Cochran, 1981; Parker *et al.*, 1982). A further example is provided by *Crepis capillaris* in which natural population plants showed no consistent interbivalent effects (Tease, 1977), while chromosome-specific desynaptic mutants from an experimental population in which certain chromosomes showed high levels of univalence consistently showed small but significant compensatory increases in the chiasma frequencies of the other bivalents (Tease and Jones, 1976). This lack of consistency as to the occurrence and direction of interchromosomal effects suggests that this is not a general phenomenon or a regular feature of chiasma control. Interbivalent effects have been variously attributed to competition between bivalents for limited amounts of a particular substance or substances required for chiasma formation (Mather, 1936b; Holliday, 1977) or, in the case of structural rearrangements, to an effect on the duration of prophase I and hence the time available for chiasma formation (Lucchesi and Suzuki, 1968).

The positional distribution of chiasmata within bivalents is, with one or two rare exceptions, never random. If it were so, chiasmata would occur with equal frequencies, on average, in all bivalent regions and the positions of two or more chiasmata occurring simultaneously in the same bivalent would be independent of one another. Neither of these conditions is found to occur in practice and these departures from randomness reflect the two principal identifiable constraints on chiasma distribution within bivalents, namely, localization and interference.

7. Chiasmata

Chiasma localization arises from the tendency for chiasmata to occur preferentially or exclusively in certain bivalent regions (Darlington, 1931). Extreme localization of chiasmata is found in some species, sometimes confined to one sex (Callan and Perry, 1977). Chiasmata may be procentrically or proximally localized near centromeres as in *Stethophyma grossum* males and *Allium fistulosum*, or more commonly distally localized near chromosome ends as in *Secale cereale* and in the L bivalents of *Chloealtis conspersa* and *Euthystira brachyptera*. The acrocentric or telocentric bivalents of some species show chiasmata localized to both centromeric and telomeric ends (= proterminal localization). In addition to these cases of extreme localization, many, probably all, species show some degree of localization in the sense that while chiasmata are not restricted to certain regions, they occur more frequently in some regions than others. This is a quantitative rather than a qualitative effect and usually requires careful measurements and plotting of frequency histograms of chiasma positions within bivalents for its detection. This type of study was first undertaken by Henderson (1963) in *Schistocerca gregaria* and subsequently by Fox (1973) in the same species, Southern (1967) and Laurie (1980) in truxaline grasshoppers, Hulten (1974) in human males, Shaw and Knowles (1976) and Coates and Shaw (1982) in Australian grasshoppers of the genus *Caledia*, and by Maudlin and Evans (1980) in mouse oocytes. All these studies revealed some degree of localization in the distribution of chiasmata, confirming the conclusion reached from studies of the distribution of genetic recombination in *Drosophila melanogaster* (Mather, 1938). For example, the L_3 bivalent of *Chorthippus brunneus* shows a tendency to distal localization, particularly in the short arm, and relatively few chiasmata occur around the centromere (Laurie, 1980) (see Fig. 5).

The second major identifiable constraint on chiasma distribution within bivalents is chiasma interference (Mather, 1933). This phenomenon was first recognized in genetic studies (Muller, 1916) and hitherto has been most thoroughly investigated and characterized at the genetic level. Genetically, interference is expressed as reduced frequencies of double crossovers compared to the frequencies expected on the basis of crossover independence, that is, crossovers appear to reduce the probability of other crossovers occurring in their vicinity. Cytologically, interference is expressed in a number of different ways. One rather indirect expression was recognized by Haldane (1931), who showed that the distribution of chiasma number per bivalent among the M bivalents of *Vicia faba* was much narrower (overdispersed) than the Poisson distribution predicted. As discussed earlier, this is partly attributable to the effect of interference in limiting the number of chiasmata which can occur in bivalents. Interference between chiasmata can also be inferred from comparisons of chiasma distributions in bivalents or bivalent arms having one or two chiasmata. It is commonly observed that these distributions differ considerably, so that single chiasmata

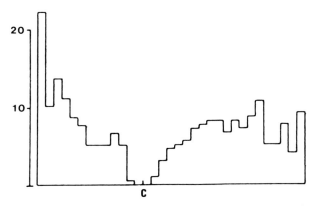

Fig. 5. Pooled histogram of chiasma distribution in the L_3 bivalent of *Chorthippus brunneus* normalized to chiasma frequency per 100 bivalents, based on 511 bivalents from 13 full sib individuals from a single family. The bivalent is divided into sections corresponding to 5% of the long arm; the position of the centromere is indicated (C).

tend to occupy more central positions in bivalent arms while two chiasmata occupy relatively more proximal and distal locations, from which the operation of interference can be reasonably inferred. This effect appears to be more or less universal but its extent is variable; the effect is quite marked in the long arm of the *Chorthippus brunneus* L_3 bivalent (Laurie, 1980), while in other cases, such as bivalent 6 of the Torresian taxon of *Caledia captiva* (Coates and Shaw, 1982), the shift in chiasma position is discernible but not very marked.

A more direct expression of chiasma interference appears in the widespread tendency for chiasmata to be spaced apart along bivalents. This was recognized in a qualitative way in early cytological studies (Mather, 1940) and has been thoroughly quantified in more recent studies of chiasma distribution (Fox, 1973; Shaw and Knowles, 1976; Laurie, 1980). In these studies, interchiasma distances were plotted in histogram form, with the following conclusions: (1) Long bivalents show a wide range of interchiasma (ix) distances. (2) Mean interchiasma ($\overline{ix}$) distances vary in proportion to the lengths of the different bivalents. (3) Bivalents show minimum interchiasma (ix min.) distances. In *Schistocerca gregaria* (Fox, 1973) ix min. is constant in different bivalents irrespective of length and amounts to 13% of the length of the longest bivalent and 33% of the shortest bivalent. In *Chorthippus brunneus*, where only one bivalent (L_3) was studied, ix min. represents 30% of the long arm; in molecular terms this is a huge distance, equal to 3×10^5 kbp of DNA. How do the two parameters $\overline{ix}$ and ix min. relate to genetic interference? Clearly ix min. can be directly related to interference as this represents a distance or region of virtually total prohibition and it is therefore reasonable to conclude that ix min. is an expression of complete interference. The extent and strength of interference over greater distances cannot be judged simply by inspection of ix histograms or from the value of $\overline{ix}$; interference might fall off rapidly or extend in a weak form for considerable distances. Newcombe (1941) attempted to measure the physical extent of interference in *Trillium erectum* bivalents by comparing observed and expected frequencies of double chiasmata in pairs of segments of known physical length. Interference was found to be very effective over short distances of 1 μm but fell off rapidly and was barely detectable over distances greater than 2 μm, this distance approximating 20% of long arms and 50% of short arms. A similar analysis has been undertaken in *Chorthippus brunneus* (Jones, 1986), where extensive information is available on the L_3 bivalent (Laurie, 1980). The long arm of the L_3 bivalent, which invariably forms either one or two chiasmata, was divided into 10 regions of equal lengths and their chiasma frequencies were determined from chiasma distribution measurements on 1467 bivalents. The expected numbers of double chiasmata were computed for all combinations of pairs of regions and these were compared to the observed numbers of double chiasmata in these regions. The coefficients of coincidence were calculated in the usual way. The existence of complete interference over distances equalling 25–30% of the arm was confirmed, since pairs of regions separated by none, one, or two other regions all show coincidence values of zero. Over longer distances, interference continued to be expressed for a further 30% of the arm but fell off gradually over this distance until chiasmata separated by five other regions showed coincidence values approximating unity, that is, no interference (Jones, 1986).

Many studies have investigated the question whether interference is operative across centromeres (reviewed by Sybenga, 1975) but with very divergent conclusions; some have found no interference, others significant positive interference, and others significant negative interference. Sybenga (1975) has reasoned that negative interference across the centromere could be the result of pairing variations. The detection of significant positive interference across centromeres in certain species, e.g., *Culex* (Callan and Montalenti, 1947) and *Paeonea* (Harte, 1956), demonstrates that centromers are *not* always effective barriers to interference. Failures to detect such interference in other species may be due to technical causes; chiasmata may be localized at the distal ends of different arms, too far apart for interference to be operative. In other cases, correlation coefficients of chiasma *numbers* in the two arms of a bivalent may be nonsignificant, but interference may nevertheless be influencing the relative positions of chiasmata in the two arms, as was found in the L_3 bivalent of *Chorthippus brunneus* (Laurie, 1980).

VIII. GENETIC CONTROL OF CHIASMA DISTRIBUTION

The characteristics of chiasma distribution patterns have been interpreted in terms of a series of constraints and controls on chiasma formation and this interpretation is supported by a substantial body of evidence for genetic controls of chiasma distribution based on investigations of naturally occurring and induced major gene mutants, comparisons of inbred lines, and selection experiments (Rees, 1961; Baker et al., 1976). These controls regulate the distribution of chiasmata at several different levels, including the numerical distribution of chiasmata between cells and within cells (e.g., Rees and Thompson, 1956) and the positional distribution of chiasmata within bivalents (Jones, 1967). To some extent, these different levels are subject to separate and independent controls (Rees and Thompson, 1956; Karp and Jones, 1982, 1983) and there is evidence for considerable autonomy of local recombination controls. For example, chromosome-specific desynaptic mutants have been described in *Hypochaeris radicata* and *Crepis capillaris* which show high levels of univalence affecting particular single chromosome pairs (Parker et al., 1976). In fungi such as *Neurospora crassa* and *Schizophyllum commune* genetic controls of recombination have been discovered which operate very locally and affect recombination levels within very short genetic intervals (reviewed by Catcheside, 1977). However, there is also evidence from the analysis of certain genotypes of rye for coordinate controls of chiasma distribution which simultaneously affect the distribution of chiasmata at the between cell, between bivalent, and within bivalent levels. Normal rye genotypes show a highly controlled pattern of chiasma distribution with very even distributions of chiasmata between bivalents and between cells, and distally localized chiasmata. An abnormal genotype on the other hand showed a severe breakdown of control at all levels, with much wider ranges of cell and bivalent chiasma frequencies which fit Poisson distributions and loss of localization within bivalents resulting in proximal and interstitial as well as distal chiasmata (Jones, 1967). A range of rye genotypes showing different degrees of loss of control also showed correlated changes in chiasma distribution at different levels (Jones, 1974). The correlated changes in interbivalent and intrabivalent chiasma distribution parameters suggest that some aspects of chiasma control at these different levels are coordinately controlled, perhaps by a single mechanism. Meiotic mutants of *Drosophila melanogaster* in which the distribution of crossovers is altered in a similar manner have been interpreted as having defects in preconditions for exchange, which determine crossover positions, and as a result they have been termed precondition mutants (Sandler et al., 1968; Baker et al., 1976). It is likely that meiotic mutants recovered from a number of other species, including *Podospora anserina*, *Zea mays*, *Pisum sativum*, *Oenothera*, *Triticum*, and *Lycopersicon esculentum*, are also defective in preconditions necessary for the control of chiasma distribution (Baker et al., 1976).

IX. MODELS OF CHIASMA CONTROL

Consideration of chiasma control in terms of a series of constraints and the characteristics of "precondition" meiotic mutants and abnormal genotypes showing defective control of crossover and chiasma distributions suggest an approach to the formulation of models of chiasma control based on site probabilities (stochastic models). Each bivalent can be considered as consisting of a large number of evenly distributed sites where chiasma formation can occur. In the absence of control, each site would have a uniform low probability of forming a chiasma. Under these conditions, chiasma distribution would be subject to no constraints and would be random at every level; that is, there would be no localization or interference within bivalents, no obligate chiasma, and the distribution of chiasmata between cells and between similar-sized bivalents would be Poissonian. This is the situation which is found, or is approached, in the abnormal genotype of rye described above and in the "precondition" meiotic mutants of *Drosophila* and other species.

A corollary of this model is that the imposition of control might be achieved through the modification of site probabilities, so that some sites have increased probabilities of chiasma formation and others have reduced probabilities; that is, the sites which eventually form chiasmata are a subset of an evenly distributed population of potential sites. This model has parallels with proposals for crossover site selection based on biochemical analyses of meiotic DNA metabolism, which regard crossover site determination as a selective elimination of an evenly dispersed set of potential sites (Stern and Hotta, 1978; Stern, 1981). An example of a simple site probability modification will illustrate both the potential of this approach and also its limitations. Suppose that of all the many potential sites for chiasma formation in each bivalent, two sites, each located distally at either end of each bivalent, have *unit* probability of chiasma formation and all other sites have *zero* probability. This relatively simple, but extreme, change in site probabilities will immediately generate distal localization, obligate chiasmata, and an overdispersed, narrow distribution of chiasmata per bivalent, in fact most of the inter- and intrabivalent chiasma distribution characteristics found in rye and many other organisms. However, this simple model is inadequate as it fails to account for a number of important characteristics of chiasma formation. The model, in its simple form, assumes that site probabilities are independent; in order to account for interference a modification of the model is required which specifies that chiasma determination at a site modifies the probabilities of chiasma determination at adjacent sites. Second, chiasmata seldom, if ever, occur uniquely at defined sites, even in cases of extreme localization (e.g., Fletcher, 1978), that is, chiasma distributions cannot be specified in the simple terms of unit or very large probabilities at certain sites. Instead, high or unit probabilities of chiasma formation appear to be the properties of regions, arms, or whole bivalents. This is not to deny the existence of local site-specific controls which undoubtedly exist also, but these should be regarded as "fine" controls (Smichen and Stamberg, 1969) superimposed on the more general region- or bivalent-based controls.

X. MECHANISMS OF CHIASMA DISTRIBUTION CONTROL

Stochastic models are a convenient approach to formulating general models of chiasma distribution control, but they do not address the question of the mechanism or mechanisms which determine where chiasmata form. To a large extent this question remains unresolved, but the nature of the controlling mechanism has been identified in a small number of species showing extreme localization of chiasmata.

Localized or incomplete pairing has been advanced as a causal mechanism of chiasma localization on the basis of light microscopical (LM) analyses of pairing in a few plant species (Darlington, 1935, 1958; Stebbins and Ellerton, 1939) and in some grasshoppers (Darlington, 1937, 1958; Henderson, 1969b). The analysis of meiotic prophase stages by light microscopy is, however, notoriously difficult, particularly in many grasshopper species which show extensive diffuse stages. Recent electron microscopical (EM) studies of synaptonemal complexes which overcome these difficulties have confirmed the association of localized pairing with localized chiasma formation in male meiosis of the grasshopper *Stethophyma grossum* and the flatworm *Mesostoma ehrenbergii ehrenbergii*. *Stethophyma grossum* ($2n = 23$ ♂) forms 11 bivalents in male meiosis, 8 of which show extreme proximal localization of a single chiasma in each bivalent (see Fig. 6), while EM analyses of synaptonemal complexes, prepared by three-dimensional reconstruction from serial sections and by the surface-spreading method, show that 8 bivalents are incompletely paired to about the extent expected if chiasma localization results from localized pairing (Fletcher, 1978; Wallace and Jones, 1978; Jones and Wallace, 1980). The mean length of SCs belonging to partially paired bivalents in surface spreads was 36 μm, about the same as the mean SC length of the fully paired very short S_{11} bivalent (Jones and Wallace, 1980). An even more extreme restriction on pairing has been described in male meiosis of *Mesostoma ehrenbergii ehrenbergii* ($2n = 10$), which regularly has two pairs of univalents at metaphase I and three bivalents each with a single very localized distal chiasma (see Fig. 7). A three-dimensional reconstruction analysis of SCs in pachytene spermatocytes of this species showed that pairing is restricted to three very short SCs, averaging only 3.42 μm in length, which are presumed to correspond to the distal chiasmate regions of the three bivalents (Oakley and Jones, 1982). The alternative explanation, proposed by

7. Chiasmata

Fig. 6. Metaphase I of male meiosis in the grasshopper *Stethophyma grossum* showing 11 autosomal bivalents and a univalent X chromosome. Note the marked proximal localization of chiasmata in the large- and medium-sized bivalents.

John (1976), that localized pairing may have evolved secondarily in species showing localized chiasma formation must also be considered possible but in *Stethophyma grossum,* at least, it seems unlikely (discussed by Wallace, 1980).

The number of authenticated cases in which chiasma localization can be reasonably attributed to localized pairing is low, and while it is likely that other examples, as yet undescribed, do exist, it seems that only a minority of cases of localized chiasmata show this association with localized pairing. A few species are intermediate, in that they show quite pronounced localization of chiasmata but only semilocalized pairing. For example, *Chloealtis conspersa* shows distally localized chiasmata in its three long metacentric bivalents and proximal or proximal/distal localization in most of its shorter telocentric bivalents, while pairing as judged from surface-spread SCs is complete in the telocentric bivalents and in the metacentric bivalents it extends to roughly half the length of each arm; occasionally the metacentric bivalents are seen to be completely paired (Moens and Short, 1983). That is, in *Chloealtis* pairing is much more extensive and less localized than the distribution of chiasmata. A similar situation appears to be

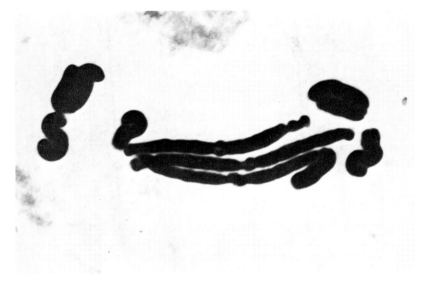

Fig. 7. Metaphase I of male meiosis in *Mesostoma ehrenbergii ehrenbergii* showing the typical configuration of three bivalents, each with a single distally localized chiasma and four univalents.

present in the grasshopper *Euthystira brachyptera* (Craig-Cameron and Jones, 1970b; Fletcher and Hewitt, 1980). However, most species, including many showing pronounced chiasma localization and many others showing less obvious localization of chiasmata, show complete pairing of homologous chromosomes at pachytene. Clearly in these cases the final extent of pairing is not a factor in the determination of chiasma positions, and other mechanisms must be sought.

The most influential proposal relating to the control of chiasma distribution in species with fully paired chromosomes at meiosis is the sequential model put forward by Mather (1936a, 1937, 1938) on the basis of the relationship between chromosome lengths and chiasma frequencies in various species and the distribution of crossover events along chromosomes in *Drosophila melanogaster*. Briefly, Mather's proposal was that chiasma determination follows a linear and temporal sequence from a fixed point in each bivalent; he proposed that the fixed starting point was the centromere. The distance from the starting point to the first chiasma he termed the differential distance (d) and the distances between the first and second chiasmata, and any subsequent chiasmata, he termed the interference distance (i). Other than proposing a sequential process of chiasma formation, Mather's original hypothesis did not advance any detailed mechanism of chiasma determination. Mather's d and i distances can, therefore, be regarded as descriptive concepts rather than implying any particular deterministic mechanism. Subsequently, a number of detailed quantitative studies of chiasma distribution have been performed based on careful measurements of chiasma positions in extended diplotene or diakinesis bivalents (Henderson, 1963; Southern, 1967; Fox, 1973; Hultén, 1974; Shaw and Knowles, 1976; Maudlin and Evans, 1980; Laurie, 1980; Coates and Shaw, 1982). These studies have consistently interpreted their findings in terms of Mather's sequential model, although all but one (Coates and Shaw, 1982) propose telomeres rather than centromeres as sequence origins. They are unanimous in placing heavy emphasis on the role of interference, probably because of the choice of species with large chromosomes and hence several chiasmata per bivalent. However, Mather (1938) was careful to point out that interference can only influence the positions of chiasmata in bivalents or bivalent arms having two or more chiasmata. Since many species frequently form only one chiasma per arm or per bivalent, it is clear that interference has little or no influence on the distribution of the majority of chiasmata in these cases.

The detailed study conducted by Fox (1973) on chiasma distribution in the locust *Schistocerca gregaria* led to the formulation of a very elaborate deterministic sequential model of control. According to this model, a chiasma determination mechanism (C.D.M.) passes along the bivalent from the fixed initiation point at a constant rate and, in response to a trigger, determines the site of formation of the first chiasma. The C.D.M. continues to pass along the chromosome, but requires a certain time to become recharged after chiasma determination, which accounts for the phenomenon of interference. A number of rather similar proposals have been made which invoke the activity of diffusible chemical substances, present in limiting amounts, which are necessary for either initiation (King, 1970) or stabilization (Holliday, 1977) of exchanges. It has also been suggested that the synaptonemal complex may be involved in producing interference either by acting as the agent for the transmission of structural or chemical information (Egel-Mitani *et al.*, 1982; Holm *et al.*, 1981) or by imposing constraints on the free isomerization of DNA strands involved in crossing-over (Perkins, 1979; Whitehouse, 1982).

Fox (1973) also found that particular bivalents could vary in length in different cells and that this length variation had an apparently direct and causal effect on chiasma frequency, such that when a bivalent was longer it had (on average) more chiasmata. Similar correlations have been noted between SC lengths and genetic crossover frequencies in maize genotypes carrying extra heterochromatic segments which affect SC lengths (Gillies *et al.*, 1973; Mogensen, 1977) and also in human meiosis where total SC length in oocytes is two to three times longer than that in spermatocytes (Holm *et al.*, 1982; B. M. N. Wallace, personal communication) while the frequency of genetic recombination is 40% higher in females compared to males. It is tempting to interpret these correlations in terms of a sequential deterministic mechanism, as proposed by Fox (1973); that is, if chiasma determination is sequential from an end, and d and i distances are relatively constant parameters (at least for particular bivalents in particular species), any increase in SC length will automatically result in more chiasmata being formed.

At a different level, there have been numerous suggestions that chiasma distribution control may depend on the pairing behavior of chromosomes during zygotene (e.g., Suomalainen, 1952; Moens, 1969; Sybenga, 1975). According to this view, features such as chiasma localization and interference are not necessarily the results of deterministic mechanisms traveling along paired pachytene bivalents, but could instead be indirect expressions of patterns of pairing behavior, such as the initiation, progression, and timing of pairing. Evidence has already been presented that in certain species showing extreme localization the distribution of chiasmata is closely related to the extent of pairing. The question therefore arises whether the pattern of pairing could have a more general influence on chiasma distribution, even in species with complete pairing. As previously noted, chiasma determination appears to be a characteristic of quite large chromosome segments, or whole arms or even whole bivalents, rather than particular sites. This collectivity is consistent with a mechanism of chiasma determination based on some general characteristic of quite large chromosome segments, such as pairing behavior. However, this is only a very general and indirect indication of a possible relationship. More direct indications of a relationship between pairing and chiasma determination come from observations on numerical and structural chromosome variants, and also on pairing behavior in normal bivalents.

Maguire (1965, 1966, 1977) found very close correlations between the pachytene pairing frequencies of certain chromosome segments in maize/*Tripsacum* translocation trisomics and in inversion heterozygotes of maize and the crossover frequencies of those same segments. For instance, in maize material heterozygous for a short paracentric inversion, a consistent 1 : 1 relation was found between the occurrence of observable homologous pairing within the region of inversion heterozygosity and the occurrence of a crossover within this region. These comparisons led Maguire (1977) to conclude "that crossover site establishment is usually associated with the initiation of synapsis." That is, it appears as if the pairing of cytologically marked segments, bounded by changes of pairing partner, leads with a very high probability to chiasma formation.

The analysis of chiasma frequencies in a series of reciprocal translocations and centric fissions in the plant *Hypochaeris radicata* has led to broadly similar conclusions (Parker *et al.*, 1982). *Hypochaeris radicata* ($2x = 8$) normally forms four bivalents at meiosis with a mean chiasma frequency of just over one chiasma per bivalent. Most reciprocal translocation heterozygotes (but not homozygotes) show *increased* chiasma frequencies which are associated with the establishment of additional, independent, paired segments at pachytene, separated from other paired segments in the translocation multiple by changes of pairing partner. In the case of centric fissions, both heterozygotes *and* homozygotes show increased chiasma frequencies which are again associated with increased numbers of effectively independent paired segments, bounded by chromosome discontinuities. One possible interpretation of these findings is that structural rearrangements result in reduced interference because of the effective separation of large chromosome segments by changes or breaks in pairing, and hence discontinuities in SCs. However, Parker *et al.* (1982) suggest that the primary factor causing increased chiasma frequencies in these situations is the altered pairing relations and behavior of these chromosomes, and that alterations in interference may be secondary consequences.

Observations on structurally normal bivalents from a variety of species also indicate that pairing behavior and chiasma formation may show at least a general correspondence. The precise nature and extent of this relationship is, however, uncertain. At one extreme we have the suggestion, already considered in relation to some numerical and structural chromosome variants, that establishment of a chiasma follows directly from each initiation of pairing. Extensive data on the numbers and distributions of pairing initiations in normal bivalents are scarce, but in some cases there appears to be some parallel between pairing initiation and chiasma determination, as in maize (Gillies, 1975). However, it is clear that an exact parallel is not generally observed and in some species the numbers of separate pairing initiations far exceed the numbers of chiasmata as in *Lilium longiflorum* (Holm, 1977), *Secale* (Albini *et al.*, 1984), and *Locusta migratoria* (Croft, 1984); in the latter example the number of independent initiations of pairing is in the region of 25–30 per nucleus while the number of chiasmata is between 13 and 17 per nucleus. In addition, the locations of pairing initiation sites and chiasmata frequently do not correspond. For example, in *Locusta migratoria* the great majority of pairing initiations appear to be terminal or nearly so (Moens, 1969; Croft, 1984) but relatively few chiasmata are similarly localized. Nevertheless, although there is little evidence to support the idea that each individual site of pairing initiation forms a chiasma, there is considerable evidence of a general correspondence between pairing initiation and progression and chiasma distribution. For instance, in many species pairing is initiated at or near chromosome ends, progressing sequentially to other regions, and these species show a corresponding tendency to terminal (i.e., distal or distal/proximal) localization of chiasmata. Even species with many interstitial pairing initiations show a tendency for terminal regions to pair first followed by the more interstitial regions (e.g., Holm, 1977). Indications of sequential chiasma determination may, therefore, simply reflect a general tendency for sequential pairing.

These considerations suggest that a general, albeit quantitative, relationship exists between pairing behavior and chiasma distribution. A suggested basis for this relationship is that the probability of chiasma formation in any region depends on the duration, extent, and intimacy of pairing, as indeed has been suggested by Rhoades (1968) from observations on the relationship between pairing behavior and crossing-over in a range of maize genotypes. Any factors, including structural rearrangements and genetic or environmental effects, which lead to more rapid, prolonged, or intimate pairing would result in increased probabilities of chiasma formation in the regions concerned. Clearly a mechanism of this type could adequately explain various patterns of chiasma distribution, but there are insufficient observational and experimental data to conclude that this is the principal determining mechanism.

REFERENCES

Albini, S. M., Jones, G. H., and Wallace, B. M. N. (1984). A method for preparing two-dimensional surface-spreads of synaptonemal complexes from plant meiocytes for light and electron microscopy. *Exp. Cell Res.* **152**, 280–285.

Allen, J. W. (1982). SCE and meiotic crossover exchange in germ cells. In "Sister Chromatid Exchange" (A. A. Sandberg, ed.), pp. 297–311. Alan R. Liss, Inc., New York.

Arana, P., Santos, J. L., and Giraldez, R. (1980). Chiasma interference and centromere coorientation in a spontaneous translocation heterozygote of *Euchorthippus pulvinatus gallicus* (Acrididae:Orthoptera). *Chromosoma* **78**, 327–340.

Baker, B. S., Carpenter, A. T. C., Esposito, M. S., Esposito, R. E., and Sandler, L. (1976). The genetic control of meiosis. *Annu. Rev. Genet.* **10**, 53–134.

Basak, S. L., and Jain, H. K. (1963). Autonomous and interrelated formation of chiasmata in *Delphinium* chromosomes. *Chromosoma* **13**, 577–587.

Belling, J. (1927). Configurations of bivalents of *Hyacinthus* with regard to segmental interchange. *Biol. Bull. (Woods Hole, Mass.)* **52**, 480–487.

Belling, J. (1928). A working hypothesis for segmental interchange between homologous chromosomes in flowering plants. *Univ. Calif. Publ. Bot.* **14**, 283–291.

Belling, J. (1931). Chiasmas in flowering plants. *Univ. Calif. Publ. Bot.* **16**, 311–338.

Callan, H. G., and Montalenti, G. (1947). Chiasma interference in mosquitoes. *J. Genet.* **48**, 119–134.

Callan, H. G., and Perry, P. E. (1977). Recombination in male and female meiocytes contrasted. *Philos. Trans. R. Soc. London, Ser. B* **277**, 227–233.

Catcheside, D. G. (1977). "The Genetics of Recombination." Edward Arnold, London.

Christensen, B. (1961). Studies on cyto-taxonomy and reproduction in the Enchytraeidae. With notes on parthenogenesis and polyploidy in the animal kingdom. *Hereditas* **47**, 287–450.

Church, K., and Wimber, D. E. (1969). Meiotic timing and segregation of H^3-thymidine labelled chromosomes. *Can. J. Genet. Cytol.* **11**, 573–581.

Coates, D. J., and Shaw, D. D. (1982). The chromosomal component of reproductive isolation in the grasshopper *Caledia captiva*. I. Meiotic analysis of chiasma distribution patterns in two chromosomal taxa and their F_1 hybrids. *Chromosoma* **86**, 509–531.

Couzin, D. A., and Fox, D. P. (1974). Variation in chiasma frequency during tulip anther development. *Chromosoma* **46**, 173–179.

Craig-Cameron, T., and Jones, G. H. (1970a). The analysis of exchanges in tritium-labelled meiotic chromosomes. I. *Schistocerca gregaria*. *Heredity* **25**, 223–232.

7. Chiasmata

Craig-Cameron, T., and Jones, G. H. (1970b). Meiosis in *Euthystira brachyptera. Ann. Stn. Biol. Besse-en-Chandesse* **5**, 275–292.
Croft, J. A. (1984). Studies on chromosome synapsis and chiasma formation. Ph.D. Thesis, University of Birmingham.
Darlington, C. D. (1929). Chromosome behaviour and structural hybridity in the Tradescantiae. *J. Genet.* **21**, 207–286.
Darlington, C. D. (1930). A cytological demonstration of "genetic" crossing-over. *Proc. R. Soc. London, Ser. B* **107**, 50–59.
Darlington, C. D. (1931). Meiosis. *Biol. Rev. Cambridge Philos. Soc.* **6**, 221–264.
Darlington, C. D. (1932). "Recent Advances in Cytology." Churchill, London.
Darlington, C. D. (1935). The internal mechanics of the chromosomes 11. Prophase pairing at meiosis in *Fritillaria. Proc. R. Soc. London, Ser. B* **118**, 59–73.
Darlington, C. D. (1937). "Recent Advances in cytology," 2nd ed. Churchill, London.
Darlington, C. D. (1958). "The Evolution of Genetic Systems." Oliver & Boyd, Edinburgh.
Egel-Mitani, M., Olson, L. W., and Egel, R. (1982). Meiosis in *Aspergillus nidulans:* Another example for lacking synaptonemal complexes in the absence of crossover interference. *Hereditas* **97**, 179–187.
Elliot, C. G. (1958). Environmental effects on the distribution of chiasmata among nuclei and bivalents and correlation between bivalents. *Heredity* **12**, 429–439.
Ennis, T. J. (1972). Meiosis in *Diabrotica* (Coleoptera: Chrysomelidae). Chiasma frequency and variation. *Can. J. Genet. Cytol.* **14**, 113–128.
Evans, E. P., Burtenshaw, M. D., and Cattanach, B. M. (1982). Meiotic crossing-over between the X and Y chromosomes of male mice carrying the sex-reversing (Sxr) factor. *Nature (London)* **300**, 443–445.
Fletcher, H. L. (1978). Localised chiasmata due to partial pairing: A 3D reconstruction of synaptonemal complexes in male *Stethophyma grossum. Chromosoma* **65**, 247–269.
Fletcher, H. L., and Hewitt, G. M. (1980). A comparison of chiasma frequency and distribution between sexes in three species of grasshoppers. *Chromosoma* **77**, 129–144.
Fox, D. P. (1973). The control of chiasma distribution in the locust, *Schistocerca gregaria* (Forskal). *Chromosoma* **43**, 289–328.
Fuge, H. (1979). Synapsis, desynapsis and formation of polycomplex-like aggregates in male meiosis of *Pales ferruginea* (Diptera:Tipulidae). *Chromosoma* **70**, 353–373.
Gillies, C. B. (1975). An ultrastructural analysis of chromosome pairing in maize. *C. R. Trav. Lab. Carlsberg* **40**, 135–161.
Gillies, C. B. (1979). The relationship between synaptonemal complexes, recombination nodules and crossing-over in *Neurospora crassa* bivalents and translocation quadrivalents. *Genetics* **91**, 1–17.
Gillies, C. B., Rasmussen, S. W., and von Wettstein, D. (1973). The synaptonemal complex in homologous and non-homologous pairing of chromosomes. *Cold Spring Harbor Symp. Quant. Biol.* **38**, 117–122.
Haldane, J. B. S. (1931). The cytological basis of genetical interference. *Cytologia* **3**, 54–65.
Harte, C. (1956). Die variabilitat der Chiasmenbildung bei *Paeonia tenuifolia. Chromosoma* **8**, 152–182.
Henderson, S. A. (1962). Temperature and chiasma formation in *Schistocerca gregaria*. II. Cytological effects at 40°C and the mechanism of heat induced univalence. *Chromosoma* **13**, 437–463.
Henderson, S. A. (1963). Chiasma distribution at diplotene in a locust. *Heredity* **18**, 173–190.
Henderson, S. A. (1969a). Chromosome pairing, chiasmata and crossing-over. *In* "Handbook of Molecular Cytology" (A. Lima de Faria, ed.), pp. 326–357. North-Holland Publ., Amsterdam.

Henderson, S. A. (1969b). Chiasma localisation and incomplete pairing. *Chromosomes Today* **2**, 56–60.

Hewitt, G. M. (1964). Population cytology of British grasshoppers. I. Chiasma variation in *Chorthippus brunneus*, *Chorthippus parallelus* and *Omocestus virudulus*. *Chromosoma* **15**, 212–230.

Hewitt, G. M. (1967). An interchange which raises chiasma frequency. *Chromosoma* **21**, 285–295.

Hewitt, G. M., and John, B. (1965). The influence of numerical and structural chromosome mutations on chiasma conditions. *Heredity* **20**, 123–135.

Hillel, J., Feldman, M. W., and Simchen, G. (1973). Mating systems and population structure in two closely related species of the wheat group. III. Chiasma frequency and population structure. *Heredity* **31**, 1–9.

Holliday, R. (1977). Recombination and meiosis. *Philos. Trans. R. Soc. London, Ser. B* **277**, 359–370.

Holm, P. B. (1977). Three-dimensional reconstruction of chromosome pairing during the zygotene stage of meiosis in *Lilium longiflorum*. *Carlsberg Res. Commun.* **42**, 103–151.

Holm, P. B., and Rasmussen, S. W. (1980). Chromosome pairing recombination nodules and chiasma formation in diploid *Bombyx* males. *Carlsberg Res. Commun.* **45**, 483–548.

Holm, P. B., and Rasmussen, S. W. (1983). Human meiosis. VII. Chiasma formation in human spermatocytes. *Carlsberg Res. Commun.* **48**, 415–456.

Holm, P. B., Rasmussen, S. W., Zickler, D., Lu, B. C., and Sage, J. (1981). Chromosome pairing, recombination nodules and chiasma formation in the basidiomycete *Coprinus cinereus*. *Carlsberg Res. Commun.* **46**, 305–346.

Holm, P. B., Rasmussen, S. W., and von Wettstein, D. (1982). Ultrastructural characterization of the meiotic prophase. A tool in the assessment of radiation damage in man. *Mutat. Res.* **95**, 42–59.

Hughes, A. (1959). "A History of Cytology." Abelard-Schuman, London.

Hultén, M. (1974). Chiasma distribution at diakinesis in the normal human male. *Hereditas* **76**, 55–78.

Janssens, F. A. (1905). Spermatogénèse dans les Batrachiens. III. Evolution des auxocytesmales du Batracoseps attenuatus. *Cellule* **22**, 379–425.

Janssens, F. A. (1909). Spermatogénèse dans les Batraciens. V. La Théorie de la Chiasmatypie, nouvelle interprétation des cinèses de maturation. *Cellule* **25**, 387–411.

Janssens, F. A. (1924). La chiasmatypie dans les insectes. *Cellule* **34**, 135–359.

John, B. (1976). Myths and mechanisms of meiosis. *Chromosoma* **54**, 295–325.

John, B., and King, M. (1982). Meiotic effects of supernumerary heterochromatin in *Heteropternis obscurella*. *Chromosoma* **85**, 39–65.

John, B., and Lewis, K. R. (1965). "The Meiotic System. Protoplasmatologia V1/F/1." Springer-Verlag, Vienna.

Jones, G. H. (1967). The control of chiasma distribution in rye. *Chromosoma* **22**, 69–90.

Jones, G. H. (1971). The analysis of exchanges in tritium-labelled meiotic chromosomes. II. *Stethophyma grossum*. *Chromosoma* **34**, 367–382.

Jones, G. H. (1974). Correlated components of chiasma variation and the control of chiasma distribution in rye. *Heredity* **32**, 375–387.

Jones, G. H. (1977). A test for early terminalisation of chiasmata in diplotene spermatocytes of *Schistocerca gregaria*. *Chromosoma* **63**, 287–294.

Jones, G. H. (1978). Giemsa C-banding of rye meiotic chromosomes and the nature of "terminal" chiasmata. *Chromosoma* **66**, 45–57.

Jones, G. H. (1986). The control of chiasma distribution. *Sympo. Soc. Exp. Biol.* **38**, 291–320.

Jones, G. H., and Tease, C. (1980). Meiotic exchange analysis by molecular labelling. *Chromosomes Today* **7**, 114–125.

7. Chiasmata

Jones, G. H., and Tease, C. (1984). Analysis of exchanges in differentially stained meiotic chromosomes of *Locusta migratoria* after BrdU-substitution and FPG staining. IV. the nature of "terminal" associations. *Chromosoma* **89**, 33–36.

Jones, G. H., and Wallace, B. M. N. (1980). Meiotic chromosome pairing in *Stethophyma grossum* spermatocytes studied by a surface-spreading and silver-staining technique. *Chromosoma* **78**, 187–201.

Kanda, N., and Kato, H. (1980). Analysis of crossing-over in mouse meiotic cells by BrdU labelling technique. *Chromosoma* **78**, 113–122.

Karp, A., and Jones, R. N. (1982). Cytogenetics of *Lolium perenne*. 1. Chiasma frequency variation in inbred lines. *Theor. Appl. Genet.* **62**, 177–183.

Karp, A., and Jones, R. N. (1983). Cytogenetics of *Lolium perenne*. 2. Chiasma distribution in inbred lines. *Theor. Appl. Genet.* **64**, 137–145.

King, M., and Hayman, D. (1978). Seasonal variation of chiasma frequency in *Phyllodactylus marmoratus* (Gray) (Gekkonidae:Reptilia). *Chromosoma* **69**, 131–154.

King, R. C. (1970). The meiotic behaviour of the *Drosophila* oocyte. *Int. Rev. Cytol.* **28**, 125–168.

Kundu, S. C., and Bogdanov, F. Y. (1979). Ultrastructural studies of late meiotic prophase nuclei of spermatocytes of *Ascaris suum*. *Chromosoma* **70**, 375–384.

Laurie, D. A. (1980). Inter-individual variation in chiasma frequency and chiasma distribution. A study of *Chorthippus brunneus* and man. Ph.D. Thesis, University of Birmingham.

Laurie, D. A., and Jones, G. H. (1981). Inter-individual variation in chiasma distribution in *Chorthippus brunneus* (Orthoptera:Acrididae). *Heredity* **47**, 409–416.

Lederberg, J. (1955). Recombination mechanisms in bacteria. *J. Cell Comp. Physiol.* **45**, Suppl. 2, 75–107.

Loidl, J. (1979). C-band proximity of chiasmata and absence of terminalisation in *Allium flavum* (Liliaceae). *Chromosoma* **73**, 45–51.

Lucchesi, J. C., and Suzuki, D. T. (1968). The interchromosomal control of recombination. *Annu. Rev. Genet.* **2**, 53–86.

Maguire, M. P. (1965). The relationship of crossover frequency to synaptic extent at pachytene in maize. *Genetics* **51**, 23–40.

Maguire, M. P. (1966). The relationship of crossing over to chromosome synapsis in a short paracentric inversion. *Genetics* **53**, 1071–1077.

Maguire, M. P. (1977). Homologous chromosome pairing. *Philos. Trans. R. Soc. London, Ser. B* **277**, 245–258.

Mather, K. (1933). The relations between chiasmata and crossing-over in diploid and triploid *Drosophila melanogaster*. *J. Genet.* **27**, 243–259.

Mather, K. (1936a). The determination of position in crossing-over. I. *Drosophila melanogaster*. *J. Genet.* **33**, 207–235.

Mather, K. (1936b). Competition between bivalents during chiasma formation. *Proc. R. Soc. London, Ser. B* **120**, 208–227.

Mather, K. (1937). The determination of position in crossing-over. II. The chromosome length–chiasma frequency relation. *Cytologia Fujii Jubilee Vol.*, pp. 514–526.

Mather, K. (1938). Crossing-over. *Biol. Rev. Cambridge Philos. Soc.* **13**, 252–292.

Mather, K. (1940). The determination of position in crossing-over. III. The evidence of metaphase chiasmata. *J. Genet.* **39**, 205–223.

Maudlin, I. (1972). Developmental variation of chiasma frequency in *Chorthippus brunneus*. *Heredity* **29**, 259–262.

Maudlin, I., and Evans, E. P. (1980). Chiasma distribution in mouse oocytes during diakinesis. *Chromosoma* **80**, 49–56.

Moens, P. (1969). The fine structure of meiotic chromosome polarization and pairing in *Locusta migratoria* spermatocytes. *Chromosoma* **28**, 1–25.

Moens, P. B., and Short, S. (1983). Synaptonemal complexes of bivalents with localized chiasmata in *Chloealtis conspersa* (Orthoptera). In "Kew Chromosome Conference II" (P. E. Brandham and M. D. Bennett, eds.), pp. 99–106. Allen & Unwin, London.

Mogensen, H. L. (1977). Ultrastructural analysis of female pachynema and the relationship between synaptonemal complex length and crossing-over in *Zea mays*. *Carlsberg Res. Commun.* **42**, 475–497.

Morgan, T. H., and Cattell, E. (1912). Data for the study of sex-linked inheritance in *Drosophila*. *J. Exp. Zool.* **13**, 79–101.

Muller, H. J. (1916). The mechanism of crossing-over. *Am. Nat.* **50**, 193–221.

Newcombe, H. B. (1941). Chiasma interference in *Trillium erectum*. *Genetics* **26**, 128–136.

Oakley, H. A., and Jones, G. H. (1982). Meiosis in *Mesostoma ehrenbergii ehrenbergii* (Turbellaria, Rhabdocoela). I. Chromosome pairing, synaptonemal complexes and chiasma localisation in spermatogenesis. *Chromosoma* **85**, 311–322.

Orellana, J., and Giraldez, R. (1981). Metaphase I bound arms and crossing over frequency in rye. *Chromosoma* **84**, 439–449.

Parker, J. S., Jones, G. H., Tease, C., and Palmer, R. W. (1976). Chromosome-specific control of chiasma formation in *Hypochoeris* and *Crepis*. In "Current Chromosome Research" (K. Jones and P. E. Brandham, eds.), pp. 133–142. North-Holland Publ., Amsterdam.

Parker, J. S., Palmer, R. W., Whitehorn, M. A. F., and Edgar, L. A. (1982). Chiasma frequency effects of structural chromosome change. *Chromosoma* **85**, 673–686.

Peacock, W. J. (1968). Chiasmata and crossing-over. In "Replication and Recombination of Genetic Material" (W. J. Peacock and R. D. Brock, eds.), pp. 242–252. Aust. Acad. Sci., Canberra.

Peacock, W. J. (1970). Replication, recombination and chiasmata in *Goniae australiasiae* (Orthoptera:Acrididae). *Genetics* **65**, 593–617.

Perkins, D. D. (1979). *Genetics* **91**, 594.

Polani, P. E., Crolla, J. A., and Seller, M. J. (1981). An experimental approach to female mammalian meiosis. Differential chromosome labelling and an analysis of chiasmata in the female mouse. In "Bioregulators of Reproduction" (H. J. Vogel and G. Jagiello, eds.), pp. 59–87. Academic Press, New York.

Rees, H. (1955). Genotypic control of chromosome behaviour in rye. I. Inbred lines. *Heredity* **9**, 93–116.

Rees, H. (1961). Genotypic control of chromosome form and behaviour. *Bot. Rev.* **27**, 288–318.

Rees, H., and Jones, G. H. (1967). Chromosome evolution in *Lolium*. *Heredity* **22**, 1–18.

Rees, H., and Naylor, B. (1960). Developmental variation in chromosome behaviour. *Heredity* **15**, 17–27.

Rees, H., and Thompson, J. B. (1956). Genotypic control of chromosome behaviour in rye. III. Chiasma frequency in homozygotes and heterozygotes. *Heredity* **10**, 409–424.

Rhoades, M. M. (1968). Studies on the cytological basis of crossing-over. In "Replication and Recombination of Genetic Material" (W. J. Peacock and R. D. Brock, eds.), pp. 229–241. Aust. Acad. Sci., Canberra.

Robertson, W. R. B. (1916). Chromosome studies. I. *J. Morphol.* **27**, 179–332.

Ross, M. H., and Cochran, D. G. (1981). Synthesis and properties of a double translocation heterozygote involving a stable ring-of-six interchange in the German cockroach. *J. Hered.* **72**, 39–44.

Rowlands, D. G. (1958). The control of chiasma frequency in *Vicia faba* L. *Chromosoma* **9**, 176–184.

Rückert, J. (1892). Zur Entwicklungsgeschichte des Ovarialeies bei Selachiern. *Anat. Anz.* **7**, 107–158.

Sandler, L., Linsley, D. L., Nicoletti, B., and Trippa, G. (1968). Mutants affecting meiosis in natural populations of *Drosophila melanogaster*. *Genetics* **60**, 525–558.

7. Chiasmata

Santos, J. L., and Giraldez, R. (1978). The effect of heterochromatin on chiasma terminalisation in *Chorthippus biguttulus* L. (Acrididae:Orthoptera). *Chromosoma* **70**, 59–66.

Schultz, J., and Redfield, H. (1951). Interchromosomal effects on crossing-over in *Drosophila*. *Cold Spring Harbor Symp. Quant. Biol.* **16**, 175–197.

Seuanez, H. N., Armada, J. L., Barroso, C., Rezende, C., and da Silva, V. F. (1983). The meiotic chromosomes of *Cebus apellus* (Cebidae:Platyrhini). *Cytogenet. Cell Genet.* **36**, 517–524.

Sharp, P. (1982). Sex chromosome pairing during meiosis in marsupials. *Chromosoma* **86**, 27–47.

Shaw, D. D., and Knowles, G. R. (1976). Comparative chiasma analysis using a computerised optical digitiser. *Chromosoma* **59**, 103–127.

Simchen, G., and Stamberg, J. (1969). Fine and coarse controls of genetic recombination. *Nature (London)* **222**, 329–332.

Solari, A. J. (1970). The behaviour of chromosomal axes during diplotene in mouse spermatocytes. *Chromosoma* **31**, 217–230.

Solari, A. J. (1974). The behaviour of the XY pair in mammals. *Int. Rev. Cytol.* **38**, 273–317.

Solari, A. J. (1980a). Chromosome axes during and after diplotene. *In* "International Cell Biology" (H. G. Schweiger, ed.), pp. 178–186. Springer-Verlag, Berlin and New York.

Solari, A. J. (1980b). Synaptonemal complexes and associated structures in microspread human spermatocytes. *Chromosoma* **81**, 315–337.

Solari, A. J., and Ashley, T. (1977). Ultrastructure and behaviour of the achiasmatic telosynaptic XY pair of the sand rat (*Psammomys obesus*). *Chromosoma* **62**, 319–336.

Sotello, R. J., Garcia, R. B., and Wettstein, R. (1973). Serial sectioning study of some meiotic stages in *Scaptericus borelli* (Grylloidea). *Chromosoma* **42**, 307–333.

Southern, D. I. (1967). Chiasma distribution in truxaline grasshoppers. *Chromosoma* **22**, 164–191.

Speed, R. M. (1977). The effects of ageing on the meiotic chromosomes of male and female mice. *Chromosoma* **64**, 241–254.

Stebbins, G. L., and Ellerton, S. (1939). Structural hybridity in *Paeonia californica* and *P. brownii*. *J. Genet.* **38**, 1–36.

Stern, H. (1981). Chromosome organisation and DNA metabolism in meiotic cells. *Chromosomes Today* **7**, 94–104.

Stern, H., and Hotta, Y. (1978). Regulatory mechanisms in meiotic crossing-over. *Annu. Rev. Plant Physiol.* **29**, 415–436.

Suomalainen, H. O. T. (1952). Localization of chiasmata in the light of observations on the spermatogenesis of certain *Neurospora*. *Ann. Zool. Soc. Zool. Bot. Fenn. Vanamo* **15**(3), 1–104.

Sybenga, J. (1975). Meiotic configurations. *In* "Monographs on Theoretical and Applied Genetics 1" (R. Frankel, ed.), pp. 1–251. Springer-Verlag, Berlin and New York.

Taylor, J. H. (1965). Distribution of tritium-labelled DNA among chromosomes during meiosis. I. Spermatogenesis in the grasshopper. *J. Cell Biol.* **25**, 57–67.

Tease, C. (1977). Genetic control of chiasma variation in *Crepis capillaris*. Ph.D. Thesis, University of Birmingham.

Tease, C. (1978). Cytological detection of crossing-over in BudR substituted meiotic chromosomes using the fluorescent plus Giemsa technique. *Nature (London)* **272**, 823–824.

Tease, C., and Jones, G. H. (1976). Chromosome specific control of chiasma formation in *Crepis capillaris*. *Chromosoma* **57**, 33–49.

Tease, C., and Jones, G. H. (1978). Analysis of exchanges in differentially stained meiotic chromosomes of *Locusta migratoria* after BrdU substitution and FPG staining. I. Crossover exchanges in monochiasmate bivalents. *Chromosoma* **69**, 163–178.

Tease, C., and Jones, G. H. (1986). In preparation.

Wallace, B. M. N. (1980). A study of chromosome synapsis in the grasshopper *Stethophyma grossum*. Ph.D. Thesis, University of Birmingham.

Wallace, B. M. N., and Jones, G. H. (1978). Incomplete chromosome pairing and its relation to chiasma localisation in *Stethophyma grossum* spermatocytes. *Heredity* **40**, 385–396.

Wenrich, D. H. (1916). The spermatogenesis of *Phrynottetix magnus*, with special reference to synapsis and the individuality of the chromosomes. *Bull. Mus. Comp. Zool.* **60**, 57–135.

Westergaard, M., and von Wettstein, D. (1970). Studies on the mechanism of cross-over. IV. The molecular organization of the synaptinemal complex in *Neotiella* (Cooke) Saccardo (Ascomycetes). *C. R. Trav. Lab. Carlsberg* **37**, 239–268.

Westerman, M. (1967). The effect of X-irradiation on male meiosis in *Schistocerca gregaria* (Forskal). I. Chiasma frequency response. *Chromosoma* **22**, 401–416.

White, M. J. D. (1973). "Animal Cytology and Evolution," 3rd ed. Cambridge Univ. Press, London and New York.

Whitehouse, C., Edgar, L. A., Jones, G. H., and Parker, J. S. (1981). The population cytogenetics of *Crepis capillaris*. I. Chiasma variation. *Heredity* **47**, 95–103.

Whitehouse, H. L. K. (1965). "Towards an Understanding of the Mechanism of Heredity." Edward Arnold, London.

Whitehouse, H. L. K. (1982). "Genetic Recombination." Wiley, New York.

Zickler, D. (1977). Development of the synaptonemal complex and the "recombination nodules" during meiotic prophase in the seven bivalents of the fungus *Sordaria macrospora* Anersw. *Chromosoma* **61**, 289–316.

8

The Synaptonemal Complex and Meiosis: An Immunocytochemical Approach

MICHAEL E. DRESSER*

Department of Anatomy
Duke University Medical Center
Durham, North Carolina 27710

> The origin of the internal phenomena of meiosis is still wholly unknown. What led to the conjugation of the chromosomes and their subsequent disjunction in orderly system, and what part this process may have played in the physiological activities of the cell can only be conjectured. It is easy to offer hypothetical teleological answers to such questions. We may say that only through these processes could the normal organization and activities of the species be maintained, the "summation of ancestral germ-plasms" prevented, the species held true to its type, or (conversely) provided with a fruitful source of variation and natural selection; but such "explanations" leave us no wiser than before. If the cytological problems here involved remain unsolved they are none the less real and offer an interesting field for further inquiry.
> E. B. Wilson, 1925

I. INTRODUCTION

The synaptonemal complex (SC) is a nuclear organelle that forms the axis of chromosomal bivalents in the pachytene stage of meiotic prophase. The SC is of remarkably similar architecture in organisms ranging from mammals to fungi, and is clearly involved with the meiotic chromosomal processes (for reviews, see Moses, 1968, 1969; Westergaard and von Wettstein, 1972; Gillies, 1975; von

---

*Present address: Cellular and Genetic Toxicology Branch, N.I.E.H.S., Research Triangle Park, North Carolina 27709

Wettstein *et al.*, 1984). The manner in which the SC forms in zygotene and the absence of meiotic crossing-over in organisms that lack SCs implicate the SC in the processes of synapsis and genetic recombination, but questions of its specific function(s) are difficult to address directly. Clearly, clues and answers will come from determining the molecular organization of the SC.

Seminal information lies in defining the components of the SC, where they come from and where they go, and their interactions in a variety of circumstances. Although the remarkable similarities of SC architecture in all sexually reproducing species so far examined presumably reflect conservation from a common progenitor, there is likely to have been some divergence in the components and perhaps even some adaptation of the components to different functions in and out of the meiotic nucleus. If so, it will be necessary to sort out the common versus the species-specific attributes of the SC, of its components, of the DNA sequences that are acted on by the SC, and of the genes that encode the components.

One means of gathering the needed information involves the use of immunocytological and immunochemical techniques. We have begun by identifying antibodies that bind the SC with the intent of using the antibodies in a number of ways: to gather information on the position of and any changes in the sites bound during meiosis, to follow the appearance and fate of the (presumed) proteins cytologically, to identify the proteins electrophoretically, to isolate the genes for these proteins, to pull the proteins and/or complexes out of crude suspensions for any of a number of manipulations *in vitro*, and the list goes on. Judging from our early results and from the conflicting reports from other laboratories, the difficulties encountered in this approach arise from oversimplifications of SC biology and from ambiguities inherent in the immunological methods. It is the intent of the following to provide a background of observations and speculations concerning SC structure and function, to raise some of the considerations that form the rationale and basis for a careful immunological approach, and to illustrate some of the points made as well as the potential power of the approach with a number of preliminary results.

II. AN INTRODUCTION TO SC BIOLOGY

Examples abound of associations of SC components outside the SC per se, of the nonuniformity of SC structure within a single nucleus, and of the continual changes in SC structure from zygotene to diplotene that must be understood if we are to move from the components and structure of the SC to the question of its role(s) in meiosis. Only a full review of these could do the subjects justice, but the major points, the questions raised by them, and the applicability of immuno-

8. Synaptonemal Complex and Immunocytochemistry

logical techniques can be illustrated by a relatively brief consideration. Many of the basic facts of SC structure omitted here can be found in the reviews cited above.

A. Extra-SC Associations of SC Components

A number of observers have identified structures apart from the paired bivalents that have morphological characteristics in common with the SC. Termed "polycomplexes" (and "multiple complexes"; see Moses, 1968), they apparently arise from the assembly of SC components into well-ordered arrays, can be quite large, and, in many species, are found consistently [for examples: in prophase and postprophase nuclei (reviewed in Moses, 1968; Wettstein and Sotelo, 1971); in later prophase nuclei of mosquitoes (Fiil and Moens, 1973); in the cytoplasm of preprophase cells of *Ascaris* (Bogdanov, 1977; Fiil *et al.*, 1977); in human spermatocytes at metaphase (Solari and Vilar, 1978); in yeast (Zickler and Olson, 1975; Horesh *et al.*, 1979; Moens and Kundu, 1982)]. The repeating structure, particularly of the larger polycomplexes, is well suited to image analysis (see Esponda, 1977).

The first steps in identification of components common to SCs and polycomplexes can be carried out immunocytologically, and the use of monoclonal antibodies in combination with optical diffraction may provide detailed, high-resolution information on the location and perhaps orientation of SC components with respect to one another. Complementary information, possibly at lower resolution, should be available by a similar analysis of SCs. Choosing between whole-mount and sectioned preparations for analysis will depend on further development of the techniques, the specific questions, and the steps necessary to provide access of antibody to substrate and to preserve antibody–epitope binding.

Schin (1965) suggested that polycomplex formation in the cricket *Gryllus* (=*Acheta*) *domesticus* "has occurred without the influence of a higher order, that is, without orientation with respect to the chromosomes, and that it accordingly is determined only through properties of its own structural elements." This interpretation indicates an essential independence of the SC components from the chromosomes in bringing about certain characteristics of SC architecture, though clearly the components depend on the chromosomes to bring about the SCs per se. In addition, it raises questions of the SC component–chromosome interactions (modeled in Moses, 1968) and of the possibility of *in vitro* associations of the components. *In vitro* assembly of SC components would provide a valuable tool for analysis; antibody-based precipitation and affinity chromatography for pulling particular proteins and complexes out of solution could provide the starting material.

Associations of SC components outside the SC, obvious in the polycomplexes, may be of more general occurrence though difficult to demonstrate by morphology alone. In mosquito oocytes, cytoplasmic accumulations of membranes similar to the nuclear envelope and with a high concentration of putative pore complexes, termed "annulate lamellae," form in close association with the polycomplexes and increase in size and number as the polycomplexes dwindle (Fiil and Moens, 1973). These authors suggested that the lateral elements (LEs) may have arisen during evolution from chromosome attachment sites at the nuclear envelope, perhaps pore annuli, and that the transverse filaments of the central region of the SC might be descendants of other components of the nuclear membrane. This proposition was made in bolder form by Englehardt and Pusa (1972) on the basis of ruthenium red-staining similarities of the pore complexes and LEs. They suggested that the nuclear half of the pore complexes detach from the nuclear envelope and align to form the LEs. The gathering of nuclear pores into large collections during meiosis, often near the SC attachment plaques, has long been recognized; the suggestion of a physical association between SCs and nuclear pores is perhaps best made by their continued connection in otherwise isolated SCs (Walmsley and Moses, 1981).

B. The Dynamic Nature of the SC

Morphological differentiations along the SC, particularly evident in the LEs, are seen regularly at the telomeres and centromeres and occur frequently at the nucleoli and elsewhere. The appearance of these differentiations depends on the species and on the preparative technique, and can change through zygotene, pachytene, and diplotene.

1. Differentiations at Centromeres and Telomeres

In Chinese hamster, the LEs at the kinetochores and telomeres appear thickened with phosphotungstic acid (PTA) stain (Moses, 1977a) but generally appear similar in width to the rest of the SC with silver stain (Dresser and Moses, 1980). In Armenian hamster, the LEs at the kinetochores appear relatively thinned with silver stain (Dresser, 1986), while in the telocentric SCs of mouse, the kinetochore ends of the LEs are thicker than the distal ends (P. A. Poorman, unpublished observations). In silver-stained preparations of the grasshopper *Melanoplus differentialis* the LEs at the subtelocentric kinetochores appear grossly thickened (Dresser, 1986) as they do in PTA-stained preparations (Solari and Counce, 1977). A similar observation in PTA-stained surface spreads of *Locusta* spermatocytes has led to the suggestion that "the kinetochore originates as a specialization of the axial element" (Counce and Meyer, 1973). Consistent

8. Synaptonemal Complex and Immunocytochemistry   249

with this proposal is the observation that some human sera that contain anti-kinetochore antibodies label the SCs (but see discussion below).

2. Other Interstitial Differentiations

Localized thickenings independent of any otherwise evident morphological distinctions can develop along the LEs and these thickenings can occur on one but not the other of the two LEs at a single place along the SC [triploid *Lilium longiflorum* (Moens, 1968, 1970); *Locusta* (Moens, 1973); *Stethophyma* nymphs (Jones, 1973); *Zea mays* (Gillies and Hyde, 1973; Gillies, 1981, 1983); *Phaedranassa* (La Cour and Wells, 1973); *Brachystola* (Church, 1976); *Sordaria humana* (Zickler and Sage, 1981)]. If it is the association of the SC components with the chromosomes that controls their assembly then clearly the interaction is different at the thickenings where, apparently, there is a localized, continued addition of LE components. The unpaired X and Y axes of many mammals (see Solari, 1970, 1974a,b; Moses, 1977b) become thickened relative to the autosomal LEs during pachytene, remarkably so in the hamster species. In mice where various rearrangements lead to persistent asynapsis of autosomal LEs, these LEs thicken to the same extent as the X and Y axes (see Fig. 2 in Moses *et al.*, 1982). These observations suggest that at least some of the thickening, particularly where symmetric across the SC, might be attributed to the modification or failure of some aspect of synapsis.

3. Changes in LE Thickness

The LEs can thicken more or less equally throughout their lengths in a variety of organisms [*Locusta* (Moens, 1973); *Drosophila* (Carpenter, 1979a); Chinese hamster (Dresser and Moses, 1980); maize (Gillies, 1981, 1983); female *Bombyx*, where it may be related to the lack of chiasma formation (Rasmussen and Holm, 1982)]. In Chinese hamster spermatocytes it is clear that this thickening is accompanied by the appearance of doubling of the LEs, which seems to proceed independently along the two LEs from an unknown number of interstitial foci (see Fig. 10b in Dresser and Moses, 1980). Similarly developed doubling is apparent in other species as well [in male mouse (P. A. Poorman, unpublished observations); *Panorpa* (Welsch, 1973); male *Bombyx* (Rasmussen and Holm, 1980)]. Curiously, in Chinese hamster, mouse, and the chiasmate females of *Panorpa*, the LEs become single again before disappearing, rendering uncertain the relationship between LE doubling and chromatid duality. In the achiasmate male *Panorpa* and chiasmate male *Bombyx* they seem to remain double. While it is possible that differences in preparation and staining may account for the observed discrepancies it seems unlikely. Notwithstanding these differences, it may be that in formed SCs the localized thickenings, doublings, and more widespread thickening along the LEs are in some way related to one another and to SC function.

4. Changes in LE Length

The relationship between the morphological differentiations discussed above and widely documented changes in length is not at all clear. The absolute lengths of the SCs can vary markedly during pachytene, while the relative lengths remain more or less constant. In *Bombyx* females (achiasmate) and males (chiasmate) the SCs lengthen from early to late pachytene while the LEs of the females accumulate material and those of the male accumulate local LE doublings (Holm and Rasmussen, 1980; Rasmussen and Holm, 1982). In *Locusta* the SCs shorten and thicken during their tenure in the nucleus (Moens, 1973). In Chinese hamster and *Drosophila melanogaster*, the SCs shorten and thicken in the first half of pachytene, then lengthen and thin during the last half [hamster (Moses *et al.*, 1977; Dresser and Moses, 1980); *Drosophila* (Carpenter, 1975, 1979a)]. Some rearrangement of SC components may account for the correlated changes. However, in various fungi where the SCs lengthen during pachytene there is no disruption of the repeat spacing of thick and thin bands along the lateral elements (Zickler, 1973).

As is the case with localized thickenings and doublings, length changes can occur independently in the two LEs of an SC. In the heterogametic sex pair of chicken oocytes, the longer Z chromosome axis shortens (and thickens) during early to middle pachytene until it nearly equals the length of the W chromosome axis (Solari, 1977). Similarly, the longer of the two LEs of the heteromorphic SC in a mouse tandem duplication heterozygote comes to equal the length of the shorter LE during pachytene, apparently thickening while unsynapsed and then thinning to equal the size of its companion LE (Poorman *et al.*, 1981; Moses and Poorman, 1981). In both examples, part of the "excess" length of the longer axis is taken up in the SC by asymmetric twisting around the shorter axis; the asymmetric twists disappear in the mouse heterozygote once the LEs become equal in length. The LEs of the autosomal SCs often differ in length by as much as 2% in normal mouse (Poorman *et al.*, 1981) and it seems possible that asynchronous changes and inequalities in LE length are common and account for the twists, and perhaps turns, seen regularly in the SCs of mouse and other species (see Moens, 1978).

5. Changes in Synaptic Association

Complex homosynaptic configurations, which result from chromosome rearrangements and departures from diploidy, are often abandoned during pachytene (or as early as zygotene) in favor of simpler, heterosynaptic configurations

8. Synaptonemal Complex and Immunocytochemistry

(Moses and Poorman, 1981; Moses et al., 1982; Holm and Rasmussen, 1984). Such changes, best referred to by the neutral term "synaptic adjustment" (Moses et al., 1978), are further evidence of the dynamic aspects of SC structure. Whether synaptic adjustment reflects mechanisms that are induced by the synaptic abnormalities or simply uncovers normally occurring activities is not obvious, but clearly it represents some turnover or reassociation of components of the central region. Contrary to the statement that, aside from this activity of the central region, SC structure is "stable . . . apart from minor stage dependent changes in length" (Holm and Rasmussen, 1984), the evidence collected above suggests that changes in SC structure outside the central region are common and may also be involved in synaptic adjustment.

6. Recombination Nodules

Recombination nodules, dense cylindrical to spherical accumulations of material associated with the central region of the SC in numerous species, are found among and along the SCs in a distribution similar to that of recombination events (reviewed in Carpenter, 1977b, 1984; von Wettstein et al., 1984). From evidence of tritiated thymidine incorporation, DNA synthesis occurs preferentially at the nodules in *Drosophila melanogaster* (Carpenter, 1981), consistent with current models for recombination that require DNA repair at the site of crossovers and gene conversions (see Carpenter, 1984).

Pachytene DNA synthesis along the SCs has been detected in spermatocytes of mouse and hamsters (Moses et al., 1981, and unpublished results) and in oocytes of mouse (Moses and Poorman, 1984) where, unfortunately, no association with recombination nodules can be determined for technical reasons. In these mammals the time course of SC-associated DNA synthesis is distinct from that of the synthesis in the rest of the nucleus, reaching a peak level during that part of pachytene characterized by SC shortening, desynapsis of the XY-SC and, in mouse, synaptic adjustment. The temporal correlations among these events suggest direct interactions, but to demonstrate and to understand these presumed interactions will require dealing with the underlying molecular mechanisms.

7. Implications

The connection between the various aspects of SC morphology outlined above and SC function will be difficult to establish, but it seems clear that any speculation as to SC function should derive first from the observations of SC biology and second from those outcomes of meiosis for which the SC may be responsible. The question of the control of crossover position and number provides a good example.

It has been proposed that the distribution of crossover sites is controlled by an undefined change along preestablished SC segments. The change would result from the establishment of a crossover within the segment and would lead to loss of affinity for, or rejection of, recombination nodules along that segment, thus preventing further crossovers in that segment (Holm and Rasmussen, 1983a,b). This model has the advantage of simplicity, but at present there is no physical evidence for a preestablished segmentation of SCs, for changes limited to defined segments or for the (proposed) role of recombination nodules in crossing over.

It is possible that the observed regional differences and changes during pachytene in SC structure are manifestations of the mechanisms involved in controlling crossover position. The correlations between LE bulge and recombination nodule position in *Sordaria* (Zickler and Sage, 1981) may be evidence for this and be particularly obvious for reasons of SC structure peculiar to *Sordaria* rather than to an unusual underlying mechanism. Given the apparent turnover in SC components, one might predict that a change in the amount of availability of one of the components, or in the relative amounts of different components, could alter the positions and/or numbers of reciprocal exchanges either by affecting SC structure directly or by altering the extent or timing of the observed changes. As a mechanism, this could account for the interchromosomal effects on exchange seen in *Drosophila* and elsewhere (Lucchesi and Suzuki, 1968), and could mimic the "diffusible substance" postulated to account for these effects (see Holliday, 1968, 1977). It will require a concerted combination of genetics, molecular genetics, and cytology to test any of the possibilities.

Insofar as nonuniformities and changes in SC structure are related to its function, a detailed understanding of the associations and conditions which lead to their appearance will provide a better understanding of meiosis. The composition of the SC, the nature of the modifications in the SC and its components that bring about structural changes, and the composition of the recombination nodules are questions that can be addressed initially by using immunocytochemistry.

III. AN IMMUNOCYTOCHEMICAL APPROACH

Until the protein components of the SC have been identified, anti-SC antibodies will have to be identified cytologically, a process most easily carried out using indirect immunofluorescence. Whole-mount, surface-spread pachytene spermatocyte nuclei are well suited as substrates for the assay, being easily prepared, readily available, and already well characterized cytologically. Nucleoli, pore annuli, poorly understood dense bodies, and heterochromatin are present and readily identified as are the attachment plaques, kinetochores, and differentiations along the SCs of the autosomes and the LEs of the sex chromo-

8. Synaptonemal Complex and Immunocytochemistry 253

somes. All of these structures are displayed at once in the light microscope by virtue of the flattening of the nuclei on drying, and the stages of nuclei from zygotene to diplotene are readily determined by clear morphological criteria. Finally, identical preparations can be analyzed with the light and electron microscopes, so that results at one level of resolution can be confirmed and extended at the other.

A. Caveats

Possible shortcomings of these preparations for application to immunocytology are the same as for any cytological preparation. Before drawing any conclusions, one must consider the potentials for (1) extraction, degradation, or altered conformation of the proteins of interest, (2) contamination or reassociation of materials during processing, and (3) inaccessibility of epitope to antibody. The last of these (which makes applicable the adage that "absence of evidence is not evidence of absence") is perhaps the most annoying to cytologists. Unfortunately, improving accessibility often means increasing the chances that the other shortcomings will come into play. Ultimately, this means that the immunological approach must be iterative and must accommodate independently derived information. [For a useful review of the principles and methods involved, see Goding (1983).]

A reliable and relatively quick assay for detecting antibodies that bind the SC has been developed using a number of "control positives" that demonstrate penetration of the antibodies into the nucleus. The assay has been used to screen antibodies from a variety of sources as well as to detect and prepare monoclonal anti-SC antibodies (manuscript in preparation; and see below). Early results obtained using this assay revealed that many sera from humans, mice, and rabbits contain anti-SC antibodies, an observation which raises another of the major stumbling blocks of the antibody-based approach. Multiple antibody specificities in a serum and the multiplicity of proteins bearing a single epitope (=binding site) account for the frequent observation that both polyclonal and monoclonal antibodies bind more than one protein. Insofar as the antibodies in, or affinity-purified from, a serum represent a collection of monoclonal antibodies, these sources introduce an increased risk that such cross-reactions will interfere with the interpretation of results, particularly when a *sensitive* method for detecting binding is used. If the method is (1) less sensitive, so that binding of multiple epitopes on a given molecule or structure is required in order to detect labeling, and *at the same time* (2) the various antibodies cross-react with different substrates, then a mixture of antibodies will be, operationally, more selective. Unless these conditions are known to hold, and until the SC components are identified and isolated in sufficient quantity to serve either as immunogens for production of adequately selective sera or as substrates for affinity purification, the simplest approach is to begin by finding and isolating monoclonal anti-SC antibodies.

B. Alternative Approaches

As a means of identifying the proteins of the SC, the biochemical approach of enrichment for SCs followed by analysis with gel electrophoresis has not yet been particularly informative. To date the most widely employed approach involves preparation of the pachytene nuclear matrix, an operationally defined assembly consisting of those elements of the nucleus that, together with the SCs, have in common their resistance to various salt extractions and digestions of RNA and DNA (Comings and Okada, 1976; Ierardi et al., 1983; Gambino et al., 1981; Li et al., 1983; Raveh and Ben-Ze'ev, 1984). Where preceded by enrichment for pachytene cells (for methods, see Meistrich, 1972, 1977; Romrell et al., 1976; Bellvé et al., 1977a,b; Shepherd et al., 1981; Grogan et al., 1981) one might expect few enough bands on gel electrophoresis of the final fraction to indicate those proteins likely to be SC components, particularly if the proteins of the pachytene nuclear matrices and SCs are the same (see Comings and Okada, 1976). However, a cytological comparison of the matrices with unextracted nuclei reveals fewer differences than might be expected on the basis of the substantial amount of extraction of RNA, DNA, and protein (for examples, see Fig. 4 in Ierardi et al., 1983; Fig. 2 in Li et al., 1983), and indeed gel electrophoresis reveals that the matrices are composed of a large number of polypeptides. It may be that the SC components remaining in the matrices form a small and distinct subset of the total fraction and are "too few to be detected by the techniques used" (Ierardi et al., 1983). Consistent with this possibility are preliminary immunocytological results that reveal labeling of the SC in the absence of labeling of other nuclear constituents (see below). It remains to be seen whether isolated SCs (Walmsley and Moses, 1981) can be separated from other debris in sufficient quantity to provide adequate starting material for straightforward gel electrophoretic analysis.

Synaptonemal complex proteins present as minor components in a sample might be readily detected and distinguished by immunochemical techniques, and any means that allows the enrichment procedures to be curtailed or circumvented will reduce the ever-present possibilities for extraction or degradation of the SC components before analysis. All of the procedures for pachytene cell enrichment require a combination of enzymatic and physical forces to sever intercellular bridges in order to form a monodisperse cell suspension, and all require a significant amount of time for separation during which autolytic changes could occur. Although the cells are "viable" by trypan blue exclusion and other criteria (see references above and O'Brien *et al.,* 1983), effects on SC composition that do not involve gross structural rearrangement would go undetected. Should some enrichment for SC components be necessary for their identification and analysis, then assaying for the continued presence of components by antibody binding following each separation and extraction step would provide more confidence in the final result.

IV. PRELIMINARY RESULTS

If it were not easy cytologically to screen defined antibodies for activity against the SC, then the fact that positive results (visible labeling) can be as inconclusive as negative results would make the attempt all too frustrating. The uncertainties, imposed by factors discussed above, can only be diminished by using different antibodies, incorporating morphological observations, and correlating cytological, biochemical, and genetic data. Nevertheless, initial screenings can provide information which indicates specific questions to be asked in the second round of analysis.

Development of the assay used in our laboratory included testing a number of previously characterized polyclonal and monoclonal antibodies to demonstrate adequate preservation of binding sites and penetration of antibody into the nucleus. Once the procedure was established, the products of cell fusion experiments for monoclonal antibody production were screened. These experiments provided a number of results that illustrate the potential of the approach for addressing questions of the SC and meiosis. The results also illustrate the potential for the methods to provide misleading information.

A. Nucleoli and Nuclear Dense Bodies

Physical association of SC and nucleolar components is reported to occur during meiosis in insects (e.g., Schin, 1965; Wolstenholme and Meyer, 1966; Wettstein and Sotelo, 1971; Rasmussen, 1975), in plants (e.g., Stockert *et al.,* 1970; Gillies and Hyde, 1973; Kehlhoffner and Dietrich, 1983), in fungi (e.g., von Wettstein, 1971, 1977; Moens and Rapport, 1971; Horesh *et al.,* 1979; Moens and Kundu, 1982), and in the slime mold *Echinostelium* (Haskins *et al.,* 1971), and apparently occurs in mammals as well (see below). In many cases there is an association of structures formed by these components with the sex chromosome(s), as is true for putative nucleoli in some mammals (for a brief review, see Solari, 1974a), for the nucleolus-related "double dense body" (ddb) of Chinese hamster (Moses, 1977a; Dresser and Moses, 1980), and for a "dense body" seen in mouse (Solari and Tres, 1967; Solari, 1969; Oud and Reutlinger, 1981).

The Chinese hamster ddb forms in association with nucleoli during zygotene (Dresser and Moses, 1980) and is free of further association until middle pachytene when the "RNA-containing" moiety (as defined by orange-red fluorescence on staining with acridine orange) fuses with material surrounding the X chromosome axis and is lost from view (Dresser and Moses, 1980). The remaining moiety, now a "single dense body," fluoresces yellow-green on staining with acridine orange, as does the polycomplex- and X chromosome-associated round body in *Gryllus domesticus* (Wolstenholme and Meyer, 1966). Antibodies in various anti-nucleolar human sera (provided by G. McCarty and F. Barada, Duke University) bind dense bodies as well as nucleoli in Chinese hamster (unpublished results), supporting the proposal that the two are related (Dresser and Moses, 1980; cf. Takanari *et al.*, 1982).

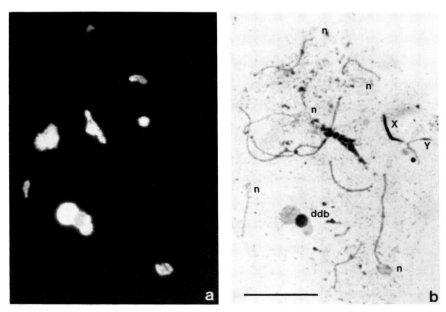

Fig. 1. Chinese hamster pachytene spermatocyte nucleus indirect immunofluorescence-labeled using serum from a rabbit immunized with the silver-staining nucleolar protein C23 (a) and then silver-stained (b). Nucleoli (n), identified by their characteristic shapes and positions (see Dresser and Moses, 1980), are less well contrasted by silver staining than usual, presumably caused by the preceding immunolabeling. The double dense body (ddb; in this case a triple body) and other structures in the nucleus are also immunolabeled, while the synaptonemal complexes (the dense, linear, silver-staining structures in b), the X and Y axes, and other nuclear material are not. (Serum provided by M. Olson, University of Mississippi.) Bar equals 10 μm.

8. Synaptonemal Complex and Immunocytochemistry

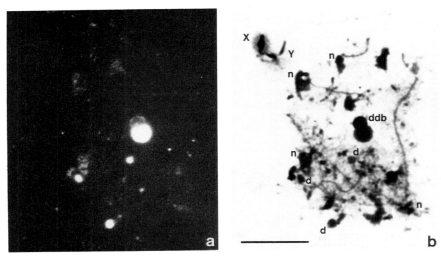

Fig. 2. Chinese hamster pachytene spermatocyte nucleus at a slightly later stage than the nucleus in Fig. 1, indirect immunofluorescence-labeled using monoclonal antibody DB1 (a) and then silver-stained (b). One moiety of the double dense body (ddb) and several unidentified dense bodies (d) are well labeled, while the nucleoli (n) at this stage are at most only slightly fluorescent. Bar equals 10 μm.

Antibodies directed against the silver-staining, nucleolar protein C23 (rabbit serum provided by M. O. J. Olson, University of Mississippi) label nucleoli and the ddb and other dense bodies in Chinese hamster nuclei in zygotene and pachytene (Fig. 1). The C23 protein has been localized to preribosomal particles (Prestayko *et al.*, 1974; Olsen *et al.*, 1974) and the ddb may represent, at least in part, the accumulation of these at some stage of maturation. Evidence that C23 binds DNA, in particular part of the nontranscribed spacer region of the robosomal DNA (Olson *et al.*, 1983), raises another possibility (not mutually exclusive with the first), i.e., that the ddb is composed, at least in part, of extranucleolar, amplified, ribosomal DNA sequences. Polycomplex formation in association with gene amplification has been suggested in *Gryllus domesticus* (Jaworska and Lima-de-Faria, 1969) and, while the observations were made on abnormal material, it may be that the association was only made more obvious, as opposed to being effected, by the abnormality.

A monoclonal antibody termed DB1 (by the author) labels the dense bodies and nucleoli at leptotene/zygotene. but in pachytene labels one moiety (sometimes both) of the ddb and *not* nucleoli (Fig. 2). The regularly labeled moiety of the ddb is that which fuses with the X chromosome. The monoclonal anti-SC antibody SC3, which binds the central region of the SC (see below; Moses *et al.*, 1986), begins to label the other, i.e., nonfusing, moiety of the ddb at early

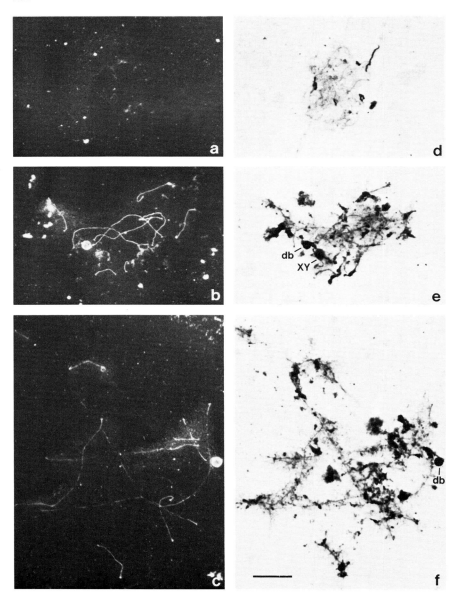

Fig. 3. Chinese hamster spermatocyte nuclei at zygotene (a,d), middle pachytene (b,e), and late pachytene (c,f) indirect immunofluorescence-labeled using monoclonal antibody SC3 in solutions buffered with phosphate (a,b,c) and then silver-stained (d,e,f). Synaptonemal complexes (SCs) are virtually unlabeled at zygotene (a), are well labeled at middle pachytene (b), and are labeled less at late pachytene (c) when the SCs begin to break down. The attachment plaques at either end of the SCs

8. Synaptonemal Complex and Immunocytochemistry

pachytene. The fluorescence of the SC3-labeled moiety increases in intensity until it is lost from view in late pachytene, sometime after the other moiety has fused with the X chromosome and disappeared (Fig. 3). These results may reflect a stage-dependent change in accessibility to antibody in the different structures but seem more likely to reveal the changing location of some components.

In mouse spermatocyte nuclei, DB1 labels the (single) dense body in early pachytene and then labels the sex body in late pachytene (unpublished results). The common labeling of the mouse dense body and one moiety of the Chinese hamster ddb with a single monoclonal antibody builds a stronger, if still circumstantial, case for the similarity in composition of these structures than can be made on the basis of morphology alone. It is too early to tell where these observations will lead, but it seems that the formation of nucleolus-related dense bodies and the association of these with SC components and with the sex chromosome(s) are regular concomitants of meiosis. Immunocytology, using these and other antibodies as probes, will better define the similarities and disparities among the various examples of dense bodies in meiosis, and in addition should identify those antibody probes likely to be useful in the biochemical characterization of the components involved.

B. Kinetochores

The location of centromeres along the LEs is apparent in most species by morphological differentiations in the axes themselves or in the surrounding chromatin, but there are cases where, for various reasons, the differentiations are obscure (for example, along the Chinese hamster X and Y chromosomes, Moses, 1977b). Anti-centromere antibodies are prevalent in the sera of humans with a relatively mild autoimmune disease ("CREST syndrome") and have been shown to bind the kinetochores (Brenner *et al.*, 1981). A number of these sera (provided by G. McCarty and F. Barada, Duke University), used to guide development of the assay, confirm the correctness of the earlier centromere localizations by labeling kinetochores where predicted, for example, in Chinese hamster and mouse (Fig. 4), and presumably could be used to localize kinetochores in any case where morphology alone is not indicative. In addition, some of the sera label the SCs (Fig. 4c,d). Labeling of the SCs could be accounted for (1) by antibodies distinct from those that bind the kinetochores or (2) by antibodies directed only against certain epitopes or components common to the two. In

---

are made evident by the widening in the fluorescence. The double dense body has just fused with the X chromosome and divided in b, making it obvious that it is the nonfusing moiety (db) that is labeled with SC3. This moiety continues to label into late pachytene (c). Other patches of label, particularly evident in b, may represent overlying debris. Bar equals 10 μm.

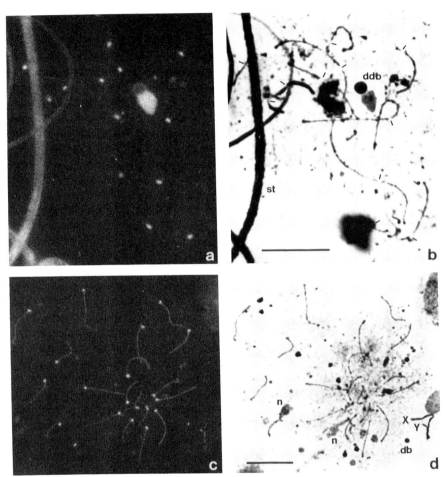

Fig. 4. Pachytene spermatocyte nuclei from Chinese hamster (a,b) and mouse (c,d) indirect immunofluorescence-labeled with human anti-centromere sera (a,c) and then silver-stained (b,d). (a,b) The positions of the kinetochores along autosomal SCs and the X and Y axes of Chinese hamster are established clearly, confirming certain earlier suggestions (see Moses, 1977b). Kinetochore positions are marked in b by small bars; e.g., see bar at upper left, adjacent to sperm tail (st). One moiety of the double dense body (ddb) and overlying sperm tails are labeled as well, a reminder of the presence of multiple antibodies in sera as well as of the possibility of "nonspecific" antibody binding. (c,d) The telomeric kinetochores along the mouse autosomal SCs and X and Y axes are demonstrated clearly. The SCs are also labeled, while other nuclear material, including the nucleoli (n), dense body (db), and unpaired X and Y axes, are not. (Sera provided by G. McCarty and F. Barada, Duke University.) Bars equal 10 μm.

8. Synaptonemal Complex and Immunocytochemistry

favor of the first possibility, anti-SC antibodies are not uncommon in human sera (unpublished results). In favor of the second possibility, there is evidence that anti-kinetochore antibodies bind several different proteins and are found in different amounts in different sera (B. R. Brinkley, personal communication; Cox *et al.*, 1983; Earnshaw *et al.*, 1984; Ayer and Fritzler, 1984; Guldner *et al.*, 1984). If these observations do support the suggestion that kinetochores and LEs are structurally related (Counce and Meyer, 1973; and see above), one would predict a correlation between the presence of SC binding and the presence of binding of particular kinetochore components.

C. Candidate Proteins

A rationale for screening antibodies against any of a large number of well-characterized proteins is not difficult to develop. However, to date such attempts in various laboratories have provided little direction and instead have revealed more about the caprices of immunocytology than about the SC.

1. Actin

On the basis of the observation that a human serum containing anti-actin antibodies labeled SC ends, De Martino *et al.* (1980) concluded that there was actin at the attachment plaques. In fact, the number of dots of fluorescence per nucleus in their figures is about half that expected, and it seems likely that instead it is the centromeres that are labeled. Nevertheless, Heyting *et al.* (1984) found that affinity-purified anti-actin antibodies (as well as those antibodies in that particular serum that did *not* bind actin) labeled both SCs and "terminal dots," and felt that confirmed the earlier report. Using various preparative techniques, Spyropoulos and Moens (1986) reported the absence of any labeling with affinity-purified anti-*Dictyostelium* actin antibodies or with phallacidin, which binds F-actin.

One rabbit anti-actin serum (provided by K. Burridge, University of North Carolina) does label SCs of mouse (Fig. 5), although in a manner similar to labeling by some "control" rabbit sera. However, there is no clear labeling with NBD-phallacidin or with a widely cross-reacting monoclonal anti-actin antibody ("JLA20," provided by J. Lin, Cold Spring Harbor) under the conditions of the assay employed here. If there is actin associated with the SC, there is not yet convincing proof of it.

2. Myosin

On the basis of the observation that a rabbit serum containing anti-myosin antibodies labeled the length of the SCs, De Martino *et al.* (1980) concluded that

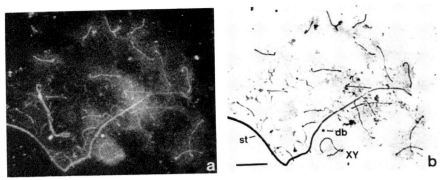

Fig. 5. Mouse pachytene spermatocyte nucleus indirect immunofluorescence-labeled using serum from a rabbit immunized with actin (a) and then silver-stained (b). Synaptonemal complexes, the dense body (db), and sperm tails (st) are labeled, as is, to a lesser extent, the chromatin (the background "haze"). Preimmune sera from a number of rabbits (not shown) show similar labeling, making it difficult to interpret this result (see Fig. 6). (Serum provided by K. Burridge, University of North Carolina.) Bar equals 10 μm.

there was myosin in the SC and proposed a model to indicate its location and possible function. This observation has been repeated (Wahrman, 1981) but other investigators, using affinity-purified antibodies, have found no labeling of the SCs (Heyting *et al.*, 1984; Spyropoulos and Moens, 1986), and results using anti-myosin antibodies in a whole serum (provided by K. Burridge, University of North Carolina) and monoclonal antibodies (provided by F. Schachat, Duke University) have also been negative (M. E. Dresser and M. J. Moses, unpublished results). The widespread occurrence of anti-SC antibodies in sera provides an alternative interpretation for the positive results reported; on the other hand, the negative results might be explained by inaccessibility of epitopes or by tissue-specific myosins with epitopes that are not recognized by some of the anti-myosin antibodies.

3. Tubulin

On the basis of SC labeling with sera directed against tubulin and microtubule-associated proteins (MAP2, tau), del Mazo and Avila (1984) concluded that these proteins were located in the SC. Antibodies in sera from rabbits immunized with sea urchin tubulin do bind the SC. However, affinity purification against pig brain tubulin (antibodies and tubulin provided by W. Voter and H. Erickson, Duke University) provides a fraction of anti-tubulin antibodies that do not label the SC and a fraction of "nonspecific" antibodies from the serum that do label the SCs (Fig. 6). Two monoclonal anti-tubulin antibodies (provided by L. I. Binder, University of Virginia) label sperm tails but not SCs. The arguments

8. Synaptonemal Complex and Immunocytochemistry

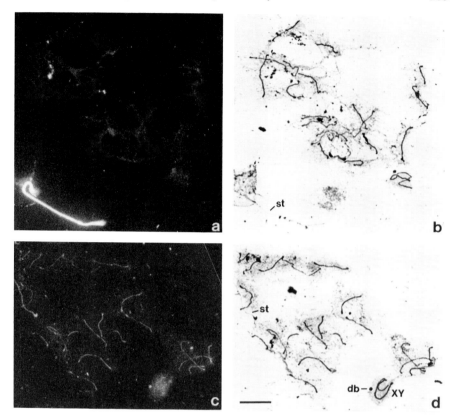

Fig. 6. Mouse pachytene spermatocyte nuclei indirect immunofluorescence-labeled using affinity-"purified" anti-tubulin antibodies from a rabbit serum (a) and with the antibodies in that serum that were not retained by the tubulin column (c) and then silver-stained (b,d). (a,b) There is slight fluorescence only of the background of the nucleus, while sperm tails (st) are well labeled as expected. (c,d) Synaptonemal complexes, the dense body (db), and sex chromatin (XY) are labeled, while labeling of a sperm tail is greatly decreased from that in a. (Serum and tubulin provided by W. Voter and H. Erickson, Duke University.) Bar equals 10 μm.

raised above can be applied here with the same conclusion—no proof exists either way for the presence or absence of tubulin in the SC.

4. Small Nuclear Ribonucleoproteins

Two monoclonal antibodies (provided by D. Pisetsky, Duke University), directed against small nuclear ribonucleoproteins that have been implicated in mRNA processing, do not bind the SCs but do illustrate the potential of the screening procedure for analyzing other activities in the nucleus. Antibody labeling (not shown here) is apparent only over the euchromatin of those cells at stages of meiotic prophase when there is nonnucleolar RNA synthesis (Utakoji, 1966; Kierszenbaum and Tres, 1974; Monesi *et al.*, 1978; unpublished results from [^{3}H]uridine uptake followed by autoradiography). It should be possible in the future to demonstrate similar time-and-place correlations between incorporated radiolabeled precursors and antibody-labeled components. The whole-mount preparations of meiotic prophase nuclei are easily staged on morphological characteristics, follow well-established patterns of synthesis, provide a good substrate for autoradiographical, cytochemical, and immunocytochemical analysis, and can be analyzed by light and electron microscopy, in combination making them particularly well suited for collecting data that can be correlated directly.

5. *Nuclear Lamins*

Morphological and immunocytological evidence indicates the disappearance of the nuclear lamina during meiosis, perhaps a necessity in order "to allow movement and condensation of the chromosomes" (see Stick and Schwarz, 1982, 1983). Unlike the situation in mitosis, where the entire nuclear envelope is broken down and the lamin proteins are found in the cytoplasm, in meiosis the nuclear membranes and pore complexes remain in place, and Stick and Schwarz reported no evidence for the lamins in the cytoplasm of either spermatocytes or oocytes. Results using various anti-lamin sera and monoclonals (provided by G. Maul, Wistar Institute) suggest the presence of some lamin(s) in the pachytene spermatocytes (Fig. 7), though not apparently as an even layer at the nuclear

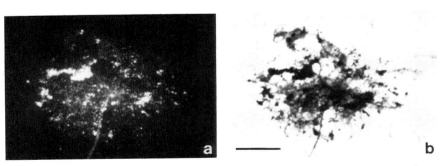

Fig. 7. Mouse pachytene spermatocyte nucleus indirect immunofluorescence-labeled using a monoclonal antibody directed against nuclear lamins (a) and then silver-stained (b). Fluorescence of speckles that seem to have coalesced in places is correlated roughly with silver stain and, given the increased density of the nucleus in b as well as the poor staining of the SCs, it may be that the antibody labeling has influenced the silver staining. (Antibody "178" provided by G. Maul, Wistar Institute.) Bar equals 10 μm.

envelope. It is possible that the antibodies used by Stick and co-workers do not bind meiosis-specific lamins or lamin-related proteins, that the specific epitopes their antibodies bind may not have been accessible, or that the lamins were removed during preparation for cytology. Alternatively, the antibodies used by us may be cross-reacting with components that are otherwise unrelated to the somatic lamins, or may reveal some contamination by proteins or structures from cells that are adjacent in the testis that have become associated with the spermatocytes during preparation for cytology.

Both the SC and nuclear lamina are implicated in DNA binding and it would not be surprising if some component(s) of the SC were related to the lamins, one of the more specific possibilities raised by the general SC–nuclear envelope relationship postulated by Fiil and Moens (1973). Biochemical evidence indicates that the usual polypeptides attributed to the nuclear lamina are not found in the nuclear matrices of mouse spermatogenic cells (Ierardi *et al.*, 1983), but it is possible that two polypeptides of electrophoretic mobility similar to that of the somatic lamins represent germline-specific lamins as seen in *Xenopus* (Benavente and Krohne, 1985). It will be interesting to see if these polypeptides are bound by anti-lamin antibodies.

D. Monoclonal Anti-SC Antibodies

An alternative to screening antibodies against candidate SC proteins is to isolate anti-SC antibodies, to define their behavior cytologically, and to use them to characterize the components that they bind. A number of anti-SC antibodies have been identified in the products of cell fusion experiments, demonstrating the success of the assay for the first steps of this approach.

A common and particularly annoying problem with this approach is that some of the antibody-producing cell lines die or stop making antibody before they can be cloned and utilized fully. Occasionally, the antibodies can still be informative. Among the most interesting of these was a class of antibodies that bind only SCs and/or the unpaired X and Y axes (Fig. 8). The results support the suggestion (problems of accessibility aside) that the composition of these structures is, at least in part, distinct from that of other constituents in the nucleus. Exclusive labeling of the unpaired sex chromosome axes may relate to any of a number of possibilities, including the failure of these elements to synapse, the absence of transcription of the associated chromatin, and the relatively increased condensation of that chromatin.

Two monoclonal antibodies, SC3 from an autoimmune mouse (fusion and initial cloning performed by R. Warren and D. Pisetsky, Duke University) and SC4 from a BALB/c mouse immunized with whole testis protein (fusion products provided by P. Saling, Duke University), are similar in their labeling pat-

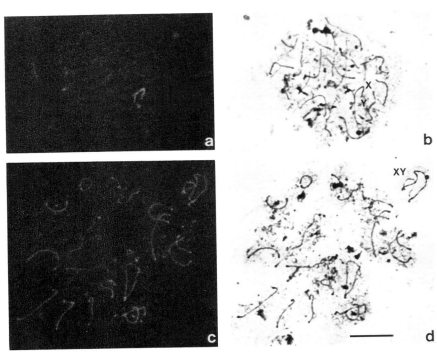

Fig. 8. Mouse pachytene spermatocyte nuclei indirect immunofluorescence-labeled with supernatants from fused cells that subsequently stopped producing the antibodies (a,c) and then silver-stained (b,d). (a) Only the unpaired X and Y axes are well-contrasted. (c) Autosomal SCs and the unpaired X and Y axes are labeled. (Mice immunized with mouse heavy membrane fraction, provided by Y. Hotta, University of California, San Diego.) Bar equals 10 μm.

terns, both cytologically and on protein blots (Dresser and Moses, 1986). Both label the central region of the SC and attachment plaques in mouse and Chinese hamster spermatocyte nuclei and both label intermediate filaments in various cell types (Figs. 3 and 9; and see Figs. 12 and 13 in Moses *et al.*, 1986; Dresser and Moses, 1983, 1984, 1986). The labeling pattern, however, can be dependent on the conditions used during preparation for cytology and during antibody binding. For example, SC3 binds interstitially along the SCs of Chinese hamster when Tris-buffered saline is used for antibody dilutions, but binds SCs, attachment plaques, and dense bodies when phosphate-buffered saline is used (Figs. 3 and 9). The differences here may be related to differences in phosphorylation and/or accessibility of a single component or of closely related components (see Sternberger and Sternberger, 1983), or they could reflect the artifactual reassociation of some component(s), depending on buffer conditions. In any case, labeling of

8. Synaptonemal Complex and Immunocytochemistry

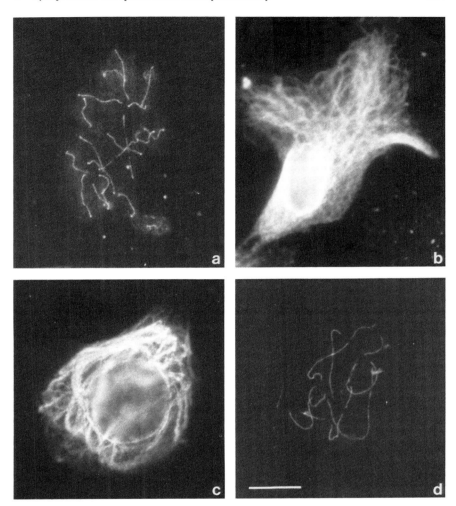

Fig. 9. (a) Mouse pachytene spermatocyte indirect immunofluorescence-labeled with monoclonal antibody SC4. Synaptonemal complexes and attachment plaques (see as thickenings at the ends of the SCs) are well labeled. (b) Cultured gerbil fibroma cell (CCL 146) labeled with SC4, showing filamentous labeling of the cytoplasm. (c) Gerbil fibroma cell treated in culture with 10 μM Colcemid for 24 hours then labeled with SC4, showing a collapse of the labeling into perinuclear whorls typical of intermediate filaments. (d) Chinese hamster pachytene spermatocyte nucleus labeled with monoclonal antibody SC3 in solutions buffered with Tris, showing label exclusively over the SCs (compare with Fig. 3b). Bar equals 10 μm.

the SCs is consistent using various buffers, fixatives, and other preparative procedures, indicating that this primary result is not artifactual.

Efforts to identify the component(s) bound by SC3 and SC4, using electrophoretically separated pachytene proteins blotted on nitrocellulose paper, have been slowed by their binding to multiple bands, a common problem with monoclonals. For example, both monoclonals label bands that contain (are?) actin and vimentin as distinguished by electrophoretic mobility and by labeling with an anti-actin monoclonal (provided by J. Lin, Cold Spring Harbor) and an anti-vimentin monoclonal (provided by F. Schachat, Duke University). Results concerning the presence of actin in the SC have been discussed above. The simplest interpretation is that the labeling of intermediate filaments in somatic cells results from the binding of SC3 and SC4 to vimentin. The failure of the anti-vimentin antibody to label SCs, while it does label filaments in Chinese hamster Sertoli cells both in spread preparations and in frozen sections, could result from lack of accessibility to the epitope or from absence or modification of the epitope. However, it seems most likely that the anti-SC monoclonals bind an SC protein that shares an epitope with actin, vimentin, and other proteins but otherwise is distinct from these proteins. Preliminary attempts to simplify the starting material, using enriched and extracted fractions of pachytene nuclei (following the scheme of Ierardi *et al.*, 1983) and enriched fractions of various spermatogenic cells and nuclei (provided by D. O'Brien, N.I.E.H.S., Research Triangle Park, N.C.), have demonstrated the marked affinity of SC3 and particularly of SC4 for a protein band that runs near but distinctly below actin in SDS-polyacrylamide gels (Dresser and Moses, 1986).

It is interesting that no anti-intermediate filament antibody, including SC3 and SC4, has yet revealed intermediate filaments in the cytoplasm of cells in meiotic prophase (for examples, see Franke *et al.*, 1979; Franz *et al.*, 1983; Fig. 4 in van Vorstenbosch *et al.*, 1984; Dresser and Moses, 1986). The appearance of the SCs, and the disappearance of both a well-formed nuclear lamina and obvious intermediate filaments, may be indicative of a fundamental difference of arrangement of these structural elements in meiotic versus somatic cells. If related, but meiosis- or germ line-specific, proteins are involved, the immunocytochemical approach provides just those analytical methods required for gathering the salient information.

V. SUMMARY

A wealth of cytological information has revealed the synaptonemal complex to be a dynamic and morphologically heterogeneous organelle that changes in structure throughout the pachytene stage of meiosis. Nothing is known of the function(s) reflected by these changes, or by the formation of the SC itself, because too little is known of the molecular mechanisms involved. The components of the SC can be identified and their behaviors characterized using immunochemical and immunocytological techniques but, because of ambiguities inherent in the methods, the results from each experiment must be interpreted carefully. Nevertheless, initial results indicate that immunocytochemistry will play an important role in the solving of meiosis.

ACKNOWLEDGMENTS

First and foremost, thanks go to M. J. Moses in whose lab and under whose tutelage all of this work was performed and prepared for publication. I am indebted as well to P. A. Poorman for her able assistance in much of the work reported here, to Todd Gambling and Rex Anderson for their technical support, to all those (listed in the text and figure legends) who provided antibodies, proteins, cell lines, and expertise, and to S. J. Counce for her support, conversations, and criticisms. This work was supported by NIH training grant GM-7171 and NIH grant CA-14236 (to M.J.M./Duke University).

REFERENCES

Ayer, L. M., and Fritzler, M. J. (1984). Anticentromere antibodies bind to trout testis histone 1 and a low molecular weight protein from rabbit thymus. *Mol. Immunol.* **21,** 761–770.

Bellvé, A. R., Cavicchia, J. C., Millette, C. F., O'Brien, D. A., Bhatnagar, Y. M., and Dym, M. (1977a). Spermatogenic cells of the prepuberal mouse. Isolation and morphological characterization. *J. Cell Biol.* **74,** 68–85.

Bellvé, A. R., Millette, C. F., Bhatnagar, Y. M., and O'Brien, D. A. (1977b). Dissociation of the mouse testis and characterization of isolated spermatogenic cells. *J. Histochem. Cytochem.* **25,** 480–494.

Benavente, R., and Krohne, G. (1985). Change of karyoskeleton during spermatogenesis of *Xenopus:* Expression of lamin L_{IV}, a nuclear lamina protein specific for the male germ line. *Proc. Natl. Acad. Sci. USA* **82,** 6176–6180.

Bogdanov, Yu. F. (1977). Formation of cytoplasmic synaptonemal-like polycomplexes at leptotene and normal synaptonemal complexes at zygotene in *Ascaris suum* male meiosis. *Chromosoma* **61,** 1–21.

Brenner, S., Pepper, D., Berns, M. W., Tan. E., and Brinkley, B. R. (1981). Kinetochore structure, duplication, and distribution in mammalian cells: Analysis by human autoantibodies from scleroderma patients. *J. Cell Biol.* **91,** 95–102.

Carpenter, A. T. C. (1975). Electron microscopy of meiosis in *Drosophila melanogaster* females. I. Structure, arrangement, and temporal change of the synaptonemal complex in wild-type. *Chromosoma* **51,** 157–182.

Carpenter, A. T. C. (1979a). Recombination nodules and synaptonemal complex in recombination-defective females of *Drosophila melanogaster*. *Chromosoma* **75,** 259–292.

Carpenter, A. T. C. (1979b). Synaptonemal complex and recombination nodules in wild type *Drosophila melanogaster* females. *Genetics* **92,** 511–541.

Carpenter, A. T. C. (1981). EM autoradiographic evidence that DNA synthesis occurs at recombination nodules during meiosis in *Drosophila melanogaster* females. *Chromosoma* **83,** 59–80.

Carpenter, A. T. C. (1984). Genic control of meiosis. *Chromosomes Today* **8,** 70–79.

Church, K. (1976). Arrangement of chromosome ends and axial core formation during early meiotic prophase in the male grasshopper *Brachystola magna* by 3D, E.M. reconstruction. *Chromosoma* **58,** 365–376.

Comings, D. E., and Okada, T. A. (1976). Nuclear proteins. III. The fibrillar nature of the nuclear matrix. *Exp. Cell Res.* **103,** 341–360.

Counce, S. J., and Meyer, G. F. (1973). Differentiation of the synaptonemal complex and the kinetochore in *Locusta* spermatocytes studied by whole mount electron microscopy. *Chromosoma* **44,** 231–253.

Cox, J. V., Schenk, E. A., and Olmsted, J. B. (1983). Human anticentromere antibodies: Distribution, characterization of antigens, and effect on microtubule organization. *Cell* **35,** 331–339.

del Mazo, J., and Avila, J. (1984). Tubulin and microtubule-associated proteins (MAPs) in the synaptonemal complex. *Chromosomes Today* **8,** 324a.

DeMartino, C., Capanna, E., Nicotra, M. R., and Natali, P. G. (1980). Immunochemical localization of contractile proteins in mammalian meiotic chromosomes. *Cell Tissue Res.* **213,** 159–178.

Dresser, M. E. (1986). In preparation.

Dresser, M. E., and Moses, M. J. (1980). Synaptonemal complex karyotyping in spermatocytes of the Chinese hamster (*Cricetulus griseus*). IV. Light and electron microscopy of synapsis and nucleolar development by silver staining. *Chromosoma* **76,** 1–22.

Dresser, M. E., and Moses, M. J. (1983). Cytological characterization of an anti-synaptonemal complex monoclonal antibody. *J. Cell Biol.* **97,** 185a.

Dresser, M. E., and Moses, M. J. (1984). Development and cytological characterization of an anti-synaptonemal complex monoclonal antibody. *Chromosomes Today* **8,** 304a.

Dresser, M. E., and Moses, M. J. (1986). In preparation.

Earnshaw, W. C., Halligan, N., Cooke, C., and Rothfield, N. (1984). The kinetochore is part of the metaphase chromosome scaffold. *J. Cell Biol.* **98,** 352–357.

Engelhardt, P., and Pusa, K. (1972). Nuclear pore complexes: "Press-stud" elements of chromosomes in pairing and control. *Nature (London), New Biol.* **240,** 163–166.

Esponda, P. (1977). Fine structure of synaptonemal-like complexes in *Allium cepa* microspores. *Protoplasma* **93,** 1–6.

Fiil, A., and Moens, P. B. (1973). The development, structure and function of modified synaptonemal complexes in mosquito oocytes. *Chromosoma* **41,** 37–62.

Fiil, A., Goldstein, P., and Moens, P. B. (1977). Precocious formation of synaptonemal-like polycomplexes and their subsequent fate in female *Ascaris lumbricoides* var. *suum. Chromosoma* **65,** 21–35.

Franke, W. W., Grund, C., and Schmid, E. (1979). Intermediate-sized filaments present in Sertoli cells are of the vimentin type. *Eur. J. Cell Biol.* **19,** 269–275.

Franz, J. K., Gall, L., Williams, M. A., Picheral, B., and Franke, W. W. (1983). Intermediate-size filaments in a germ cell: Expression of cytokeratins in oocytes and eggs of the frog *Xenopus. Proc. Natl. Acad. Sci. U.S.A.* **80,** 6254–6258.

Gambino, J., Eckhart, R. A., and Risley, M. S. (1981). Nuclear matrices containing synaptonemal complexes from *Xenopus laevis. J. Cell Biol.* **91,** 63a.

Gillies, C. B. (1975). Synaptonemal complex and chromosome structure. *Annu. Rev. Genet.* **9,** 91–109.

Gillies, C. B. (1981). Electron microscopy of spread maize pachytene synaptonemal complexes. *Chromosoma* **83,** 575–591.

Gillies, C. B. (1983). Ultrastructural studies of the association of homologous and non-homologous parts of chromosomes in the mid-prophase of meiosis in *Zea mays. Maydica* **28,** 265–287.

8. Synaptonemal Complex and Immunocytochemistry

Gillies, C. B., and Hyde, B. B. (1973). Intranucleolar bodies in maize pachytene microsporocytes. *Hereditas* **74**, 137–140.

Goding, J. W. (1983). "Monoclonal Antibodies: Principles and Practice." Academic Press, Orlando.

Grogan, W. M., Farnham, W. F., and Sabau, J. M. (1981). DNA analysis and sorting of viable mouse testis cells. *J. Histochem. Cytochem.* **29**, 738–746.

Guldner, H. H., Lakomek, H.-J., and Bautz, F. A. (1984). Human anti-centromere sera recognise a 19.5kD non-histone chromosomal protein from HeLa cells. *Clin. Exp. Immunol.* **58**, 13–20.

Haskins, E. F., Hinchee, A. A., and Cloney, R. A. (1971). The occurrence of synaptonemal complexes in the slime mold *Echinostelium minutum* de Bary. *J. Cell Biol.* **51**, 898.

Heyting, C., Dietrich, A. J. J., Koperdraad, F., and Redeker, E. J. W. (1984). Do synaptonemal complexes contain actin and myosin? *Chromosomes Today* **8**, 316a.

Holliday, R. (1968). Genetic recombination in fungi. In "Replication and Recombination of Genetic Material" (W. J. Peacock and R. D. Brock, eds.), pp. 157–174. Aust. Acad. Sci., Canberra.

Holliday, R. (1977). Recombination and meiosis. *Philos. Trans. R. Soc. London, Ser. B* **277**, 359–370.

Holm, P. B., and Rasmussen, S. W. (1980). Chromosome pairing, recombination nodules and chiasma formation in diploid *Bombyx* males. *Carlsberg Res. Commun.* **45**, 483–548.

Holm, P. B., and Rasmussen, S. W. (1983a). Human meiosis. VI. Crossing over in human spermatocytes. *Carlsberg Res. Commun.* **48**, 385–413.

Holm, P. B., and Rasmussen, S. W. (1983b). Human meiosis. VII. Chiasma formation in human spermatocytes. *Carlsberg Res. Commun.* **48**, 415–456.

Holm, P. B., and Rasmussen, S. W. (1984). The synaptonemal complex in chromosome pairing and disjunction. *Chromosomes Today* **8**, 104–116.

Horesh, O., Simchen, G., and Friedmann, A. (1979). Morphogenesis of the synapton during yeast meiosis. *Chromosoma* **75**, 101–115.

Ierardi, L. A., Moss, S. B., and Bellvé, A. R. (1983). Synaptonemal complexes are integral components of the isolated mouse spermatocyte nuclear matrix. *J. Cell Biol.* **96**, 1717–1726.

Jaworska, H., and Lima-de-Faria, A. (1969). Multiple synaptinemal complexes at the region of gene amplification in *Acheta*. *Chromosoma* **28**, 309–327.

Jones, G. H. (1973). Modified synaptinemal complexes in spermatocytes of *Stethophyma grossum*. *Cold Spring Harbor Symp. Quant. Biol.* **38**, 109–115.

Kehlhoffner, J.-L., and Dietrich, J. (1983). Synaptonemal complex and a new type of nuclear polycomplex in three higher plants: *Paeonia tenuifolia, Paeonia delavayi,* and *Tradescantia paludosa. Chromosoma* **88**, 164–170.

Kierszenbaum, A. L., and Tres, L. L. (1974). Nucleolar and perichromosomal RNA synthesis during meiotic prophase in the mouse testis. *J. Cell Biol.* **60**, 39–53.

La Cour, L. F., and Wells, B. (1973). Deformed lateral elements in synaptonemal complexes of *Phaedranassa viridiflora. Chromosoma* **41**, 289–296.

Li, S., Meistrich, M. L., Brock, W. A., Hsu, T. C., and Kuo, M. T. (1983). Isolation and preliminary characterization of the synaptonemal complex from rat pachytene spermatocytes. *Exp. Cell Res.* **144**, 63–72.

Lucchesi, J. C., and Suzuki, D. T. (1968). The interchromosomal control of recombination. *Annu. Rev. Genet.* **2**, 53–86.

Meistrich, M. L. (1972). Separation of mouse spermatogenic cells by velocity sedimentation. *J. Cell Physiol.* **80**, 299–312.

Meistrich, M. L. (1977). Separation of spermatogenic cells and nuclei from rodent testes. *Methods Cell Biol.* **15**, 15–54.

Moens, P. B. (1968). Synaptinemal complexes of *Lilium tigrinum* (triploid) sporocytes. *Can. J. Genet. Cytol.* **10**, 799–807.

Moens, P. B. (1970). The fine structure of meiotic chromosome pairing in natural and artificial *Lilium* polyploids. *J. Cell Sci.* **7**, 55–63.

Moens, P. B. (1973). Quantitative electron microscopy of chromosome organization at meiotic prophase. *Cold Spring Harbor Symp. Quant. Biol.* **38**, 99–107.

Moens, P. B. (1978). Ultrastructural studies of chiasma distribution. *Annu. Rev. Genet.* **12**, 433–450.

Moens, P. B., and Kundu, S. C. (1982). Meiotic arrest and synaptonemal complexes in yeast ts spo-10 (*Saccharomyces cerevisiae*). *Can. J. Biochem.* **60**, 284–289.

Moens, P. B., and Rapport, E. (1971). Synaptic structures in the nuclei of sporulating yeast, *Saccharomyces cerevisiae*. *J. Cell Sci.* **9**, 665–677.

Monesi, V., Geremia, R., D'Agostino, A., and Boitani, C. (1978). Biochemistry of male germ cell differentiation in mammals: RNA synthesis in meiotic and postmeiotic cells. *Curr. Top. Dev. Biol.* **12**, 11–35.

Moses, M. J. (1968). Synaptinemal complex. *Annu. Rev. Genet.* **2**, 363–412.

Moses, M. J. (1969). Structure and function of the synaptonemal complex. *Genetics* **61**, Suppl., 1–51.

Moses, M. J. (1977a). Synaptonemal complex karyotyping in spermatocytes of the Chinese hamster (*Cricetulus griseus*). I. Morphology of the autosomal complement in spread preparations. *Chromosoma* **60**, 99–125.

Moses, M. J. (1977b). Synaptonemal complex karyotyping in spermatocytes of the Chinese hamster (*Cricetulus griseus*). II. Morphology of the XY pair in spread preparations. *Chromosoma* **60**, 127–137.

Moses, M. J., and Poorman, P. A. (1981). Synaptonemal complex analysis of mouse chromosomal rearrangements. II. Synaptic adjustment in a tandem duplication. *Chromosoma* **81**, 519–535.

Moses, M. J., and Poorman, P. A. (1984). Synapsis, synaptic adjustment and DNA synthesis in mouse oocytes. *Chromosomes Today* **8**, 90–103.

Moses, M. J., Slatton, G. H., Gambling, T. M., and Starmer, C. F. (1977). Synaptonemal complex karyotyping in spermatocytes of the Chinese hamster (*Cricetulus griseus*). III. Quantitative evaluation. *Chromosoma* **60**, 345–375.

Moses, M. J., Poorman, P. A., Russell, L. B., Cacheiro, N. L., Roderick, T. H., and Davisson, M. T. (1978). Synaptic adjustment: Two pairing phases in meiosis? *J. Cell Biol.* **79**, 123a.

Moses, M. J., Dresser, M. E., and Poorman, P. A. (1981). DNA synthesis associated with synaptonemal complexes in meiotic prophase. *J. Cell Biol.* **91**, 70a.

Moses, M. J., Poorman, P. A., Roderick, T. H., and Davisson, M. T. (1982). Synaptonemal complex analysis of mouse chromosomal rearrangements. IV. Synapsis and synaptic adjustment in two paracentric inversions. *Chromosoma* **84**, 457–474.

Moses, M. J., Dresser, M. E., and Poorman, P. A. (1986). Composition and role of the synaptonemal complex. *Symp. Soc. Exp. Biol.* **36**, 245–270.

O'Brien, D. A., Gerton, G. L., and Millette, C. F. (1983). Protein synthesis by isolated prepuberal mouse spermatogenic cells in culture. *J. Cell Biol.* **97**, 11a.

Olson, M. O. J., Prestayko, A. W., Jones, C. E., and Busch, H. (1974). Phosphorylation of proteins of ribosomes and nucleolar preribosomal particles from Novikoff hepatoma ascites cells. *J. Mol. Biol.* **90**, 161–168.

Olson, M. O. J., Rivers, Z. M., Thompson, B. A., Kao, W.-Y., and Case, S. T. (1983). Interaction of nucleolar phosphoprotein C23 with cloned segments of rat ribosomal deoxyribonucleic acid. *Biochemistry* **22**, 3345–3351.

Oud, J. L., and Reutlinger, A. H. H. (1981). The behaviour of silver-positive structures during meiotic prophase of male mice. *Chromosoma* **81**, 569–578.

Poorman, P. A., Moses, M. J., Russell, L. B., and Cacheiro, N. L. A. (1981). Synaptonemal

8. Synaptonemal Complex and Immunocytochemistry

complex analysis of mouse chromosomal rearrangements. I. Cytogenetic observations on a tandem duplication. *Chromosoma* **81**, 507–518.
Prestayko, A. W., Klomp, G. R., Schmoll, D. J., and Busch, H. (1974). Comparison of proteins of ribosomal subunits and nucleolar preribosomal particles from Novikoff hepatoma ascites cells by two-dimensional polyacrylamide gel electrophoresis. *Biochemistry* **13**, 1945–1951.
Rasmussen, S. W. (1975). Synaptonemal polycomplexes in *Drosophila melanogaster*. *Chromosoma* **49**, 321–331.
Rasmussen, S. W., and Holm, P. B. (1980). Mechanics of meiosis. *Hereditas* **93**, 187–216.
Rasmussen, S. W., and Holm, P. B. (1982). The meiotic prophase in *Bombyx mori*. *Insect Ultrastruct.* **1**, 61–85.
Raveh, D., and Ben-Ze'ev, A. (1984). The synaptonemal complex as part of the nuclear matrix of the Flour Moth, *Ephestia kuehniella*. *Exp. Cell Res.* **153**, 99–108.
Romrell, L. J., Bellvé, A. R., and Fawcett, D. W. (1976). Separation of mouse spermatogenic cells by sedimentation velocity. A morphological characterization. *Dev. Biol.* **49**, 119–131.
Schin, K. S. (1965). Core-strukturen in den meiotischen und post-meiotischen kernen der spermatogenese von *Gryllus domesticus*. *Chromosoma* **16**, 436–452.
Shepherd, R. W., Millette, C. F., and DeWolf, W. C. (1981). Enrichment of primary pachytene spermatocytes from the human testis. *Gamete Res.* **4**, 487–498.
Solari, A. J. (1969). Changes in the sex chromosomes during meiotic prophase in mouse spermatocytes. *Genetics* **61**, Suppl., 113–120.
Solari, A. J. (1970). The spatial relationship of the X and Y chromosomes during meiotic prophase in mouse spermatocytes. *Chromosoma* **29**, 217–236.
Solari, A. J. (1974a). The behaviour of the XY pair in mammals. *Int. Rev. Cytol.* **38**, 273–317.
Solari, A. J. (1974b). The relationship between chromosomes and axes in the chiasmatic XY pair of the Armenian hamster (*Cricetulus migratorius*). *Chromosoma* **48**, 89–106.
Solari, A. J. (1977). Ultrastructure of the synaptic autosomes and the ZW bivalent in chicken oocytes. *Chromosoma* **64**, 155–165.
Solari, A. J., and Counce, S. J. (1977). Synaptonemal complex karyotyping in *Melanoplus differentialis*. *J. Cell Sci.* **26**, 229–250.
Solari, A. J., and Tres, L. (1967). The localization of nucleic acids and the argentaffin substance in the sex vesicle of mouse spermatocytes. *Exp. Cell Res.* **47**, 86–96.
Solari, A. J., and Vilar, O. (1978). Multiple complexes in human spermatocytes. *Chromosoma* **66**, 331–340.
Spyropoulos, B., and Moens, P. B. (1986). *Can. J. Genet. Cytol.* (in press).
Sternberger, L., A., and Sternberger, N. H. (1983). Monoclonal antibodies distinguish phosphorylated and nonphosphorylated forms of neurofilaments *in situ*. *Proc. Natl. Acad. Sci. U.S.A.* **80**, 6126–6130.
Stick, R., and Schwarz, H. (1982). The disappearance of the nuclear lamina during spermatogenesis: An electron microscopic and immunofluorescence study. *Cell Differ.* **11**, 235–243.
Stick, R., and Schwarz, H. (1983). Disappearance and reformation of the nuclear lamina structure during specific stages of meiosis in oocytes. *Cell* **33**, 949–958.
Stockert, J. C., Gimenez-Martin, G., and Sogo, J. M. (1970). Nucleolus and synaptonemal complexes in pachytene meiocytes of *Allium cepa*. *Cytobiologie* **2**, 235–250.
Takanari, H., Pathak, S., and Hsu, T. C. (1982). Dense bodies in silver-stained spermatocytes of the Chinese hamster: Behavior and cytochemical nature. *Chromosoma* **86**, 359–373.
Utakoji, T. (1966). Chronology of nucleic acid synthesis in meiosis of the male Chinese hamster. *Exp. Cell Res.* **42**, 585–596.
van Vorstenbosch, C. J. A. H. V., Colenbrander, B., and Wensing, C. J. G. (1984). Cytoplasmic filaments in fetal and neonatal pig testis. *Eur. J. Cell Biol.* **34**, 292–299.

von Wettstein, D. (1971). The synaptinemal complex and four-strand crossing over. *Proc. Natl. Acad. Sci., U.S.A.* **68,** 851–855.
von Wettstein, D. (1977). The assembly of the synaptinemal complex. *Philos. Trans. R. Soc. London, Ser. B* **277,** 235–243.
von Wettstein, D., Rasmussen, S. W., and Holm, P. B. (1984). The synaptonemal complex in genetic segregation. *Annu. Rev. Genet.* **18,** 331–413.
Wahrmann, J. A. (1981). Synaptonemal complexes—Origin and fate. *Chromosomes Today* **7,** 105–113.
Walmsley, M., and Moses, M. J. (1981). Isolation of synaptonemal complexes from hamster spermatocytes. *Exp. Cell Res.* **133,** 405–411.
Welsch, B. (1973). Synaptonemal Complex und Chromosomenstruktur in der achiasmatischen Spermatogenese von *Panorpa communis* (Mecoptera). *Chromosoma* **43,** 19–74.
Westergaard, M., and von Wettstein, D. (1972). The synaptinemal complex. *Annu. Rev. Genet.* **6,** 71–110.
Wettstein, R., and Sotelo, J. R. (1971). The molecular architecture of synaptonemal complexes. *Adv. Cell Mol. Biol.* **1,** 109–152.
Wilson, E. B. (1925). "The Cell in Development and Heredity," 3rd ed. Macmillan, New York.
Wolstenholme, D. R., and Meyer, G. F. (1966). Some facts concerning the nature and formation of axial core structures in spermatids of *Gryllus domesticus*. *Chromosoma* **18,** 272–286.
Zickler, D. (1973). Fine structure of chromosome pairing in ten *Ascomycetes*. Meiotic and premeiotic (mitotic) synaptonemal complexes. *Chromosoma* **40,** 401–416.
Zickler, D., and Olson, L. (1975). The synaptonemal complex and the spindle plaque during meiosis in yeast. *Chromosoma* **50,** 1–23.
Zickler, D., and Sage, J. (1981). Synaptonemal complexes with modified lateral elements in *Sordaria humana:* Development of and relationship to the "recombination nodules." *Chromosoma* **84,** 305–318.

9

The Rabl Orientation: A Prelude to Synapsis

CATHARINE P. FUSSELL

Department of Biology
The Pennsylvania State University
Abington, Pennsylvania 19001

I. INTRODUCTION

Synapsis is the key meiotic event because it initiates chromosome reduction to the haploid number. Synaptic failure of even one pair of homologues usually leads to inviable animal gametes and plant spores. An unsolved problem of meiosis is how homologues find and recognize their partner. It is accepted that leptotene chromosomes are not paired, and by pachytene they are (Rhoades, 1961). Chromosome observations during leptotene and zygotene in most cases virtually defy cytological analysis because of the fine skein of chromatin. Other than the fact that homologues do synapse, even in algae and fungi with zygotic meiosis (meiosis immediately following the union of gametes) the mechanism for bringing homologues together has not been solved experimentally. Suggestions abound (for references to this literature, see Wilson, 1925; Rhoades, 1961; Stern and Hotta, 1969; John, 1976; Dover and Riley, 1977; Church, 1981), however, there is little solid evidence to support one suggestion over the other.

In the course of proposing a model of chromosome arrangements in somatic interphase–prophase cells I pointed out that the model has implications for meiosis (Fussell, 1984). This chapter explores the connection between the proposed model of somatic chromosome arrangement and early meiotic stages. The meiotic model is based on the Rabl Orientation.

II. THE RABL ORIENTATION IN MITOTIC CELLS

A. Historical

The Rabl Orientation derives its name from Carl Rabl's observations (1885) that chromosomes retain their telophase grouping from the end of mitosis until the following late prophase. In favorable material, where interphase chromosomes do not uncoil to the usual extent, Rabl observed a telophase configuration (Fig. 1). At the following prophase chromosomes reappeared in the same position and number as the previous telophase (Wilson, 1925). Rabl found that the centromeres of prophase cells cluster in a small region of the nucleus (Fig. 2a,b) while telomeres are generally associated with the nuclear envelope at the opposite side. In both interphase and prophase Rabl found chromosomes oriented with a "pole" toward which the centromeres converge and an "antipole" opposite (Rabl, 1885; Wilson, 1925).

Rabl's studies were developed further by Boveri. During cell division in blastomere nuclei of *Ascaris megalocephala* Boveri observed that during telophase the nuclear membrane forms lobes around the ends of the long chromosomes. These lobes persist through interphase. In the following prophase telomeres always reappeared in these lobes and maintained this position until the breakdown of the nuclear envelope. Although the number and position of nuclear lobes varied among cells, the arrangement was always the same in sister cells and the lobes were mirror images of each other (Figs. 3 and 6). Boveri showed that these various interphase–telophase groupings derived from different chromosome arrangements at metaphase and were maintained until the next division. Boveri's explanation for these observations was that telophase chromosome position remains the same until the following prophase (Wilson, 1925).

Although Rabl's and Boveri's work clearly shows that interphase and prophase chromosomes have a telophase configuration, or Rabl Orientation, the concept was not incorporated into cytological canon. This is not surprising, considering the relatively narrow base of the evidence and the technical difficulty of demonstrating interphase chromosome arrangements.

B. Evidence

1. Interphase

Following Rabl and Boveri's work, a number of cytological reports confirmed a Rabl Orientation. Among these are Sutton (1902), Metz (1916), Heitz (1932), Vanderlyn (1948), and Carlson (1956), the latter on living grasshopper neuroblasts. The patterns of chromosome breaks and reunion following ionizing

9. The Rabl Orientation: A Prelude to Synapsis 277

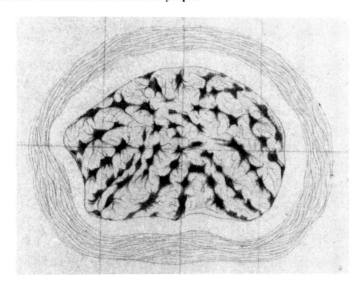

Fig. 1. Interphase nucleus with chromosomes still visible. (From Rabl, 1885.)

radiations (Sax, 1940; Evans, 1961; Kumar and Natarajan, 1966) are interpreted as reflecting a Rabl Orientation in interphase cells. C-banding, which selectively stains centromeres and telomeres in a number of plant and animal species, demonstrates that telomeres are distributed in interphase cells in a Rabl arrangement (Stack and Clarke, 1974; Fussell, 1977; Ghosh and Roy, 1977). The electron micrographs of Church and Moens (1976) show interphase telomeres attached to the nuclear membrane at one side of the nucleus. Many of the reports of telomere distribution in interphase cells have been on rapidly dividing meristematic or embryonic cells. Comparison of telomere distribution in eight differentiated tissues of *Allium cepa* shows that telomeres maintain a typical Rabl arrangement in differentiated as well as meristematic cells (Fussell, 1983).

A majority of reports observe that centromeres tend to occupy a small area of the nucleus. This has been demonstrated by fluorescence staining (Ellison and Barr, 1972; Ellison and Howard, 1981), by C-banding (Ghosh and Roy, 1977; Tanaka and Tanaka, 1977; Korf *et al.*, 1982; Evans and Filion, 1982), and by serial electron micrographs (Church and Moens, 1976; Lafontaine and Luck, 1980; Bennett *et al.*, 1981). Exceptions to a Rabl centromere arrangement are the C-band reports of Hsu *et al.* (1971) on mouse, Schmid and Krone (1976) on several species of *Urodeles,* and immunofluorescence antibody staining on cultured mammalian cells (Moroi *et al.*, 1981; Brenner *et al.*, 1981). In the C-band reports the patterns of centromere distribution varied among tissues, but was constant for any one. Both light and electron micrographs of antibody

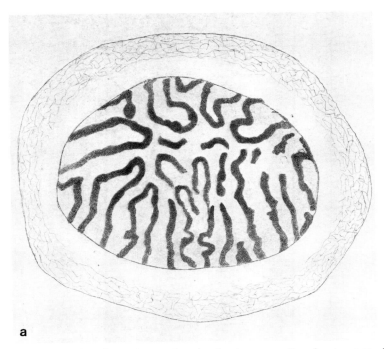

Fig. 2. (a) Polar view of a prophase nucleus showing the congregation of centromeres within a limited area. Many telomeres are associated with the nuclear membrane. (b) Side view of a prophase

techniques show that interphase centromeres are randomly distributed (Moroi et al., 1981), except in late telophase and early G_1 (Brenner et al., 1981). Thus the evidence for centromeres having a Rabl arrangement in interphase cells is mixed. Perhaps the apparent contradictions arise from a spreading out of centromeres in response to, or initiating, changes in cell functions. Some electron microscope studies indicate that centromeres are not attached to the nuclear membrane (Lafontaine and Luck, 1980; Brenner et al., 1981; Jensen, 1982), while others find that they are (Moroi et al., 1981).

A Rabl arrangement of telomeres and centromeres in the same cell is shown by autoradiographs of *Allium cepa* root tip cells (Fussell, 1975). In these experiments roots were pulse-labeled with high specific activity tritiated thymidine, then fixed at various times. Autoradiographs of chromosomes at the very end of the S phase are labeled mainly in their telomere and centromere regions (Fig. 4). Late labeled G_2 cells have a characteristic pattern of a cluster of label at one side of the nucleus with 12 to 16 small regions of label at the opposite side (Fig. 5). Prophase of the late labeled population of cells has a similar distribution of label, but it can be seen that the large cluster is the centromeres and the smaller clusters

9. The Rabl Orientation: A Prelude to Synapsis 279

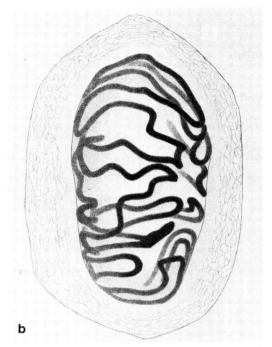

b

nucleus. Most centromeres are within a restricted area at one side of the nucleus. Telomeres are at the opposite side (From Rabl, 1885.)

the telomeres (Fig. 6). *Allium cepa* has 32 telomeres. Inasmuch as all telomeres late label, and there are about one-half the number of groups of labeled telomeres in both interphase (Fig. 5) and prophase (Fig. 6) as telomeres, it appears telomeres are paired or associate to some degree. C-banding also shows that interphase telomeres associate to some extent (Stack and Clarke, 1973; Godin and Stack, 1975; Fussell, 1977; Ghosh and Roy, 1977; Ashley, 1979). The pulse-labeled *A. cepa* autoradiographs show clearly that interphase chromosomes in G_2 and throughout the following interphase maintain a Rabl Orientation (Fussell, 1975).

2. Prophase

The Rabl Orientation is both an interphase and prophase phenomenon (Figs. 2 and 6). Renewed interest in chromosome arrangements in prophase cells was aroused by Kitani's (1963) study of prophase configurations in eight species of plants. Kitani (1963) found that prophase chromosomes in all species have their centromeres at one pole of the nucleus and telomeres at the opposite side. This

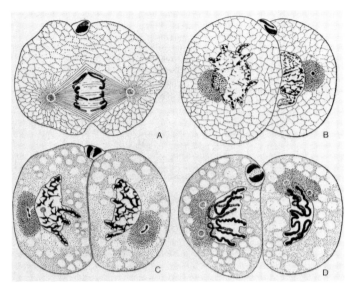

Fig. 3. Boveri's drawings of *Ascaris* cleavage. (A) Anaphase of first cleavage. (B) Lobed nuclei at the two-cell stage. Lobes have formed around the ends of the arms of the long chromosomes. (C) Reappearance of chromosome ends in the lobes during the following prophase. (D) Late prophase showing ends of long chromosomes in the nuclear lobes. (From "The Cell in Development and Heredity," Third Edition, by Edmund B. Wilson. Copyright 1925 by Macmillan Publishing Company.)

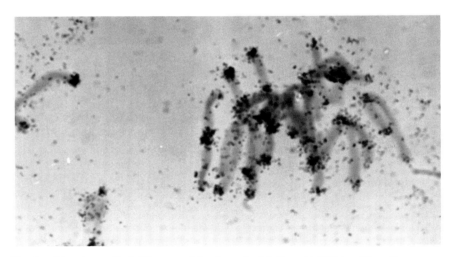

Fig. 4. Metaphase cell of *Allium cepa* 5 hr after pulse-labeling with [^{3}H]thymidine. Centromeres and telomeres of chromosomes are preferentially labeled. (From Fussell, 1975.)

9. The Rabl Orientation: A Prelude to Synapsis

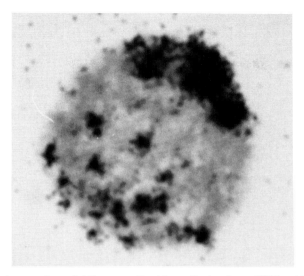

Fig. 5. Interphase nucleus of *Allium cepa* fixed immediately after a [^{3}H]thymidine pulse. The larger mass of label is the centromeres. The smaller groups of label at the opposite side are the telomeres. (From Fussell, 1975.)

orientation is maintained until chromosomes start to align on the metaphase plate (Kitani, 1963). Tanaka (1981) reports that prophase nuclei of nonsquashed root tip cells of *Haplopappus gracilis* (2n = 4) and *Crepis capillaris* (2n = 6) have their centromere regions localized in a small area close to the nuclear envelope. Telomeres are at the opposite side. Observations of Wagenaar (1969) show not only that prophase chromosomes of eight species of plants have a Rabl Orientation, but that their telomeres are paired. Similar observations on prophase chromosomes of *Ornithogalum virens* (2n = 6) are reported by Ashley (1979). Telomere associations are not unique to prophase, but are an extension of associations found in interphase by autoradiography (Fussell, 1975) and C-banding (Stack and Clarke, 1973; Godin and Stack, 1975; Fussell, 1977; Ghosh and Roy, 1977; Ashley, 1979).

The Rabl Orientation persists through prophase until the nuclear membrane breaks down at prometaphase. Time-lapse cinemaphotography shows that as prophase chromosomes coil and become thicker they do not change positions (Bajer and Molè-Bajer, 1956). At prometaphase, when the nuclear membrane breaks down and spindle fibers attach to centromeres (Harris, 1965; Molè-Bajer, 1958), chromosomes begin to move. Their movements are irregular and complex, but the net effect is that centromeres move from a clustered region at one side of the nucleus to the metaphase plate (Bajer and Molè-Bajer, 1956; McIntosh *et al.*, 1975; Roos, 1976). Chromosomes move independently of each other

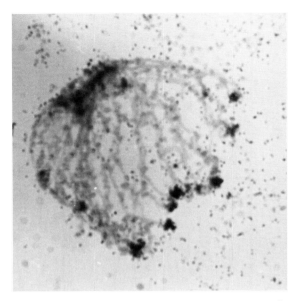

Fig. 6. Prophase nucleus of *Allium cepa* 5 hr after a pulse treatment with [^{3}H]thymidine. This confirms that the chromosomes have a Rabl Orientation in both interphase and prophase. (From Fussell, 1975.)

during prometaphase (Luykx, 1970) and are individually attached to spindle fibers (Begg and Ellis, 1979). During prometaphase movements chromosome order could be altered.

3. Polytene Chromosomes

The arrangements and positions of polytene chromosomes of *Drosophila melanogaster* salivary gland chromosomes are being studied by Sedat's group (Agard and Sedat, 1983; Mathog *et al.*, 1984). Cells are stained with a DNA-specific fluorescent dye. "Optical sections" taken through each nucleus are computerized and positions of the chromosomes are traced from the stack of images for each nucleus. The banding patterns unequivocally identify each chromosome. The results from 14 analyzed nuclei show that polytene chromosomes of *D. melanogaster* have a Rabl Orientation. The centromeres, long known to be fused in the chromocenter, are connected to the nuclear envelope. Centromere orientation is not random, but is located toward the exterior surface of the gland. A similar, nonrandom orientation of the chromocenter toward the topological outside of early embryos of *Drosophila virilis* previously was reported by Ellison and Howard (1981). Sedat's group (Mathog *et al.*, 1984) finds that telomeres are on the opposite side of the nucleus from the chromocenter. Each chromosome arm was folded or coiled into a rough right-hand helix and occupied its own topological space. Arms were not coiled or twisted about one another, but were associated with the nuclear membrane at many points (Mathog *et al.*, 1984).

From optical density values of Feulgen-stained HeLa S-3 cells at the G_1–S border, Belmont *et al.* (1984) also find DNA distributed in a narrow layer next to the nuclear envelope as well as bordering the nucleolus. For a review of the concept that interphase chromatin is distributed in some specific way and each chromosome occupies its own domain see Hancock and Boulikas (1982).

If further analysis of polytene chromosome position continues to support the Sedat group findings, and polytene chromosomes are comparable to those of somatic diploid cells, then this work (Agard and Sedat, 1983; Mathog *et al.*, 1984) adds support to the proposition that a Rabl Orientation is the standard arrangement of chromosomes in interphase and prophase cells.

In summary, observations over the last 100 years, from Rabl's initial proposal in 1885 to the present, overwhelmingly confirm that interphase and prophase chromosomes maintain their telophase configuration, i.e., a Rabl Orientation. The Rabl Orientation can be considered the standard arrangement of chromosomes during mitotic interphase and prophase just as the highly visible chromosomes of metaphase and prophase are of their stages.

C. Models

My model of mitotic interphase–prophase chromosome arrangement is based on a Rabl Orientation (Fussell, 1984). In light of experimental evidence since Rabl's (1885) proposal, refinements are suggested. These are that telomeres tend to associate, more or less in pairs, becoming attached to the nuclear envelope when it reforms following cell division. Telomere associations and distribution on the nuclear envelope are considered to be a function of arm lengths. Consequently, chromosome arms of similar lengths will tend to associate. However, due to the independent movement of chromosomes during prometaphase (Luykx, 1970), and perhaps into anaphase, chromosome arrangements may change at each mitosis as Boveri observed (Wilson, 1925). If telomere associations are a function of arm length, then pairing patterns between chromosome arms will be restricted, e.g., very long and very short arms would be unlikely to pair, while short-arms would. In essence the model imposes on a Rabl arrangement a variable chromosome order. The order may change at each mitosis during prometaphase, but telomere associations would be restricted to arms of similar lengths— homologous or nonhomologous (Fussell, 1984).

Other models of interphase chromosome arrangements are proposed by Ashley and Pocock (1981) and Bennett (1982, 1983). Chromosome arrangements in interphase nuclei are reviewed by Avivi and Feldman (1980) and Comings (1980). Both the Ashley and Pocock (1981) and Bennett (1982, 1983) models suggest chromosomes maintain a specific order from cell generation to cell generation within haploid genomes. A difficulty with both models is the lack of known mechanisms for maintaining strict chromosome order during prometaphase, and perhaps through anaphase, especially in light of the fact that chromosomes move indepedently of each other during prometaphase (Bajer and Molè-Bejar, 1956; McIntosh et al., 1975). For a more comprehensive critique see Fussell (1984).

III. MEIOSIS AND THE RABL ORIENTATION

A. Rabl Orientation and Synapsis: Overview

Accepting the Rabl Orientation as the normal, standard chromosome configuration in mitotic interphase–prophase cells has a number of ramifications (Fussell, 1984). One of these concerns meiotic synapsis. A Rabl arrangement orients chromosomes in such a way that they are halfway home to synapsis as they enter meiosis. Rabl-oriented chromosomes in premeiotic cells at a minmum would be aligned in the same way: centromeres at one side of the nucleus and telomeres fanned out on the opposite side. Such a centromere–telomere axis would position homologous arm segments opposite one another within the nucleus. If telomere distribution on the nuclear envelope is a function of arm length, then telomeres of similar arm lengths would lie in the same circumferal band. An orderly, polarized arrangement of mitotic Rabl-oriented chromosomes would set the stage for synapsis and minimize chromosome movements during homologous pairing. This implies that during the interphase between the last mitosis and the beginning of meiosis chromosomes undergo little or no rearrangement, e.g., a Rabl Orientation is maintained.

B. Premeiotic Mitosis

What evidence is there that Rabl-oriented chromosomes in mitotic cells carry over into meiosis and become the starting point for eventual synapsis? Other than the fact that mitotic divisions just prior to meiosis last longer (Riley and Flavell, 1977), by most accounts chromosomes progress through these mitotic stages normally. In some cases chromosomes may be more uncoiled and metaphase chromosomes somewhat longer (Rhoades, 1961). The final premeiotic telophase deposits chromosomes in a Rabl configuration (Wilson, 1925). As the telophase chromosomes uncoil and enter interphase they presumably retain a Rabl configuration as they do in mitotic cells.

Although overall premeiotic mitosis progresses normally through the standard stages, there have been a number of reports that homologues "prealign" during this period, that is, homologues not previously paired become associated (see Maguire, 1984, for references). Others suggest that organisms, in addition to Diptera, normally have homologues associated (see Avivi and Feldman, 1980). Cytological evidence for one or the other of these views is given by Sutton (1902), Smith (1942), Brown and Stack (1968); and Maguire (1983); see Church (1981) for other references. For accounts of associations between homologous chromosome regions, such as heterochromatic blocks (Maguire, 1972), see Dover and Riley (1977). However, there are also a number of cytological studies, often on the same material, which find no evidence of presynaptic chromosome pairing (Walters, 1970; Palmer, 1971; Darvey and Driscoll, 1972; John, 1976; for other references, see Dover and Riley, 1977; Church, 1981). A strong argument against premeiotic pairing has been advanced by John (1976).

Clearly, light microscope evidence for premeiotic alignment is contradictory. Several explanations arise. Cytological evidence for homologous associations are often narrowly based on only a few chromosomes in a limited number of cells (Maguire, 1983). Another possibility is that Rabl-oriented chromosome arm associations, homologous or nonhomologous, may be mistaken for alignment of homologues. Finally, the degree of normal somatic pairing will influence the results. As Wilson (1925) points out, metaphase plates show that in some species all chromosomes show somatic pairing, e.g., Diptera, in others there is a tendency toward homologous pairing, and in other cases there is none. In species with a bias toward somatic homologue pairing this could be misinterpreted in premeiotic mitosis as "prealignment." Church (1981) reports that within the Liliaceae there are species with specific differences in centromere and nucleolar associations. As determined by electron microscopy, *Ornithogalum virens* (2n = 6) centromeres in premeiotic interphase tend to pair with their homologue, while *Lilium* centromeres are less likely to associate. Thus chromosome associations termed "prealigned" may be normally arranged Rabl Orientations, or some degree of somatic association, occurring in all cells, not just premeiotic interphase.

The ambiguous light microscope data on chromosome "prealignment" are offset by the clear results of serial-sectioned electron micrographs. In the interphase interval between the last mitosis and meiotic prophase in *Allium fistulosum,* Church and Moens (1976) find centromeres clustered in one area close to the nuclear membrane. In some cells centromeres were single and in others they were associated with other centromeres judged to be nonhomologous. At this stage, centromeres are not attached to the nuclear envelope. At the opposite side of the nucleus Church and Moens (1976) find heterochromatic masses considered to be telomeres. The number of telomere groups was less than the number of telomeres, indicating associations which Church and Moens (1976) consider to be nonspecific.

In the male grasshopper *Brachystola magna,* Church (1977) finds that the heterochromatic chromocenters of all autosomes are attached to the nuclear envelope. They are nonrandomly restricted to one-half the nuclear envelope. Since the attachment sites of the nucleolar organizing autosomes were well separated, Church (1977) concludes that homologous chromosome ends are not adjacent on the nuclear envelope during premeiotic interphase and thus not paired.

In a similar study, centromere positions in *Lilium speciosum* microsporocytes in mid- and late premeiotic interphase were studied by reconstruction of serial-sectioned electron micrographs (Del Fosse and Church, 1981). Centromeres, clustered in a small region of the nucleus, showed little or no evidence of movement during premeiotic interphase. On the basis of the number of centromeres observed, Del Fosse and Church (1981) conclude that homologous centromeres are not associated. No evidence for direct attachment of centromeres to the nuclear envelope was observed (Del Fosse and Church, 1981).

Bennett *et al.* (1979a) have also looked at the distribution of centromeres in premeiotic interphase nuclei by means of reconstructions of serial-sectioned electron micrographs. In electron micrographs, centromeres can be identified by their distinctive morphology. Bennett *et al.* (1979a) looked at five cereals: wheat, rye, triticale, barley, and oats. A total of 19 different ploidies, varieties, and species were analyzed. The data in all 19 cases show that in premeiotic interphase centromeres are clustered in a small volume toward one pole of the nucleus. Triticale "Rosner" has 42 chromosomes. Groups of 17–18 centromeres were observed, while rye ($n = 7$) had two to three centromeric bodies. Bennett *et al.* (1979a) suggest that this represents homologous pairing. However, in both cases the number of centromere clumps is less than half the diploid chromosome number, suggesting that at least some nonhomologous pairing is involved.

A C-band light microscopic study of rye telomere position in premeiotic interphase shows that telomeres are distributed in one-half the nucleus, probably on the side opposite the centromeres (Thomas and Kaltsikes, 1976).

In summary, electron micrograph studies and the C-band report show that in premeiotic interphase cells, centromeres and telomeres have a Rabl Orientation. The data do not resolve whether homologous telomeres or centromeres are paired at this stage. Church (1981) and associates suggest that homologues do not regularly pair, while Bennett *et al.* (1979a) suggest that centromeres pair. In any event, the data, although not extensive, strongly support the concept that chromosomes retain a Rabl Orientation during the mitotic–meiotic interphase. If this is the case, then leptotene chromosomes will have a Rabl Orientation.

C. Early Meiotic Prophase

Leptotene is the first meiotic stage with cytologically visible chromosomes. As the unpaired chromosomes begin to condense and shorten, the nuclear volume increases. Centromeric chromosome regions are close together, and chromosomes have a Rabl Orientation, reflecting a carryover of the final mitotic telophase configuration (Rhoades, 1961). Classical descriptions of leptotene indicate that chromosomes in many animals develop a bouquet stage: telomeres are attached to the side of the nuclear envelope nearest the centrosome and the arms loop out toward the other side (Rhoades, 1961). A somewhat similar configuration, synizesis, frequently is found in plants, although the chromosomes usually clump into a dense mat of threads. Synizesis is considered the counterpart of the bouquet stage by some (Kaufmann, 1925) and a fixation artifact by others (see Rhoades, 1961). During leptotene, telomeres move toward each other while attached to the nuclear membranes. Chromosome arms of the clustered telomeres continue to loop out in a bouquet arrangement. As Moses (1968) points out, telomere movement is restricted to one plane. Homologous pairing marks the beginning of zygotene. At this point, telomeres have moved into a small area of the nuclear envelope (Rhoades, 1961).

More recent evidence supports this sequence of events. For example, all seven rye chromosomes have large blocks of heterochromatin on most telomeres which C-band (Thomas and Kaltsikes, 1976). C-banded rye premeiotic interphase cell telomeres have a Rabl arrangement. As cells move into early meiotic prophase, telomeres aggregate into a single, large heterochromatic mass next to the nuclear envelope. The heterochromatic mass lasts well into zygotene and begins to break up before the beginning of pachytene (Thomas and Kaltsikes, 1976). Both telomeres and centromeres of rye can be identified in electron micrographs. Bennett *et al.* (1979a) in serial-sectioned reconstructed rye nuclei find the same series of events as Thomas and Kaltsikes (1976): in premeiotic interphase, telomeres are widely separated at one side of the nucleus, and during leptotene telomeres move along the nuclear envelope and aggregate in a bouquet arrangement. Clustered centromeres are at the opposite side of the nucleus (Bennett *et al.*, 1979a).

The most satisfactory evidence for events during leptotene are three-dimensional reconstructions of serial sectional nuclei. Rasmussen and Holm (1980) reviewed their results of 30 reconstructed nuclei. The species studied are the basidiomycete *Schizophyllum commune, Lilium longiflorum, Drosophila melanogaster, Bombyx mori,* and *Homo sapiens.* Although there are some differences of detail among the species, the overall sequence of events is the same. In leptotene, telomeres are attached to the nuclear envelope more or less in a polar fashion. A bouquet arrangement is established by the beginning of zygotene after the telomeres have congregated, presumably by moving along the inner nuclear membrane. In cases where individual homologues could be identified no evidence for presynaptic alignments, except in *D. melanogaster,* was found. The chromosome regions where pairing began varied. In *Bombyx* and *H. sapiens* pairing usually started at the telomeres, while in *Lilium* a number of interstitial initiations occurred, many in regions well removed from the telomeres. Homologous pairing and synaptonemal complex formation are quite asynchronous (Rasmussen and Holm, 1980).

Another report of interest is that of Zickler's (1977) on the ascomycete *Sordaria macrospora* (n = 7), which undergoes zygotic meiosis. Each chromosome has distinctive characteristics. Consequently, the position of each leptotene chromosome could be determined from reconstructed serial-sectional electron micrographs. Seven leptotene nuclei were completely reconstructed, from just after karyogamy to just before zygotene. In early leptotene there is no sign of homologue alignment. Some telomeres initially were not attached to the nuclear envelope, but well before mid-leptotene were. In the earliest leptotene reconstruction that Zickler (1977) illustrates, I estimate that 19 of the 28 telomeres are in one-half of the nucleus (Fig. 7). Two telomeres are anchored to the nucleolus which lies in the other half of the nucleus. Of the nine telomeres in the nucleolar half, all had the other telomere in the other half (Fig. 7). Thus during early to mid-leptotene telomeres are *not* randomly distributed. Zickler (1977) could not identify centromeres. As leptotene progressed homologues tended to become parallel to each other. By late leptotene all homologues had two telomeres closely paired on the nuclear envelope (Fig. 8). At the beginning of zygotene, all homologues were roughly paired. The synaptonemal complex starts to form close to the nuclear envelope, although some pairing was initiated interstitially (Zickler, 1977).

A similar pattern of telomere distribution and "movement" has been described by others from three-dimensional reconstructions of electron micrographs. In these cases it was not possible to identify all individual chromosomes during early prophase. Moens (1973) finds that polarized, unpaired telomeres of locust spermatocytes are attached to the nuclear membranes at leptotene. While still attached, the ends move, bringing homologous chromosomes closer together in a classic bouquet arrangement. Rasmussen (1976) reports essentially the same sequence of events during leptotene and zygotene in female *Bombyx mori.* Gillies (1975) in a review states that maize chromosomes have a bouquet arrangement in leptotene and zygotene.

The surface-spreading technique for electron microscopy also provides infor-

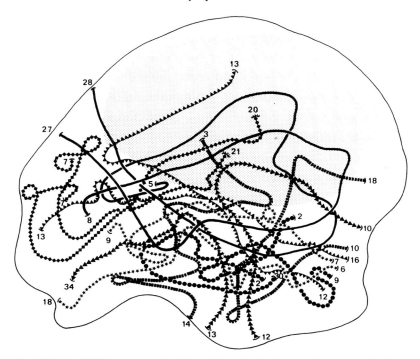

Fig. 7. Zickler's (1977) reconstruction of a serial-sectional electron microscope early leptotene nucleus of *Sordaria macrospora*. The gray area is the nucleolus. The numbers are the section numbers. Homologues are indicated by different line symbols, e.g., solid line, stars, circles, etc. Note that the majority of the telomeres lie in the lower half of the nucleus and that all chromosomes have at lease one telomere in this region of the nucleus. (From Zickler 1977, Fig. 7.)

mation on meiotic chromosome arrangements, although *in situ* chromosome arrangements are disrupted and reduced to two dimensions. In early zygotene of Chinese hamster, prior to synapsis, Moses (1977) finds chromosome ends widely separated but attached to pieces of the nuclear envelope. This result points up the strong attachment of telomeres to the nuclear envelope (Moses, 1977), and also the disruptive nature of the technique on chromosome arrangements. Solari and Counce (1977) used a more gentle spreading technique on grasshopper spermatocytes. In early zygotene they found that telomeres are usually clustered in one region. The difference in their results from Moses (1977) is probably due to Solari and Counce's (1977) improved technique, which greatly reduces the extent of nuclear spreading.

The autosomes of chicken oocytes also have the familiar pattern of telomere attachment to the nuclear envelope, the telomeres moving during leptotene to form a bouquet arrangement (Solari, 1977).

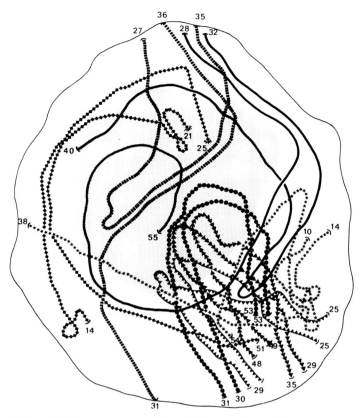

Fig. 8. Zickler's (1977) reconstruction of a late leptotene nucleus of *Sordaria macrospora*. Note the bouquetlike configuration of the smaller chromosomes and that at least two telomeres of each homologue are quite close together. (From Zickler 1977, Fig. 9.)

Stack and Soulliere's (1985) light and electron microscope-sectioned and surface-spread study of synapsis in *Rhoeo spatheceae* tells the same story. At leptotene one to three chromocenters are at one side of the nucleus and telomeres at the other. During late leptotene/very early zygotene, the chromocenters, still at one side of the nucleus, fuse into one or two masses. The telomeres at the opposite side move closer together and roughly align in pairs. Synapsis appears to start close to the ends of the chromosomes, which are attached to the nuclear membrane and polarized in a bouquet arrangement (Stack and Soulliere, 1984).

Information from the studies just cited show that synapsis very frequently starts at chromosome ends as previously observed (Wilson, 1925; McClintock, 1933; Rhoads, 1961) and continues in a zipperlike fashion. However, it is also

9. The Rabl Orientation: A Prelude to Synapsis

clear that synapsis may start interstitially (Moses, 1968; Moens, 1973; Zickler, 1977) and that the pattern of initiation of pairing is species-specific and varies (Rasmussen and Hold, 1980).

Both classical cytological studies (Rhoades, 1961) and electron micrograph observations show that chromosome behavior during leptotene and zygotene is essentially standard in higher plants and animals. Leptotene chromosomes in the earliest stage have a Rabl Orientation (see above for evidence). It seems highly likely that the leptotene Rabl arrangement is set up and carried over from the last premeiotic mitosis. If telomeres of interphase and prophase chromosomes are distributed on the nuclear envelope according to the length of their arms, then early leptotene telomeres presumably would be distributed in the same way. Telomeres on short chromosome arms would be close to their centromere, and telocentric chromosomes would have one telomere very close to their centromere position. Consequently, depending on arm lengths, leptotene telomeres could be distributed from the centromeric pole to the opposite side. The distribution of telomeres in species with diverse arm lengths would tend to mask a Rabl Orientation.

During leptotene telomeres move along the nuclear membrane to form a bouquet arrangement. Telomeres distributed according to arm length may facilitate bringing homologous telomeres close together as they move into a bouquet arrangement. This would hold particularly if the rate of telomere movement is constant for all telomeres. Congregation of telomeres in a limited area of the nuclear membrane, especially if sorted by arm lengths, would expedite homologue recognition. As synapsis proceeds, chromosomes tend to lose their bouquet arrangement, and by pachytene are often dispersed (for reviews of these events, see Wilson, 1925; Rhoades, 1961; Moses, 1968; Westergaard and von Wettstein, 1972; Church, 1981).

D. Zygotic Meiosis

Karyogamy of haploid nuclei followed by meiosis occurs in many algae and fungi. Seemingly, this presents synaptic problems. However, mechanically the solution is straightforward. If Rabl-arranged chromosomes in both nuclei are oriented in the same way, i.e., centromeres pointing in the same direction, then karyogamy will result in both sets of chromosomes having the same orientation within a single nucleus. In fact, this is what happens during karyogamy in *Neurospora crassa* (Singleton, 1953) and *Sordaria fimicola* (Carr and Olive, 1958). In the conjugate division of these fungi the spindles of the two haploid nuclei lie almost parallel. Consequently, Rabl-oriented chromosomes in the crozier curve have their centromeres oriented in the same direction. With chromosomes retaining this orientation, nuclei fuse, placing both centromeres on the same side of the nucleus (Singleton, 1953). Following nuclear fusion in the ascomycetes, chromosomes contract, becoming almost as short as metaphase chromosomes. At the end of contraction, the two sets of chromosomes lie at opposite sides of the nucleus. Chromosomes synapse while contracted, then elongate into a standard pachytene configuration (McClintock, 1945; Singleton, 1953).

In algae and fungi with zygotic meiosis various cellular mechanisms no doubt are involved in orienting the two sets of Rabl-oriented chromosomes in the same direction prior to karyogamy. In the ascomycete-type both haploid nuclei, lying in a common cytoplasm, divide side by side depositing nuclei with similarly oriented Rabl-arranged chromosomes close together. Karyogamy follows. In cases of *cell* fusions, chromosomes in each nucleus could be polarized in the same direction prior to cell fusion, or after cell fusion, but before karyogamy. In the latter case, similar Rabl orientation could be achieved by nuclear rotation, a common phenomenon. Parallel alignment of the Rabl-oriented chromosomes in both nuclei prior to karyogamy would bring the chromosome sets together having the same orientation: centromeres at one side and telomeres toward the other. (The same process probably takes place during fertilization in higher plants and animals, thereby establishing a Rabl Orientation in the fertilized egg.)

The Rabl Orientation most likely contributes to orderly synapsis in organisms with zygotic meiosis as it does during meiosis of diploid organisms.

E. Colchicine and Temperature Effects on Synapsis

Colchicine or other spindle-inhibiting treatments given just before meiosis produce asynapsis, univalents, and reduced chiasmata (Levan, 1939; Driscoll *et al.*, 1967; Church and Wimber, 1971; Bayliis and Riley, 1972; Shepard *et al.*, 1974; Thomas and Kaltsikes, 1977; Dover and Riley, 1977; Bennett *et al.*, 1979b; Puertas *et al.*, 1984). The most studied species are *Lilium* and wheat. In both it is possible to determine premeiotic and early meiotic stages fairly precisely. Dover and Riley (1977) summarize the data for wheat. Pollen mother cells of *Triticum aestivum* are sensitive to colchicine from the end of the last premeiotic mitosis through approximately the first third of meiotic G_1, which is well before the start of meiotic DNA synthesis. No synaptic irregularities occurred from colchicine treatment given up to 30 hr before the start of leptotene or during leptotene. The temperature-sensitive period for wheat occurs after the colchicine-sensitive period, and lasts until the beginning of meiotic DNA synthesis (Dover and Riley, 1977). In *Lilium* the colchicine-sensitive period starts in premeiotic interphase and extends into early zygotene (Shepard *et al.*, 1974) or mid to late zygotene (Bennett *et al.*, 1979b). There is no explanation why the boundaries of the colchicine-sensitive period differ between *Lilium* and wheat.

Dover and Riley (1977) suggest that colchicine affects the prealignment of homologues during the last premeiotic anaphase by interfering with the spindle. Shepard et al. (1974) working with *Lilium* suggest that colchicine interferes with synapsis in some way. Bennett et al. (1979b) do not think synapsis in *Lilium* is being disrupted by colchicine acting on spindle fibers. They suggest colchicine may be acting on some cellular process such as transport of substances. Others, working with wheat, suggest colchicine disrupts homologous pairing, which occurs throughout the life of the organism (Feldman, 1966; Feldman et al., 1966; reviewed in Sears, 1976).

Hotta and Shepard (1973) found a colchicine-binding protein in the nuclear envelope of *Lilium* which appears in meiotic prophase (reviewed in Stern and Hotta, 1973). The normal level of this protein during meiotic prophase declines in cells treated with colchicine. These authors are inclined to the view that colchicine interferes with the pairing process by disrupting pairing events associated with the nuclear envelope (Stern and Hotta, 1973).

Bennett et al. (1974) report a fibrous material in premeiotic interphase cells of *Triticum aestivum* linking chromosomes to the nuclear envelope. In follow-up experiments, Bennett and Smith (1979) report that when colchicine is given during the colchicine-sensitive period fibrillar material is absent. However, the fibrillar material is not considered to be tubulin, because colchicine treatment for up to 5 days does not destroy it when present. The conclusion Bennett and Smith (1979) reach is that colchicine may affect the transport of the fibrillar material from the tapetum or cytoplasm into the nucleus.

Experiments testing whether colchicine-sensitive material anchors telomeres to the nuclear envelope have been carried out on *Allium cepa* root tips (Fussell, 1987). Only telomeres of *A. cepa* C-band lie in one hemisphere of the nucleus (Fussell, 1977). *Allium cepa* root tips were treated with colchicine for 16 hr and either fixed or centrifuged for up to 100,000 g for 15 min, other roots were only centrifuged. The results show that colchicine alone did not affect telomere distribution. Control roots had an average of 18.60 telomeres in one hemisphere and 4.03 in the other. Colchicine-treated roots averaged 18.03 telomeres in one hemisphere, and 4.97 in the other. Also, neither centrifugation alone nor colchicine followed by centrifugation changed the basic telomere distribution. However, the number of telomeres in the major hemisphere decreased with increasing force of centrifugation. For instance, the average number of telomeres per nucleus following centrifugation at 50,000 g for 15 min was 17.48 in the major hemisphere and 4.43 in the other. Roots treated with colchicine then centrifuged under the same conditions had 17.13 telomeres in the major hemisphere and 3.40 in the minor. The decrease in the average number of telomeres in the major hemisphere after centrifugation is considered to be due to centrifugal force, not colchicine, because the differences between the number of telomeres in the major hemispheres are not statistically significant. The same holds for telomeres in the other hemisphere. From these experiments it is concluded that a colchicine-sensitive structure is not binding root tip *A. cepa* telomeres to the nuclear envelope (Fussell, 1987).

Aside from the indisputable fact that colchicine and temperature treatments during premeiotic meiosis, and in some species through much of zygotene, disrupt synapsis, there is no consensus on the mode of action. The discovery of a colchicine-binding protein in nuclear membranes (Hotta and Shepard, 1973) at least locates a component which would seem to be involved. However, the *Allium cepa* root tip experiments (Fussell, 1987) suggest that the colchicine-binding protein is not anchoring telomeres to the nuclear envelope in this species. Perhaps mitotic and meiotic nuclei differ in this respect.

IV. SUMMARY

This chapter attempts to crystallize information on chromosome positions leading up to synapsis. Experimentally this is a difficult period to study because chromosomes are in interphase. Furthermore, early leptotene chromosomes are a mass of fine threads which yield little information on homologue positions. Consequently chromosome positions during this period have been inferred largely from the final premeiotic mitosis and from the action of colchicine and temperature treatments on synapsis. Observations on chromosome associations in the last mitosis are contradictory. The modes of action of colchicine and temperature are very unclear.

This review of the data suggests that mitotic Rabl-oriented chromosomes carry over into meiosis, setting the stage for synapsis. The rough alignment of Rabl-oriented chromosomes would minimize movement required for synapsis. In brief, the evidence for such a sequence of events is as follows: (a) Interphase and prophase somatic cells have a Rabl Orientation with roughly paired telomeres attached to the nuclear envelope. Telomere distance from the centromeric pole is a function of arm length (Fussell, 1984). (b) The somatic Rabl Orientation is maintained through premeiotic interphase until leptotene (Church and Moens, 1976; Church, 1977; Thomas and Kaltsikes, 1976; Bennett *et al.*, 1979a; Del Fosse and Church, 1981). (c) Leptotene chromosomes have a Rabl Orientation (Moens, 1973; Thomas and Kaltsikes, 1976; Zickler, 1977; Bennett *et al.*, 1979; Rasmussen and Holm, 1980; Stack and Soulliere, 1984; for others, see above). (d) The movement of telomeres along the nuclear envelope during leptotene into a bouquet configuration (essentially a Rabl Orientation with telomeres congregated in a small area) has long been known (Wilson, 1925; Rhoades, 1961; Moses, 1968).

Descriptions of the transition from mitosis to meiosis frequently imply that

9. The Rabl Orientation: A Prelude to Synapsis

chromosomes are randomly dispersed and must undergo extensive movement in order to synapse. A Rabl Orientation eliminates this need. Nevertheless, homologues must move close enough for the synaptonemal complex to form. Electron micrograph serial sections suggest that only a small region somewhere along the chromosome is required to initiate synapsis. The distribution of telomeres on the nuclear envelope by arm length would roughly sort homologous telomeres into the same band of the nuclear envelope. As telomeres move along the nuclear membrane into a bouquet arrangement, their presorting may facilitate the alignment of homologous chromosome segments.

Maintenance of a Rabl Orientation during the mitotic–meiotic transition keeps chromosome movements required to effect synapsis at a minimum. However, it does not solve the problem of the mechanism for moving homologues into synaptic positions.

REFERENCES

Agard, D. A., and Sedat, J. W. (1983). Three-dimensional architecture of a polytene nucleus. *Nature (London)* **302,** 676–681.

Ashley, T. (1979). Specific end-to-end attachment of chromosomes in *Ornithogalum virens. J. Cell Sci.* **38,** 357–367.

Ashley, T., and Pocock, N. (1981). A proposed model of chromosomal organization in nuclei at fertilization. *Genetica (The Hague)* **55,** 161–169.

Avivi, L., and Feldman, M. (1980). Arrangement of chromosomes in the interphase nucleus of plants. *Hum. Genet.* **55,** 281–295.

Bajer, A., and Molè-Bajer, J. (1956). Cine-micrographic studies of mitosis in endosperm. II. Chromosome, cytoplasmic and brownian movements. *Chromosoma* **7,** 558–607.

Bayliss, M. W., and Riley, R. (1972). Evidence of premeiotic control of chromosome pairing in *Triticum aestivum. Genet. Res.* **20,** 201–212.

Begg, D. A., and Ellis, G. W. (1979). Micromanipulation studies of chromosome movement. I. Chromosome-spindle attachment and the mechanical properties of chromosomal spindle fibers. *J. Cell Biol.* **82,** 528–541.

Belmont, A., Kendall, F. M., and Nicolini, C. (1984). Three-dimensional intranuclear DNA organization *in situ:* Three states of condensation and their redistribution as a function of nuclear size near the G_1–S border in HeLa S-3 cells. *J. Cell Sci.* **65,** 123–138.

Bennett, M. D. (1982). Nucleotypic basis of the spatial ordering of chromosomes in eukaryotes and the implications of the order for genome evolution and phenotypic variation. *In* "Genome Evolution" (G. A. Dover and R. B. Flavell, eds.), pp. 239–261. Academic Press, New York.

Bennett, M. D. (1983). The spatial distribution of chromosomes. *In* "Kew Chromosome Conference II" (P. E. Brandham and M. D. Bennett, eds.), pp. 71–79. Allen & Unwin, London.

Bennett, M. D., and Smith, J. B. (1979). The effect of colchicine on fibrillar material in wheat meiocytes. *J. Cell Sci.* **38,** 33–47.

Bennett, M. D., Stern, H., and Woodward, M. (1974). Chromatin attachment to nuclear membrane of wheat pollen mother cells. *Nature (London)* **252,** 395–396.

Bennett, M. D., Smith, J. B., Simpson, S., and Wells, B. (1979A). Intranuclear fibrillar material in cereal pollen mother cells. *Chromosoma* **71,** 289–332.

Bennett, M. D., Toledo, L. A., and Stern, H. (1979B). The effect of colchicine on meiosis in *Lilium speciosum* cv. "Rosemede." *Chromosoma* **72,** 175–189.

Bennett, M. D., Smith, J. B., Ward, J., and Jenkins, G. (1981). The relationship between nuclear DNA content and centromere volume in higher plants. *J. Cell Sci.* **47,** 91–115.

Brenner, S., Pepper, D., Berns, M. W., Tan, E., and Brinkley, B. R. (1981). Kinetochore structure, duplication, and distribution in mammalian cells: Analysis by human autoantibodies from scleroderma patients. *J. Cell Biol.* **91,** 95–102.

Brown, W. V., and Stack, S. M. (1968). Somatic pairing as a regular preliminary to meiosis. *Bull. Torrey Bot. Club* **95,** 369–378.

Carlson, J. G. (1956). On the mitotic movement of chromosomes. *Science* **124,** 203–206.

Carr, A. J. H., and Olive, L. S. (1958). Genetics of *Sordaria fimicosa*. II. Cytology. *Am. J. Bot.* **45,** 142–150.

Church, K. (1977). Chromosome ends and the nuclear envelope at premeiotic interphase in the male grasshopper *Brachystola magna* by 3D, E. M. reconstruction. *Chromosoma* **64,** 143–154.

Church, K. (1981). The architecture of and chromosome movements within the premeiotic interphase nucleus. *In* "Mitosis/Cytokinesis" (A. M. Zimmerman and A. Forer, eds.), pp. 83–102. Academic Press, New York.

Church, K., and Moens, P. B. (1976). Centromere behavior during interphase and meiotic prophase in *Allium fistulosum* from 3-D, E. M. reconstruction. *Chromosoma* **56,** 249–263.

Church, K., and Wimber, D. E. (1971). Meiosis in *Ornithogalum virens* (Liliaceae). II. Univalent production by preprophase cold treatment. *Exp. Cell Res.* **64,** 119–124.

Comings, D. E. (1980). Arrangement of chromatin in the nucleus. *Hum. Genet.* **53,** 131–143.

Darvey, N. L., and Driscoll, C. J. (1972). Evidence against somatic association in hexaploid wheat. *Chromosoma* **36,** 140–149.

Del Fosse, F. E., and Church, K. (1981). Presynaptic chromosome behavior in *Lilium*. I. Centromere orientation and movement during premeiotic interphase in *Lilium speciosum* cv. Rosemede. *Chromosoma* **81,** 701–716.

Dover, G. A., and Riley, R. (1977). Inferences from genetical evidence on the course of meiotic chromosome pairing in plants. *Philos. Trans. R. Soc. London, Ser. B* **277,** 313–326.

Driscoll, C. J., Darvey, N. L., and Barber, H. N. (1967). Effect of colchicine on meiosis of hexaploid wheat. *Nature (London)* **216,** 687–688.

Ellison, J. R., and Barr, H. J. (1972). Quinacrine fluorescence of specific chromosome regions. Late replication and high A:T content in *Samoaia leonensis*. *Chromosoma* **36,** 375–390.

Ellison, J. R., and Howard, G. C. (1981). Nonrandom position of the A-T rich DNA sequences in early embryos of *Drosophila virilis*. *Chromosoma* **83,** 555–561.

Evans, H. J. (1961). Chromatid aberrations induced by gamma irradiation. I. The structure and frequency of chromatid interchanges in dipoid and tetraploid cells of *Vicia faba*. *Genetics* **46,** 257–275.

Evans, K. J., and Filion, W. G. (1982). The distribution of chromatin in the interphase nucleus of *Zabrina pendula*. *Can. J. Genet. Cytol.* **24,** 583–591.

Feldman, M. (1966). The effect of chromosomes 5B, 5D, and 5A on chromosomas pairing in *Triticum aestivum*. *Proc. Natl. Acad. Sci. U.S.A.* **55,** 1447–1453.

Feldman, M., Mello-Sampayo, T., and Sears, E. R. (1966). Somatic association in *Triticum aestivum*. *Proc. Natl. Acad. Sci. U.S.A.* **56,** 1192–1199.

Fussell, C. P. (1975). The position of interphase chromosomes and late replicating DNA in centromere and telomere regions of *Allium cepa* L. *Chromosoma* **50,** 201–210.

Fussell, C. P. (1977). Telomere associations in interphase nuclei of *Allium cepa* demonstrated by C-banding. *Exp. Cell Res.* **110,** 111–117.

Fussell, C. P. (1983). Telomere arrangement in differentiated interphase cells of *Allium cepa:* A

function of chromosome arm lengths at anaphase-telophase. *Can. J. Genet. Cytol.* **25,** 478–486.
Fussell, C. P. (1984). Interphase chromosome order: A proposal. *Genetica* **62,** 193–201.
Fussell, C. P. (1987). In preparation.
Ghosh, S., and Roy, S. C. (1977). Orientation of interphase chromosomes as detected by Giemsa C-bands. *Chromosoma* **61,** 49–55.
Gillies, C. B. (1975). Synaptonemal complex and chromosome structure. *Annu. Rev. Genet.* **9,** 91–109.
Godin, D. E., and Stack, S. M. (1975). Heterochromatic connectives between the chromosomes of *Secale cereale*. *Can. J. Genet. Cytol.* **17,** 269–273.
Hancock, R., and Boulikas, T. (1982). Functional organization in the nucleus. *Int. Rev. Cytol.* **79,** 165–214.
Harris, P. (1965). Some observations concerning metakinesis in sea urchin eggs. *J. Cell Biol.* **25,** 73–77.
Heitz, E. (1932). Die Herkunft der Chromocentren. Dritter Beitrag zur Kenntnis der Beziehung zwischen Kernstruktur und qualitativer Verschiedenheit der Chromosomen in ihrer Langsrichtung. *Planta* **18,** 571–636.
Hotta, Y., and Shepard, J. (1973). Biochemical aspects of colchicine action on meiotic cells. *Mol. Gen. Genet.* **122,** 243–260.
Hsu, T. C., Cooper, J. E. K., Mace, M. L., Jr., and Brinkley, B. R. (1971). Arrangement of centromeres in mouse cells. *Chromosoma* **34,** 73–87.
Jensen, C. G. (1982). Dynamics of spindle microtubule organization: Kinetochore fiber microtubules of plant endosperm. *J. Cell Biol.* **92,** 540–558.
John, B. (1976). Myths and mechanisms of meiosis. *Chromosoma* **54,** 295–325.
Kaufmann, B. P. (1925). The existence of double spiral chromatin bands and a "bouquet" stage in *Tradescantia pilosa* Lehm. *Am. Nat.* **59,** 190.
Kitani, Y. (1963). Orientation, arrangement and association of somatic chromosomes. *Jpn. J. Genet.* **38,** 244–256.
Korf, B. R., Gershey, E. L., and Diacumakos, E. G. (1982). Centromeres are arranged in clusters throughout the Muntjac cell cycle. *Exp. Cell Res.* **139,** 393–396.
Kumar, S., and Natarajan, A. T. (1966). Kinetics of two-break chromosome exchanges and the spatial arrangement of chromosome strands in interphase nucleus. *Nature (London)* **209,** 796–797.
Lafontaine, J. G., and Luck, B. T. (1980). An ultrastructural study of plant cell (*Allium porrum*) centromeres. *J. Ultrastruct. Res.* **70,** 298–307.
Levan, A. (1939). The effect of colchicine on meiosis in *Allium. Hereditas* **25,** 9–26.
Luykx, P. (1970). Celluar mechanisms of chromosome distribution. *Int. Rev. Cytol., Suppl.* **2,** 1–173.
McClintock, B. (1933). The association of non-homologous parts of chromosomes in the midprophase of meiosis in *Zea mays. Z. Zellforsch. Mikrosk. Anat.* **19,** 191–237.
McClintock, B. (1945). *Neurospora*. I. Preliminary observations on the chromosomes of *Neurospora crassa. Am. J. Bot.* **32,** 671–678.
McIntosh, J. R., Cande, W. Z., Snyder, J. A. (1975). Structure and physiology of the mammalian mitotic spindle. *Soc. Gen. Physiol. Ser.* **30,** 31–76.
Maguire, M. P. (1972). Role of heterochromatin in homologous chromosome pairing: Evaluation of the evidence. *Science* **176,** 543–544.
Maguire, M. P. (1983). Chromosome behavior at premeiotic mitosis in maize. *J. Hered.* **74,** 93–96.
Maguire, M. P. (1984). The mechanism of meiotic homologue pairing. *J. Theor. Biol.* **106,** 605–615.

Mathog, D., Hochtrasser, M., Gruenbaum, Y., Saumweber, H., and Sedat, J. (1984). Characteristic folding pattern of polytene chromosomes in *Drosophila* salivary gland nuclei. *Nature (London)* **308**, 414–421.

Metz, C. W. (1916). Chromosome studies on the Diptera. II. *J. Exp. Zool.* **21**, 213–279.

Moens, P. B. (1973). Quantitative electron microscopy of chromosome organization at meiotic prophase. *Cold Spring Harbor Symp. Quant. Biol.* **38**, 99–107.

Molè-Bajer, J. (1958). Cine-micrographic analysis of c-mitosis in endosperm. *Chromosoma* **9**, 332–358.

Moroi, Y., Hartman, A. L., Nakane, P. K., and Tan, E. M. (1981). Distribution of kinetochore (centromere) antigen in mammalian cell nuclei. *J. Cell Biol.* **90**, 254–259.

Moses, M. J. (1968). Synaptinemal complex. *Annu. Rev. Genet.* **2**, 363–412.

Moses, M. J. (1977). Synaptonemal complex karyotyping in spermatocytes of the chinese hamster (*Cricetulus griseus*). I. Morphology of the autosomal complement in spread preparations. *Chromosoma* **60**, 99–125.

Palmer, R. G. (1971). Cytological studies of a meiotic mutant and normal maize with reference to premeiotic pairing. *Chromosoma* **35**, 233–246.

Puertas, M. J., de la Peña, A. Estades, B., and Merino, F. (1984). Early sensitivity to colchicine in developing anthers of rye. *Chromosoma* **89**, 121–126.

Rabl, C. (1885). Uber zelltheilung. *Morphol. Jahrb.* **10**, 214–330.

Rasmussen, S. W. (1976). The meiotic prophase in *Bombyx mori* females analyzed by three-dimensional reconstructions of synaptonemal complexes. *Chromosoma* **54**, 245–293.

Rasmussen, S. W., and Holm, P. B. (1980). Mechanisms of meiosis. *Hereditas* **93**, 187–216.

Rhoades, M. M. (1961). Meiosis. In "The Cell" (J. Brachet and A. E. Mirsky, eds.), Vol. 3, pp. 1–75. Academic Press, New York.

Riley, R., and Flavell, R. B. (1977). A first view of the meiotic process. *Philos. Trans. R. Soc. London Ser. B* **277**, 191–199.

Roos, U.-P. (1976). Light and electron microscopy of rat kangaroo cells in mitosis. III. Patterns of chromosome behavior during prometaphase. *Chromosoma* **54**, 363–385.

Sax, K. (1940). An analysis of x-ray induced chromosomal aberrations in *Tradescantia*. *Genetics* **25**, 41–68.

Schmid, M., and Krone, W. (1976). The relationship of a specific chromosomal region to the development of the acrosome. *Chromosoma* **56**, 327–347.

Sears, E. R. (1976). Genetic control of chromosome pairing in wheat. *Annu. Rev. Genet.* **10**, 31–51.

Shepard, J., Boothroyd, E. R., and Stern, H. (1974). The effect of colchicine on synapsis and chiasma formation in microsporocytes of *Lilium*. *Chromosoma* **44**, 423–437.

Singleton, J. R. (1953). Chromosome morphology and the chromosome cycle in the ascus of *Neurospora crassa*. *Am. J. Bot.* **40**, 124–144.

Smith, S. G. (1942). Polarization and progression in pairing. II. Premeiotic orientation and the initiation of pairing. *Can. J. Res., Sect. D* **20**, 221–229.

Solari, A. J. (1977). Ultrastructure of the synaptic autosomes in the ZW bivalent in chicken oocytes. *Chromosoma* **64**, 155–165.

Solari, A. J., and Counce, S. J. (1977). Synaptonemal complex karyotyping in *Melanoplus differentialis*. *J. Cell Sci.* **26**, 229–250.

Stack, S. M., and Clarke, C. R. (1973). Differential Giemsa staining of the telomeres of *Allium cepa* chromosomes: Observations related to chromosome pairing. *Can. J. Genet. Cytol.* **15**, 619–624.

Stack, S. M., and Clarke, C. R. (1974). Chromosome polarization and nuclear rotation in *Allium cepa* roots. *Cytologia* **39**, 553–560.

Stack, S. M., and Soulliere, D. L. (1984). The relation between synapsis and chiasma formation in *Rhoeo spatheceae*. *Chromosoma* **90**, 72–83.

9. The Rabl Orientation: A Prelude to Synapsis

Stern, H., and Hotta, Y. (1969). DNA synthesis in relation to chromosome pairing and chiasma formation. *Genetics* **61**(1), Suppl., 27–39.

Stern, H., and Hotta, Y. (1973). Biochemical controls of meiosis. *Annu. Rev. Genet.* **7**, 37–66.

Sutton, W. S. (1902). On the morphology of the chromosome group in *Brachystola magna*. *Biol. Bull. (Woods Hole, Mass.)* **4**, 24–39.

Tanaka, N. (1981). Studies on chromosome arrangement in some higher plants. III. *Haplopappus gracilis* (2n = 4) and *Crepis capillaris* (2n = 6). *Cytologia* **46**, 545–559.

Tanaka, N., and Tanaka, N. (1977). Behavior of the differentially stained kinetochores during the mitotic cell cycle in some *Polygonatum* species (Liliaceae). *Cytologia* **42**, 765–775.

Thomas, J. B., and Kaltsikes, P. J. (1976). A bouquet-like attachment plate for telomeres in leptotene of rye revealed by heterochromatin staining. *Heredity* **36**, 155–162.

Thomas, J. B., and Kaltsikes, P. J. (1977). The effect of colchicine on chromosome pairing. *Can. J. Genet. Cytol.* **19**, 231–249.

Vanderly, L. (1948). Somatic mitosis in the root tip of *Allium cepa*—A review and reorientation. *Bot. Rev.* **14**, 270–318.

Wagenaar, E. B. (1969). End-to-end chromosome attachments in mitotic interphase and their possible significance to meiotic chromosome pairing. *Chromosoma* **26**, 410–426.

Walters, M. S. (1970). Evidence on the time of chromosome pairing from the preleptotene spiral stage in *Lilium longiflorum* "Croft." *Chromosoma* **29**, 375–418.

Westergaard, M., and von Wettstein, D. (1972). The synaptonemal complex. *Annu. Rev. Genet.* **6**, 71–110.

Wilson, E. B. (1925). "The Cell in Development and Heredity," 3rd ed. Macmillan, New York.

Zickler, D. (1977). Development of the synaptonemal complex and "recombination nodules" during meiotic prophase in the seven bivalents of the fungus *Sordaria macrospora* Auersw. *Chromosoma* **61**, 289–316.

IV
Chemistry of Meiosis

10

The Biochemistry of Meiosis

HERBERT STERN AND YASUO HOTTA

Department of Biology
University of California, San Diego
La Jolla, California 92093

I. INTRODUCTION

Biochemical or molecular events directly relevant to meiosis begin no later than the premeiotic S phase and terminate on completion of the second meiotic division. Many molecular events that occur during meiosis are not relevant to the process but are essential to gametogenesis. This is particularly prominent in animal meiocytes, and much less so in those of plants where meiosis is followed by a haploid generation in which gametogenesis occurs. A major difficulty in identifying those metabolic events in a prophase spermatocyte that are concerned with meiosis is the large fraction of events that function in spermiogenesis. Although discussions of gametogenesis almost always include the meiotic process, it must be understood that the only necessary relation between them is that meiosis must occur at some time following zygote formation and preceding gametogenesis. In some organisms—yeast, for example—many haploid cell generations may intervene between meiosis and the conversion of cells into the equivalent of gametes.

Meiosis is primarily and necessarily involved in the reductional division of chromosome sets. Closely, but not necessarily, associated with it is a limited crossing-over between homologous chromosomes. Underlying both events is a pairing of homologous chromosomes. Such pairing is indispensable to a normal reductional division and it is also highly important for the occurrence of recombination. Pairing without crossing-over, however, may and does occur, for example, in male *Drosophila* and female *Bombyx*. In meiocytes, where crossing-over is a normal event of division, failure of any bivalent to undergo at least one crossover is likely to be associated with abnormal disjunction at the first meiotic division. Inasmuch as this chapter is addressed to organisms in which crossing-over is a regular feature of meiosis, chromosome pairing and recombination will be treated as interrelated components of a single program.

A. Approach

Although a number of important cytoplasmic events occur during meiosis, the process is primarily concerned with chromosome behavior in which replication, synapsis, recombination, and disjunction are the principal events. Given our present knowledge of meiosis, a coherent picture of all the biochemical processes governing chromosome behavior from S phase to telophase is beyond reach. The biochemical account of meiosis being presented is necessarily partial. Most of the information available applies to the prophase stages, the bulk of it being derived from studies of a single genus, *Lilium*. Little is known about the molecular mechanisms governing the disjunction of chromosomes, or of the many critical events that precede the initiation of meiotic prophase.

There are two principal aspects in our treatment of the subject. One is purely descriptive; it provides an account of the known metabolic activities that occur during meiotic development and that are distinctive of meiosis. The other is functional; it attempts to identify mechanisms that coordinate metabolic activities with chromosome behavior. In doing so, it must be kept in mind that although the switch from mitotically proliferating cells to meiocytes is essentially a process of cytodifferentiation, the switch has certain distinctive features. Commonly, cytodifferentiation involves the introduction of a set of stabilized changes in the morphology and function of a cell. By contrast, in the life history of a meiocyte, there are no stabilized changes. Every phase of chromosome behavior is transient and so are the processes associated with that behavior. Among the various mechanisms that govern cytodifferentiation, those that introduce stabilized functional or structural characteristics are unlikely to be involved in meiotic differentiation. Every meiosis-specific change is transient. Novel proteins and novel enzymes are short-lived in their appearance. It would seem that meiocytes have an elaborate mechanism to turn on essential processes at specific times and, in so doing, to allow for their quick expiration.

Studies to date do not permit any generalizations about mechanisms underlying meiocyte differentiation. Two regulatory characteristics are nevertheless apparent. One characteristic is the rapid synthesis of certain key proteins, including enzymes, just prior to or during early prophase, all of them returning to their initial low level by the end of prophase, usually before diplotene (see Fig. 2). A second characteristic is the nuclease susceptibility of meiotically active sites of transcription. Thus far, all chromosomal regions that have been identified as being active during meiosis have been found to be significantly more accessible to nucleases than the same regions in mitotically dividing cells (Hotta et al., 1985c). Moreover, as discussed later, the intervals of accessibility are limited to the periods when the sites in question are active. Meiosis thus involves an activation of certain meiosis-specific chromosomal regions, the chromatin in those regions being transiently modified. Some of the regions thus affected code for proteins having catalytic or structural functions. The other regions are addressed to the regulation of chromosome behavior.

B. A Brief Perspective

The plan of this presentation is to treat the biochemical events of meiosis in terms of three developmental phases. The first is the prezygotene phase which includes at least the latter part of premeiotic DNA replication and covers events up to the initiation of zygotene. The second phase covers the events associated with chromosome synapsis, thus covering zygotene alone. The third phase involves all of pachytene and is primarily concerned with the regulation of crossing-over. No sharp boundaries can be drawn between events comprising one phase and those assigned to its neighbor. The large number of cells required for biochemical analysis makes it difficult to obtain sharply separated meiotic stages. Moreover, and more importantly, it will become evident that events occurring during one phase may also continue into the next one, some events anticipating chromosome behavior at the next stage. The theme of the entire presentation is the existence of specific DNA regions that have, or are presumed to have, a unique functional role in meiosis, one that does not include conventional transcription. It is to these special DNA regions that the action of the meiosis-specific structural and regulatory genes is directed.

II. THE PREZYGOTENE PHASE

A. DNA Replication

In the few organisms in which the duration of the premeiotic S phase has been measured, it has been found to be longer than any of the S phases in somatic tissues (Callan, 1972). Although the long duration can be accounted for by the reduced number of initiation sites for replication, the role of S phase attenuation is unknown. It may be supposed that certain nuclear changes need to occur in preparation for meiosis, but if so they remain hidden. In lilies, at least, cells do not become irreversibly committed to meiosis until very late in S phase or just after its completion (Ninnemann and Epel, 1973; Ito and Takegami, 1982). The distinctive and functionally significant characteristic of premeiotic DNA replication is its incompleteness. In lilies and, from indirect evidence, also in mouse, a small component of the genome is not replicated until the zygotene stage (Hotta and Stern, 1971a; Stern and Hotta, 1985). Such zygotene DNA ("zygDNA") constitutes between 0.1 and 0.2% of the genome. It occurs in segments that are 4–10 kbp in length and that are interspersed among diverse regions of all the chromosomes of the complement. Their precise location is unknown but, judging from *in situ* hybridizations, they are broadly distributed. From C_0t analysis, it appears that zygDNA segments consist primarily of DNA sequences with a very low copy number but probably include a small fraction of repeated sequences (Hotta and Stern, 1975).

It is evident that the replication of zygDNA sequences is suppressed during the premeiotic S phase. It is also evident that in cells subjected to conditions causing a reversion to mitosis, zygDNA replication occurs before mitosis is initiated (Stern and Hotta, 1969). The unreplicated zygDNA segments are almost certainly present in duplex form throughout the first phase. It is therefore probable that sister chromatids are tightly apposed because the interstitial segments of zygDNA consist of one strand from each of the two chromatids (see Fig. 1). The function of such apposition is unknown but it has long been recognized cytologically. Indeed, the light microscope image of meiotic prophase was the basis of the first, but erroneous, theory of meiosis known as the "Precocity Theory" (Darlington, 1937). According to that theory, chromosomes paired precociously and did not undergo replication until after pairing was completed. Delayed, rather than simply late, replication is an unusual event in cell behavior and it is reasonable to assume that the delay itself has a particular relevance to meiosis. The nature of such relevance is discussed below. To the extent that this assumption is correct, the mechanism responsible for the delay must be important to the biochemical program of the process. What follows is a description of that mechanism.

B. The L-Protein

The key factor in suppressing the replication of zygDNA is a protein of approximately 70,000 Da. It is a nuclear component that is extractable with deoxycholate but not with nonionic detergents and, presumably, is a component of the nuclear membrane (Hotta *et al.*, 1984). It will be referred to as the leptotene or "L-protein." The constitution of the nuclear membrane that houses the protein is in itself of interest with respect to the prezygotene phase of meiosis. Nuclear membranes of prezygotene meiocytes have a distinctive composition when compared with membranes similarly prepared from somatic nuclei (Hotta and Stern, 1971b). The meiotic nuclei have a "heavy" nuclear membrane fraction that is absent from somatic nuclei. This membrane fraction is the seat of another important meiotic protein, the "R-protein," that is intimately connected with the process of synapsis. This will be discussed later but it may be significant that the L-protein is located close to another meiosis-specific protein, both being housed in a meiosis-specific nuclear lipoprotein. There are undoubtedly other important components that are formed during this interval but the appearance of these proteins just before the initiation of meiotic prophase is significant inasmuch as their synthesis represents a set of molecular events that precede meiosis and are critical to its operation (Hotta and Stern, 1971b; Hotta *et al.*, 1984).

Nuclei isolated from leptotene cells and incubated in the presence of standard precursors synthesize little or no DNA depending on the extent to which the preparations are free of S phase nuclei (Hotta *et al.*, 1984). Nuclei isolated from cells that have entered zygotene and similarly incubated synthesize zygDNA. Significantly, leptotene nuclei can be made to synthesize zygDNA if first treated with deoxycholate, a treatment that removes the L-protein. Preleptotene nuclei can be similarly stimulated to synthesize zygDNA. Nuclei thus rendered capable of synthesis can be blocked in their capacity to do so by treating them with L-protein. The L-protein has no appreciable effect on overall synthesis of DNA in S phase nuclei; it acts specifically on zygDNA sequences. Using the inhibition of zygDNA synthesis as an assay for L-protein, it is apparent that deoxycholate extracts contain the protein only from about the time that the premeiotic S phase is nearly completed until some time in late zygotene. The apparent time of L-protein formation coincides with the time at which the cells become irreversibly committed to meiosis. Based on extractability, there is no L-protein present during most of the S phase. Such behavior invites the speculation that zygDNA sequences are regularly replicated near the end of S phase during mitotic cycles, the timing of their replication being regulated by a factor other than the L-protein. The formation of L-protein close to the end of the premeiotic S phase is a special feature of the premeiotic cell and is addressed to the suppression of zygDNA synthesis in order to accommodate the needs of the meiotic process. Thus, in terms of chromosome behavior, the suppression of zygDNA synthesis near the end of the S phase may be viewed as the first molecular event in the chromosomes that compels the entry of potential meiocytes into meiosis. Such an interpretation requires that zygDNA play a critical role in meiosis at the time of its replication.

Given the specificity of L-protein action, it is to be expected that the protein would have a binding affinity for some sequence or group of sequences in zygDNA. This is indeed the case. Total nuclear DNA from which zygDNA has been removed shows negligible binding to double-strand (ds) DNA. No binding specificity is observed with respect to single-strand (ss) DNA; all the meiotic proteins found to act on DNA bind to ssDNA. If zygDNA is not removed from the nuclear DNA, binding does occur although at a low saturation level. On the other hand, if purified zygDNA is used, binding is very strong, bulk DNA being a very poor competitor. L-Protein does not bind to mitochondrial, chloroplast, bacterial, or phage DNA. By contrast, significant, although low, binding occurs with all eukaryotic DNAs tested (Hotta *et al.*, 1984). This makes it probable that the protein, or its binding region, is conserved across the phylogenetic spectrum of eukaryotic species. The approximate size of the binding region has been estimated from the behavior of cloned segments of zygDNA. In general, and for reasons that are yet obscure. zygDNA is refractory to cloning and among the 17 available clones, the four having large zygDNA inserts (1500–2500 bp) were tested for binding. Two of the four tested bound zygDNA. When the bound complexes were exposed to DNase it appeared that no more than 90 bp were protected by the protein. The location of the binding regions in the segments of zygDNA as they occur *in situ* is undetermined but clearly, if the segments span 4–10 kbp, each having one binding site, those sites would account for about 1% of the zygDNA or about $10^{-3}\%$ of total nuclear DNA in the case of *Lilium*.

C. Summary of Events during the Prezygotene Phase

The principal events thus far identified in the prezygotene stages of meiosis are the modification of the nuclear membrane, the synthesis of L-protein, and the suppression of zygDNA replication. Incomplete replication of the genome is a distinctive feature of preleptotene and leptotene metabolism. Modification of the nuclear membrane is displayed by the presence of a heavy membrane fraction that is absent from somatic nuclei and in which the L-protein is located. The L-protein has not been found in somatic nuclei during any phase of the cell cycle. A subset of sequences in zygDNA has a specific binding affinity for the L-protein and it is inferred that the specificity of binding is responsible for the specificity of L-protein inhibition of zygDNA synthesis.

III. THE ZYGOTENE PHASE

From the standpoint of intrachromosomal events the biochemical activities during the zygotene stage are very intense. Some of the activities are directly concerned with chromosome synapsis and others are concerned with the processes of recombination that are most prominent during the pachytene stage. The discussion in this section will be limited to the processes that seem to be directly related to synapsis. The other processes, even though initiated during zygotene, will be considered as part of the pachytene phase in relation to recombination. It should be understood, of course, that chromosome synapsis, although related to recombination, is different from molecular synapsis. The former involves a general positioning of chromosomes whereas the latter is an immediate part of the recombination event. Meiotic synapsis of chromosomes involves a gross alignment of homologous regions by a mechanism in which, at most, only a small fraction of the total DNA is involved, and for which a non-DNA structure is required to effect stabilization. How the gross chromosomal alignment relates to the gene alignment required for recombination remains unanswered.

The significant events of the zygotene phase include the replication and transcription of zygDNA and a specific action of L-protein. The assignment of zygDNA to a role in chromosome synapsis is based entirely on circumstantial evidence. The temporal correlation between zygDNA replication and homologue synapsis is suggestive of a causal relation between the two events (Hotta and Stern, 1971a). The inhibition of synapsis and of synaptonemal complex assembly by inhibiting zygDNA replication points to a dependence of synapsis on that replication (Roth and Ito, 1967). The localization of zygDNA replication in the regions of the synaptonemal complex in both mouse (Moses *et al.*, 1984) and lily (Kurata and Ito, 1978) argues for a direct role of such replication in the synaptic process. It is possible to construct models to accommodate these views and this is done below but without any direct evidence for their validity.

A. zygDNA Replication

The initiation of zygDNA replication occurs close to the beginning of zygotene and probably coincides with the initiation of chromosome pairing. In lily, the pairing extends over several days and, presumably, different zygDNA segments are replicated at different times, homologous segments being replicated simultaneously. The gradual decrease in L-protein during zygotene may be taken to reflect the gradual decrease in number of unreplicated segments (Hotta *et al.*, 1984). One feature of the replicative process during zygotene is distinctive and probably significant to meiosis. In all cases, the ends of the zygDNA segments remain unreplicated and completion of their replication does not occur until after pachytene (Hotta and Stern, 1976). The length of the unreplicated ends has not been determined, but the presence of gaps rather than unligated ends has been inferred from the fact that the discontinuities between the zygDNA segments and bulk DNA cannot be repaired by ligase alone but require polymerase activity. The role of the discontinuities that persist until close to the time of chromosome disjunction is unknown. All the zygDNA segments are thus affected and it is very likely that they function in some way to facilitate chromosome disjunction following crossing-over. One possibility is that the unreplicated end-regions maintain sister chromatid adhesion and thus prevent bivalents from dissociating at diplotene.

It may be supposed that zygDNA behaves as a functional axis of the chromosome, the completion of zygDNA replication signaling completion of chromosome replication. If this were so a special role for zygDNA in aligning chromosomes for synapsis would be easy to visualize. It would also account for the apparent lack of L-protein during most of S phase by supposing that completion of axial replication in mitotic cycles occurs at the end of the S phase, the event furnishing a signal for the initiation of the G_2 phase. In meiosis, on the other hand, pairing and recombination must occur before axial replication can be completed. Such rationalization of the persistent discontinuities in zygDNA segments is appealing; its weakness lies in an absence of evidence for its occurrence.

How the replication of zygDNA might function in chromosome alignment is considered below. It is important, however, to emphasize that zygDNA replication per se is not a sufficient condition for alignment. It occurs in mitotic cells without inducing somatic pairing, it occurs in meiotic cells that lack homologous pairing, and it also occurs in cells that have been rendered asynaptic by colchicine treatment. It is evident that ongoing alignment requires a coordination with zygDNA replication. The alignment is obviously inconsequential unless accompanied by a stabilizing mechanism. The relation between the two processes is another unknown, but it is not a mandatory one. The formation of a synaptonemal complex (SC) between chromosomes can occur in the absence of homologous alignment and the evidence for synaptic adjustment (Moses *et al.*, 1984) indicates that SC formation can occur well after the occurrence of synapsis at zygotene. It thus appears that in normal meiotic cells zygDNA replication occurs under conditions that permit the replication to effect alignment and also to stabilize the alignment by completing formation of the SC.

B. zygDNA Transcription

In the speculative model discussed above, zygDNA was assigned a structural role in chromosome behavior. The speculation derives principally from the fact that the replication of this DNA is delayed until the initiation of chromosome pairing. The phenomenon of late replication may, however, be viewed from an entirely different standpoint. It is generally true that in meiosis, the larger the genome size the lower the frequency of recombination per DNA nucleotide. The possibility thus exists that the reduction is apparent rather than real if one assumes that meiotic recombination is restricted to the portion of the genome coding for structural genes (Holliday, 1984). The complexity of zygDNA in lilies is probably sufficient to provide all the information necessary for the plant's organization and function. The unusual replicative behavior of zygDNA during meiosis could be attributed to a mechanism by which replication is tied to

10. The Biochemistry of Meiosis

recombination, zygDNA sequences being thus selectively poised for crossing-over. If this were the case, zygDNA should be transcribed in somatic cells; indeed it should be the exclusive site, or nearly so, of transcription in all tissues. The evidence for such a role is negative. Poly(A^+) RNA transcripts from somatic cells do not hybridize to any significant extent with zygDNA. However, poly(A^+) RNA transcripts from meiocytes clearly do (Hotta *et al.*, 1985a). A significant fraction of zygDNA (no less than 1–2%) is transcribed during meiotic prophase. The concentration profile of "zygRNA" transcripts with respect to meiotic stage is strongly suggestive of either a role in effecting pairing or in one immediately resulting from the occurrence of pairing. zygRNA rises abruptly beginning in late leptotene and reaches a peak at mid-zygotene. The fall is as steep as the rise so that by mid-pachytene it is no more than 20–25% of the peak level. In lilies, the peak level is close to 40% of total poly(A^+) RNA. The turnover of zygRNA is obviously rapid so that if it functions at all, it must do so during the interval beginning with zygDNA replication and terminating by mid-pachytene. Moreover, its transcription is clearly linked to the occurrence of homologous pairing. The level of zygRNA is very low at zygotene in a hybrid that is deficient in homologous pairing. Thus far, it has not been possible to determine whether the transcription of zygDNA is initiated prior to or following the initiation of chromosome pairing. Regardless of the precise order, it is apparent that pairing activity is coordinated with transcription of zygDNA. Given the high complexity of zygDNA it is conceivable that the sequence composition of the zygRNA pool is constantly changing as the meiocytes progress through zygotene. This possibility needs to be examined. At present, the established feature of meiotic metabolism is the coupling of the delayed zygDNA replication with its transcription and the association of that coupling with chromosome synapsis.

It is perhaps not surprising that the meiotic behavior of zygDNA should have its counterpart in meiocytes from a broad array of species. Despite the difficulty in identifying and isolating zygDNA from other species it is apparent that in mouse spermatocytes zygRNA reaches a peak during meiotic prophase and is at very low levels before or after that interval (Hotta *et al.*, 1985a). Moreover, there is sufficient sequence homology between lily and mouse zygDNA so that cross-hybridization occurs under conditions of moderate stringency. Indeed, although zygRNA does not hybridize with DNA from cytoplasmic organelles or from prokaryotes, it does hybridize with the DNA from all eukaryotes thus far tested. The high degree of sequence conservation across the phylogenetic spectrum makes it reasonable to suppose that interspersed zygDNA-like sequences are a common property of eukaryotic chromosomes whose function, in part at least, is coupled with synapsis at meiosis. In this sense, zygDNA constitutes a meiosis-specific set of DNA sequences.

C. The L-Protein Activity

The role of L-protein in suppressing prezygotene replication of zygDNA and its specificity of binding to one or more subsets of zygDNA sequences has already been discussed. This protein, however, is known to have at least one more function with respect to zygDNA. Plasmids with zygDNA inserts that have an L-protein binding site respond distinctively when complexed with L-protein in the presence of ATP (Hotta *et al.*, 1984). Under these conditions a nick (or very small gap) is introduced into one of the strands. There is no evidence of additional nucleolytic activity nor is there evidence for any activity by the protein in the absence of binding. When denatured, a duplex plasmid thus nicked may be resolved into a single-strand circle and a single-strand linear fragment of approximately the same length as the circular strand. The nicking action is specific to the L-protein binding region. If, as seems probable on the basis of the inhibitory action of the L-protein, each zygDNA segment has a binding site, the site-specific nicking probably functions in the process of synapsis.

Evidence for a particular role of L-protein nicking has not been obtained although various roles such as initiation of transcription or replication may be imagined. We favor the role in which the nick makes possible the formation of a single-strand tail, which is then available for hybridization with a corresponding complementary single-strand tail from the homologous chromosome. The hybridization would effect alignment but the heteroduplex would be quickly dissociated as replication of the zygDNA segment occurs. During this brief interval, an SC segment would have to be formed to stabilize the alignment. The location of the L-protein in the nuclear membrane and the simultaneity of replication in homologous zygDNA segments would facilitate the process of matching. It must be assumed, of course, that some factor in this mechanism triggers the nicking in the L-protein-complexed homologous zygDNA segments. The thrust of this scheme is to emphasize that the mechanism of homologous alignment is different from the mechanism of pairing stabilization and that the components involved in alignment need not be components of the stabilized structure. The proposed role of zygDNA in alignment does not imply the presence of zygDNA cross-bridges in the synaptonemal complex.

One additional aspect of L-protein activity requires consideration. A rationale should be provided for its suppression of zygDNA replication. By itself, the alignment just described could operate at the termination of the premeiotic S phase when L-protein appears, and therefore need not be delayed until the end of leptotene. This does not occur because effective alignment requires stabilization and some or all of the SC components are not yet present. The suppressive action of the L-protein must therefore be coordinated with the synthesis of the components essential to SC formation. The appearance of the L-protein at the time that commitment to meiosis becomes irreversible would thus tie in with the initiation of synthesis of some or all of the components of the SC. The evidence for interference with chiasma formation by perturbing meiocytes at this stage of the process is consistent with this model of events.

IV. THE PACHYTENE PHASE: DNA

All the biochemical information now available points to the occurrence of at least the ultimate events of recombination at pachytene (Carpenter, 1984; Stern and Hotta, 1984). Some of the evidence also makes it probable that these critical events occur during early to mid-pachytene as has been concluded from EM studies (Moses *et al.*, 1984). Regardless of the precise timing of events during or just preceding pachytene, behavior of the chromosomes during that interval is characterized by the localization of recombination-related activities at sites that have a distinctive sequence organization. They also have a distinctive chromatin composition. Activities at the different sites are effected by factors that regulate chromatin composition and by proteins that catalyze recombination-related processes. What should emerge from the description that follows is a picture of chromosomes equipped with specific DNA sequences that are sites of pachytene activity aimed at recombination. Activities at these sites are regulated, on the one hand, by the characteristics of chromatin organization and, on the other hand, by the catalytic properties of the different proteins that are present at significant levels only during the zygotene–pachytene stages. It would appear that chromatin organization specifies the chromosomal sites of action while the proteins specify the nature of that action.

A. Chromosomal Sites of Activity

Meiocytes that are pulse-labeled with either $^{32}P_i$ or [^{3}H]thymidine at pachytene become labeled at specific DNA sites generally referred to as "pachytene DNA" or "PDNA" (Hotta and Stern, 1974). In contrast, with the labeling that occurs under identical conditions in zygotene cells, the labeled DNA represents regions that have undergone repair replication. There is thus a sharp shift in DNA metabolism as meiotic cells enter pachytene from the zygotene stage. At the latter stage, replication is semiconservative, whereas at pachytene it is of the repair type. In mouse, as in lily, pachytene cells have been found to have a very high capacity for DNA repair (Sega, 1974; Stern and Hotta, 1978). Apart from the occurrence of a programmed repair activity at pachytene, the only other evidence pointing to its relevance to recombination is its absence in situations involving a lack of chiasma formation. This is the case for the achiasmatic lily hybrid, Black Beauty, and also for meiocytes treated with colchicine during prezygotene stages (Hotta and Stern, 1974). In studies of Black Beauty it has been unambiguously demonstrated that introduction of tetraploidy into these otherwise diploid cells provides for homologous synapsis, thus restoring chiasma formation and also repair replication (Hotta et al., 1979).

The selectivity of the sites undergoing repair replication is made evident from C_0t analyses of the DNA prepared from pachytene-labeled cells. Most of the label is found in DNA fragments that constitute a middle-repeat fraction of DNA (Bouchard and Stern, 1980). Whereas generally such repetitive DNA displays a great deal of sequence divergence, it has been found that among lilies the extent of sequence divergence in PDNA is of the same order as that of the *Escherichia coli* genome. PDNA, despite its middle repetitive nature, is highly conserved, a property that in itself argues for an essential role in meiosis. In a relatively broad study, these sequences have been found in dicots as well as monocots. By contrast, no cross-homology could be found for other middle-repeat sequences including those non-PDNA sequences that display a relatively low degree of divergence (Friedman et al., 1982). It has been proposed that the sites housing PDNA are potential sites for initiating recombination. The proposal implies that just as chromosomes have specific sites for synaptic alignment, they also have specific sites for initiating meiotic recombination. On these grounds, meiotic recombination is not randomly distributed, a point frequently made on the basis of genetic and cytological evidence. Nonrandom distribution is also made apparent on examining satellite DNA prepared from pachytene-labeled cells. In both mouse, which has about 10% of its DNA as satellite, and in the Arabian oryx, which has 40% or more of its DNA as satellite, the satellite DNA does not become radioactive when the pachytene cells are pulse-labeled (Hotta and Stern, 1978a; Stern and Hotta, 1980). Because satellite DNA is derived mainly from constitutive heterochromatin the observation is significant inasmuch as crossing-over is rare in heterochromatic regions.

The structural organization of PDNA is distinctive, quite apart from the conserved sequence arrangements, a feature that is probably relevant to its role in crossing-over. Because of the properties of PDNA chromatin (to be discussed below) it has been possible to isolate chromatin fragments housing PDNA sequences. Depending on the isolation procedure, the PDNA fragments recovered from the chromatin fragments after deproteinization are of unequal size but nevertheless fall into two size classes (Hotta and Stern, 1984). The smaller size class ranges from 150 to 300 bases in length and the larger size class ranges from 800 to 3000 bases in length. For reasons that will become apparent later, the large size class will be referred to as PDNA while the smaller one will be referred to as PsnDNA (see Fig. 1). There is a very close relationship between PDNA and PsnDNA. Each end of a PDNA segment consists of a PsnDNA sequence. The internal segment differs from PsnDNA in at least two respects. It has a higher complexity than PsnDNA and undergoes little, if any, repair replication during pachytene. The C_0t behavior of the PsnDNA label indicates that PsnDNA belongs to the highly conserved middle-repeat class; the unlabeled internal regions of the PDNA segments have not been analyzed.

B. Events Associated with PsnDNA Repair Replication

The immediate cause of pachytene repair replication is a nicking of DNA. This is evident upon analyzing the DNA from different meiotic stages by zonal centrifugation. Whereas duplex DNA displays similar profiles at all the stages, the denatured form displays a striking difference at pachytene. At this stage, the DNA molecules resolve into two groups when total nuclear DNA is centrifuged in an alkaline glycerol gradient. About half the DNA sediments at a mean value of about 105 S while the other half sediments at about 62 S (Hotta and Stern, 1974). The 62 S component is not present at other stages nor is it present in achiasmatic cells. At all other stages, the sedimentation profile is not bimodal; the mean sedimentation rate is about 105 S. There is thus at pachytene a clear relation between the introduction of nicks, but no double-strand cuts, and replication repair. Termination of repair activity coincides with the disappearance of single-strand gaps in the PDNA regions. It is important to note that the termini of the zygDNA segments are not repaired during pachytene; those gaps appear to be protected from the otherwise intense repair activity that characterizes the pachytene stage. Such programmed single-strand nicking that occurs only in meiocytes undergoing crossing-over is the basis for the view that the PDNA regions are potential sites for initiating crossovers.

Virtually all the PsnDNA regions are nicked during pachytene. When an alkaline gradient of pachytene DNA is probed with PsnDNA, almost all of the hybridization occurs in the 62 S region of the gradient (Hotta and Stern, 1984). Probing of similar DNA gradients from other stages shows all PsnDNA sequences to be in the 105 S region. The sedimentation behavior of DNA from pachytene cells makes it highly probable that the size of the 62 S fragments reflects the spacing between the PsnDNA regions. By digesting these fragments with exonuclease I it has been found that the PsnDNA sequences are clustered at the 5' ends. If the 62 S fragments (mean length = 160 kb) all display a clustering of sequences at the 5' ends, it follows that in the native DNA duplexes, a nick at the 5' end of a PDNA region in one strand should be accompanied by a nick at the 5' end of that region in the complementary strand. Nicking is thus not only confined to particular regions of the genome but it is also polar in position. It has also been possible to isolate 160-kbp duplexes from pachytene cells and in this way to demonstrate that virtually all the PsnDNA sequences are present only at the ends of the duplex fragments (Hotta and Stern, 1984). The principal conclusion to be drawn from all these studies is that pachytene nicking is part of a complex program regulated in both its temporal and spatial aspects. Such regulation obviously requires the operation of a specific set of mechanisms.

C. Site Specificity in Pachytene Nicking

The mechanisms responsible for the regulation of pachytene nicking must account both for its timing and for its selective action at PsnDNA sites. They must also account for the dependence of nicking on the prior occurrence of homologous synapsis. What has emerged from a series of studies is that neither timing, nor site selection, nor dependence on homologous pairing can be accounted for by the properties and behavior of the endonuclease that performs the nicking (Hotta *et al.*, 1979; Hotta and Stern, 1981). Timing and site specificity are regulated by chromatin structure. The PsnDNA sequences in which nicking occurs are housed in chromatin that is modified to render the PsnDNA regions accessible to the endonuclease. In the absence of such modification nicking does not occur. Despite the abundance of nuclease activity at pachytene, a feature discussed below, the genome is resistant to nuclease activity except at PsnDNA sites and, even then, only if homologous pairing has occurred.

The components of that regulation are a small RNA molecule which will be referred to as PsnRNA and a nonhistone protein which will be referred to as PSN-protein (Hotta and Stern, 1981). During zygotene the PsnDNA segments of the PDNA regions begin to be altered in composition. The alteration involves a replacement of the histones in those regions by Psn-protein and the PsnRNA. The combination renders those regions highly sensitive to DNase II, a property that has made it possible to isolate the chromatin housing PsnDNA and also, though less efficiently, that housing the entire PDNA segment. Using appropriately isolated pachytene nuclei as substrates, it can be shown that those prepared from the achiasmatic lily hybrid (Black Beauty), unlike those prepared from chiasmatic forms, are resistant to DNase II action or to that of the meiotic endonuclease. If, however, the resistant nuclei are preincubated with PsnRNA, Psn-protein, ATP, and a partially purified extract from the cytoplasm of chiasmatic cells, the treatment renders the nuclei susceptible to the action of either enzyme. Moreover, the chromatin released from the achiasmatic nuclei by DNase II treatment mainly houses PsnDNA sequences. Thus, the components of PDNA chromatin prepared from chiasmatic forms are capable of so altering achiasmatic nuclei as to render PsnDNA accessible to the action of meiotic endonuclease. Cells of the achiasmatic hybrid have low levels of PsnRNA and of Psn-protein. It may be inferred that the absence of nicking in the hybrid is directly due to the absence of the components essential to chromatin modification.

The site specificity of chromatin modification most probably derives from

PsnRNA. This small nuclear RNA is homologous with the PsnDNA sequences and, presumably, directs the sites at which histone replacement occurs (Hotta and Stern, 1981). The RNA has a specific binding affinity for Psn-protein, the protein having no specific affinity for the PsnDNA. Because there are at least several hundred different families of PsnDNA sequences (Bouchard and Stern, 1980), there must be as many sequence varieties of PsnRNA. One model which might account for the regulation of PDNA regions at pachytene is the operation of master-transcriptional PsnRNA genes that are activated by about mid-zygotene. The PsnRNA transcripts in combination with Psn-protein find the homologous PDNA regions and in doing so each of the moderately repeated PsnDNA sequences undergoes a replacement of its associated histones with PsnRNA and Psn-protein. The brief but probably intense transcription of each of the PsnRNA sequences at master sites may well flood the nucleoplasm and thus effect a concerted change in all the PsnDNA regions at the same time. The gene transcribing the PsnRNA thus controls the timing of PsnDNA accessibility while the PsnRNA itself controls the sites of chromatin change. What remains entirely unknown is the manner in which the absence of homologous pairing prevents (or fails to stimulate) the production of PsnRNA and Psn-protein.

D. Summary of DNA Events

The different pieces of evidence on pachytene DNA metabolism, limited though they are, clearly indicate that during the interval while chromosomes are completing homologous pairing and while they remain stably paired, an elaborate set of mechanisms is activated. These mechanisms are directed at coordinately exposing certain families of moderately repeated sequences within the PDNA segments to the action of recombination-related enzymes. The primary action of this group of enzymes (see Section V,A) is to introduce nicks that become extended to gaps within those repeated sequences. Susceptibility to enzyme action is effected by a protein that displaces the histones and the site specificity of the displacement appears to result from the action of an ATP-requiring mechanism in which families of small nuclear RNAs (PsnRNA) are the critical agents. These RNAs are found only in the nucleus during the zygotene–pachytene stages, and have not been detected at other times in the meiocytes or in any of the somatic cells. The components are thus exclusively meiotic. Whether the PDNA segments function exclusively in meiosis is unknown, but they certainly have a distinctive meiotic function in that they appear to serve as potential initiating sites for recombination. The number of such sites is in extreme excess over the number of chiasmata formed but this in itself may simply reflect the way in which meiosis ensures the regularity of the relatively few crossovers that do occur. At the molecular level, we have no knowledge of how the ultimate crossovers are sited; they are clearly localized in the recombination nudules (Carpenter, 1984).

V. MEIOSIS-SPECIFIC RECOMBINOGENIC PROTEINS

Even a superficial survey of the appearance and disappearance of proteins during meiosis makes it obvious that a whole set of proteins relevant to recombination becomes active during the early through mid-prophase stages—leptotene, zygotene, and pachytene. Members of this set have not been found during preleptotene, including the premeiotic S phase, nor have they been found following completion of pachytene. The apparently exclusive syntheses of these proteins beginning in early to late leptotene, their levels reaching a peak in late zygotene and early pachytene, give strong emphasis to the view that the enzymatic aspects of recombination are functions of the prophase stages. They also reinforce the long held cytogenetic evidence that chromosome synapsis is a precondition to crossing-over. Whether this organization of recombinational events holds for all meiotic systems is a question that runs well beyond the available evidence. Exceptions aside, the general structural and ultrastructural characteristics of meiosis point to its high degree of conservation and thus to the probability that the molecular organization is also highly conserved.

In considering the biochemistry of meiosis from the standpoint of recombinogenic proteins, a distinction needs to be made between the many proteins that are essential to recombination and present in a variety of tissues and those that are specifically induced during the meiotic process. Those in the first category will not be considered here because they provide no special insights into meiotic organization. They include enzymes like DNA ligase, polynucleotide kinase, topoisomerase, and DNA polymerase. It is of some general interest that DNA polymerase β is the predominant enzyme in mouse spermatocytes and that a similar form is predominant in lily microsporocytes (Sakaguchi et al., 1980). The reason for this replacement of the α by the β form is not known. Of much greater interest are two categories of proteins that are transient components of meiocytes during prophase and whose properties are appropriate for recombination. The first of these categories is composed of proteins that are likely to be involved in the recombination process, and the second consists of activities that are directly involved in effecting recombination. In all cases, the proteins are exclusive to meiotic prophase and in some cases, at least, have distinguishing characteristics from proteins with similar activities that are present in somatic tissues.

A. Recombination-Related Proteins (see Fig. 2)

Three proteins have been described that affect DNA structure, each of them not having identical counterparts in mitotically proliferating tissues. These are an endonuclease, a DNA-unwinding protein, and a protein that catalyzes DNA reassociation.

1. Endonuclease

Beginning at zygotene the activity of endonuclease rises steeply reaching its peak in early pachytene (Howell and Stern, 1971). The enzyme acts only on double-strand DNA (dsDNA) into which it introduces nicks. The nicks formed by the enzyme have 3'-phosphoryl and 5'-hydroxyl termini. In most if not all cases, the nicks are extended into single-strand gaps although the enzyme involved has not been identified. The evidence for gaps being present is that the repair of DNA discontinuities in isolated pachytene nuclei requires DNA polymerase in addition to DNA ligase. Moreover, no repair occurs unless phosphatase is used to remove the 3'-phosphoryl residues. The predominance of 3'-phosphoryl over 5'-phosphoryl residues in pachytene nuclei, but not in those from other stages, indicates that it is indeed the endonuclease that is responsible for introduction of the programmed DNA discontinuities at pachytene (Stern and Hotta, 1977). There is no evidence for any particular sequence preference; nicking activity appears to be random. As previously discussed, although the enzyme is essential to the introduction of nicks, its presence does not assure nicking. In the absence of chromatin modification, the meiocytes appear to be unaffected by the presence of the enzyme. This relation is most apparent in the achiasmatic hybrids. In the cultivar Black Beauty there is a small amount of pachytene nicking commensurate with the proportion of chiasmata formed, and yet the profile of endonuclease activity is identical with that occurring in normal chiasmatic forms.

2. The DNA-Unwinding Protein

The DNA-unwinding protein (U-protein) has been described in meiocytes of lilies and also of mammals (Hotta and Stern, 1978b). It attacks either nicked duplex DNA or the ends of linear duplexes and unwinds a limited stretch of the duplex if provided with ATP. The mechanism that limits the unwinding has not been resolved but it is neither inactivation of the enzyme nor a lack of ATP. If the enzyme attacks the end of a duplex it unwinds about 50 bp of DNA; if it unwinds at a nick, the unwinding extends to somewhere between 400 and 500 bp. From the standpoint of recombination, the unwinding provides a sufficient length of tail to permit heteroduplex formation. It is of interest that the U-protein binds to DNA at a 3′-hydroxyl terminus. It will not bind to 5′-hydroxyl or 5′-phosphoryl termini. Thus, for the protein to be active in the meiocyte at pachytene, a phosphatase is essential to remove the 3′-phosphoryl termini resulting from endonuclease action. A phosphatase is present in the meiocytes but its regulation is unknown. Because a 3′-hydroxyl terminus is essential to both repair and recombination, we presume that the formation of 3′-phosphoryl termini by endonuclease nicking must be linked in some way to a regulation of recombination activity.

3. The Reassociation Protein

From a purely physiological standpoint, the reassociation protein (R-protein) is the most interesting of the three proteins in this group. In many respects it resembles the gene 32-protein, but it appears to be more active under *in vitro* conditions in catalyzing the reassociation of naturally occurring DNAs (Hotta and Stern, 1971b, 1979). Its intracellular association is also highly distinctive. The R-protein binds strongly to single-strand DNA (ssDNA) and has very little binding affinity for dsDNA. Thus, as it catalyzes ssDNA reassociation its binding affinity for the substrate falls steeply. It can act to unwind duplex DNA but only if the DNA is suspended in a medium in which the duplex is already unstable, as in the absence of any divalent cations. Unlike the other two proteins, the R-protein is found almost entirely in the same fraction of the nucleus as the L-protein, the heavy component of the nuclear membrane. The R-protein also differs in its behavior from the other two proteins in that its level is very much affected by the presence or absence of homologous pairing. In absence of the latter, the level of R-protein is very low. Whether this is due to a rapid destruction of the protein or to a reduced synthesis has not been determined (Hotta *et al.*, 1979).

The structure of the R-protein molecule in relation to its function is also distinctive. The nature of its interaction with DNA depends on its state of phosphorylation (Hotta and Stern, 1979). If the native protein is dephosphorylated with phosphatase, it binds as strongly to dsDNA as to ssDNA. Under these conditions it is incapable of catalyzing ssDNA reassociation. If it is phosphorylated to excess with a heterologous cyclic AMP-dependent protein kinase, the protein loses binding affinity for both forms of DNA. On the other hand, an ATP-dependent protein kinase that is present in the meiocytes can restore the native activity if applied to the dephosphorylated protein. That kinase has no measurable effect on the native protein. It is unlikely that the native kinase would have no regulatory role on the behavior of the R-protein but at present there is no evidence for such a role. It is conceivable that the reassociation activity of the R-protein is entirely related to synapsis. This would be consistent with its associa-

10. The Biochemistry of Meiosis

tion with the L-protein and with its responsiveness to the conditions for homologous pairing. Such a role would nevertheless make it an important factor in meiotic crossing-over.

B. Proteins Catalyzing Recombination

The transient but steep increase in recombinogenic activity of extracts from cells in meiotic prophase follows the same pattern as those of the three proteins discussed above. These proteins are, however, distinguished by their direct involvement in recombinational activities. The recently accumulated evidence on this aspect of meiotic biochemistry consists of two different systems. One is the recA-like system and the other is partially purified extracts that effect recombination when challenged with complementary mutant plasmids (Hotta et al., 1985b). In both cases there are significant differences between the activities in meiotic and the corresponding ones in somatic cells.

1. RecA-like Proteins

The genetic evidence for a central role of the E. coli recA protein in recombination makes it virtually certain that the presence of a similar protein in meiocytes would signify its role in meiotic recombination (Radding, 1978). The search for such a protein has resulted in some significant findings. In both mouse and lily, two types of recA-like proteins, referred to as "rec proteins," are present. One type, the "s-rec" protein, has been found exclusively in somatic tissues. The other type, the "m-rec" protein, has been found exclusively in meiotic cells. The m-rec proteins have molecular weights in the neighborhood of 45,000 (measured in sodium dodecyl sulfate gels) and those of the s-rec proteins are 75,000 (mouse) and 70,000 (lily). Of striking interest, and also of significance, is the response of these proteins to differences in temperature. The s-rec proteins of mouse are optimally active under in vitro conditions at about 37°C. Mouse m-rec protein that has been prepared from spermatocytes is optimally active at 33°C. Activity drops considerably at 37°C, a property that applies to all activities of the m-rec protein. The comparison is significant because spermatogenesis in the testis is normal at 33°C but abnormal at 37°C. Parallel behavior occurs in the case of the lily rec proteins. In this case the evidence is clear that incubating m-rec lily protein at temperatures above 25°C results in much reduced activity whereas incubating the s-rec protein prepared from vegetative tissues at temperatures in the neighborhood of 33°C has no significant effect on activity. The catalytic properties of the enzymes in vitro are directly related to their in vivo activities with respect to temperature optima; they are indeed different proteins.

The activities that have been measured in vitro are essentially the same as those that have been measured with the *E. coli rec*A protein. In all cases, the m-*rec* proteins are much more active than the s-*rec* proteins whether compared on a molar or on a weight basis. Qualitative differences have not been observed. The two principal assays used were D-loop formation and strand transfer. In the D-loop assay, the enzyme catalyzes the invasion of ssDNA into a supercoiled duplex circle. The reaction requires that the ssDNA be homologous with at least a portion of the duplex DNA and that ATP be present to drive the reaction. The topology of the DNA substrates is such that the single strand cannot totally displace either member of the closed duplex; one of the strands becomes partially displaced by the invading single strand resulting in what resembles a D-loop configuration. The second reaction involves the transfer of a single strand from a linear duplex to a complementary ssDNA circle. In such a case an entire strand can be transferred thus forming an open duplex circle. The principal value of these tests has been the demonstration of a high degree of similarity between m-*rec* and *rec*A proteins. Release by the *rec* protein of one member of a DNA duplex and its association with a different complementary DNA strand is a central event in recombination. Whether there is a functional interaction between the m-*rec* protein and either the R- or U-protein is still undetermined; there is certainly some overlap in functions.

Apart from the fact that the m-*rec* protein is exclusive to meiotic cells and that the s-*rec* protein is absent from them, evidence for a role of this protein in meiosis is derived from the profile of m-*rec* activity during meiosis (see Fig. 2). There is little, if any, m-*rec* protein during the premeiotic S phase, but after the cells enter meiosis, the content of m-*rec* protein increases until it reaches a peak level at pachytene. If the behavior of m-*rec* protein is plotted in terms of activity, the same profile is observed. It is low during the premeiotic S phase and begins to increase prominently after the cells have entered meiotic prophase. The peak of activity is reached during early pachytene. Thus, the behavior of the *rec* proteins in the meiocytes adds to already strong evidence for a concentration of the enzymatic components of recombination during pachytene. The prezygotene events appear to be addressed to developing the mechanisms for stabilized chromosome synapsis, a precondition for homologous recombination but not part of the recombination mechanism itself.

2. Recombination Activity in Cell Extracts

From the standpoint of determining the stage at which meiocytes are poised to effect recombination, the most direct approach currently available is to assay cell extracts for the capacity to effect recombination between mutant plasmids. This has been done with two mutants each carrying a mutation in a different end-region of the gene for tetracycline resistance such that recombination between the two results in tetracycline resistance (Hotta *et al.*, 1985b). The rate of recombination can then be assayed by transforming a culture of $recA^-$ *E. coli*. What has emerged from these studies can be summarized as follows. Extracts of meiocytes in general have a much higher recombination activity than those of somatic cells. The highest activity among somatic tissues of mouse is displayed by extracts of bone marrow and these are about 20-fold lower than extracts from a mixture of mouse spermatocytes. In all three organisms analyzed—yeast (*Saccharomyces cerevisiae*), mouse, and lily—entry into meiosis is accompanied by a major increase in the recombinogenic activities of the extracts. The smallest increase is 100-fold in the case of yeast and the largest is 700-fold in the case of lily. In mouse, the increase is about 500-fold. The stage of meiosis at which this occurs cannot be determined in the case of yeast. In mouse spermatocytes the increase occurs during meiotic prophase but precise assignment of stage is not possible. In lily microsporocytes, the situation is clear. The steep rise in recombination activity does not begin until the cells have entered zygotene. Maximum activity is reached in early pachytene and the activity falls by late pachytene. Whatever recombinogenic activity exists during the premeiotic interval, S phase included, it is about $\frac{1}{700}$ of that found in cells at early pachytene. The question of when meiocytes are poised for effecting recombination is answered unambiguously; it occurs at a time when the chromosomes are stably synapsed.

A few characteristics of the recombination activity are worth noting even though a number of important molecular features of the system are still under study. One of the plasmids must be retained in supercoiled form in order to obtain appreciable recombination values. The maximum rates obtained are those in which recombination is measured between one of the mutant plasmids in supercoiled form and the other as linear duplex. No significant levels of recombination activity are obtained with both mutant plasmids in linear duplex form but highly significant levels are obtained with both plasmids in supercoiled form. In the three species tested, the profiles of recombination activity during meiosis were the same regardless of whether one or both mutants were in supercoiled form. Where both were supercoiled the activities at all stages were slightly lower. The *in vitro* system of the meiocytes that mediates recombination spans a spectrum of activities ranging from endonucleolysis to heteroduplex formation and ligation of exchanged strands. Extensive DNA replication is not involved in the process because deoxynucleotide triphosphates are not required. Molecular differences, if any, between the activities of meiotic and somatic cells have not been identified. The one clear piece of information is that meiocytes have considerably greater capacities for recombination than somatic cells.

A pertinent question with respect to meiotic recombination is whether any of the recombination-related enzymes play a role in the activities of the extracts. This question cannot be fully answered, of course, until the components of the extracts have been identified. Only components that are present in limiting amounts would show an effect on addition to the extract. Even so, it is evident that m-*rec* or *rec*A protein stimulates extract activity; mouse spermatocyte extracts are stimulated 2.5-fold whereas lily meiocyte extracts are stimulated 1.5-fold. The strongest stimulation is given by U-protein; twofold in the case of lily and yeast and threefold in the case of mouse. R-Protein has no significant effect.

One important question that merits discussion is the relationship between synapsis and the regulation of the levels of recombination activity. In the case of endonuclease and U-protein, failure of homologous synapsis has no effect on the profiles of their activities during meiosis. R-Protein, on the other hand, is highly responsive to the occurrence of synapsis (Hotta *et al.*, 1979). What has emerged from the studies of recombination activity in meiotic extracts is the evidence that the rise in activity is coordinated with, but is not regulated by, pairing. In lily, the extracts from cells in meiotic prophase display virtually identical profiles of recombination activity regardless of whether they are prepared from chiasmatic or achiasmatic forms. Diploid a/a, α/α, or a/α strains of yeast all show the same patterns of change in level of recombination activity when transferred to sporulation medium. The signal that induces meiosis also induces development of the recombination system; they are coordinated but not interdependent.

C. Summary of Events during the Pachytene Phase

From a biochemical standpoint, the pachytene phase is dominated by two different sets of events. One set relates to the provision of recombinogenic enzymes and the other set relates to the localization of their activities within the chromosomes. We have no information on the molecular factors that determine the ultimate crossover sites. These are extremely few compared with the number of what we presume to be the potential sites for initiating recombination. Such sites are restricted to chromosome regions sharing many families of highly conserved and moderately repeated sequences. Virtually all these sequences are nicked during pachytene, no less than two nicks occurring per PDNA segment. Within the segments, nicks occur at the 5'-end of the PsnDNA sequences. In most, if not all cases, each nick is extended to a gap. The localization of nicking in PsnDNA regions is probably due to PsnRNA. The accessibility of these regions to endonuclease is affected by the replacement of histones with Psn-protein. Such replacement does not occur in the absence of homologous pairing. The enzymatic activities of recombination are carried out in part by enzymes that are common to a variety of tissues and in part by enzymes that seem to be unique to meiosis. The latter are not present to any significant extent during the premeiotic S phase, all of them reaching peak levels during late zygotene and early pachytene. The critical evidence for recombination occurring during meiotic prophase is the capacity of pachytene cell extracts to catalyze genetic recombination.

VI. REGULATION OF MEIOTIC EVENTS

A little is known about the conditions under which cells become committed to meiosis, but virtually nothing is known about the developmental mechanisms that render cells capable of entering meiosis. The simplest situation studied is that of yeast (*Saccharomyces cerevisiae*) in which nitrogen deprivation sets off a chain of events that leads to meiosis in the diploid a/α forms. Even so, the components of that chain are unknown although the commitment to meiosis appears to occur during the premeiotic S phase. The situation is more complex in multicellular organisms. In the case of animals, the germ line is the sole source of meiocytes, while in plants meristematic cells retain the capacity to differentiate into meiocytes. Whether the synthesis of specific proteins during premeiotic cell generations anticipates the ultimate entry of cells into meiosis is unknown, but it is clear that the commitment of cells to meiosis is accompanied by meiosis-specific transcriptions, most of which occur prior to leptotene but some of which occur during meiotic prophase. The transcription of zygRNA and of PsnRNA are examples already discussed. Presumably, message RNA for the various meiosis-specific proteins already described are transcribed either during preleptotene or during early prophase.

Some direct evidence for meiosis-specific transcription of messages coding for proteins comes from the still incomplete analyses of cDNA clones. Among the 1300 clones prepared from meiocyte poly(A)$^+$ RNA, 49 were judged to be meiosis-specific (Appels *et al.*, 1982). The low proportion is illustrative of the fact that the vast majority of transcripts in meiocytes are also present in mitotic tissues. A striking feature of the meiosis-specific cDNA clones is the presence of an internal segment whose sequence is repeated in the genome and is common to the 13 clones that represent the most abundant transcripts in the meiocytes. A subsequent and more thorough analysis of the 49 clones (R. Bouchard, unpublished) revealed that 31 rather than 13 contained inserts that were homologous with the repeated sequence present in the group first analyzed. Moreover, only the cDNA sequences containing the repeated sequence could be shown to hybridize exclusively with poly(A)$^+$ RNA transcripts of meiocytes. Comparisons between the different clones and sequence analyses of some permit the following conclusions: one segment within each of the clones shows a divergence in sequence homology of about 12% and an adjacent region displays a divergence of about 18%. Each of the clones appears to code for a protein of 10–12 kDa that has a high proportion of acidic amino acids. The exclusive presence of the transcripts in meiocytes, their numerous transcriptional sites in the genome, and their high degree of sequence conservation make it probable that the proteins thus derived have a significant meiotic function. The finding of similar sequences in the maize genome (R. A. Bouchard, unpublished) adds emphasis to

TABLE I

Meiosis-Specific RNA Transcripts[a]

Stage	Type of RNA	Proposed function
1. Preleptotene–pachytene	poly(A)$^+$ RNA	Acidic protein; synapsis (?)
2. Leptotene–zygotene	zygRNA	Synapsis
3. Zygotene–pachytene	PsnRNA	Chromatin modification

[a] Each of the RNAs listed is discussed in the text. (1) This RNA group is identified from cDNA clones (Appels *et al.*, 1982). The sequence of the repeat component in this group has been determined by R. Bouchard (unpublished); the sequence points to a highly acidic protein. Its role as part of the synaptonemal complex is being considered. (2) zygRNA is transcribed from zygDNA sequences; the timing of transcription is suggestive of a role in synapsis. (3) PsnRNA is presumed to be the factor that specifies the sites at which chromatin is modified during late zygotene and pachytene.

that conjecture. A summary of the information we have accumulated on meiosis-specific RNA transcripts is given in Table I.

Taken together, the biochemical and molecular evidence now accumulated about the meiotic process points to a number of distinctive features that are unique to it. Broad generalizations are premature because the evidence derives primarily from lilies and to a much lesser extent from mouse spermatocytes. Some speculative generalizations are nevertheless in order. At the molecular level meiotic organization can be described in terms of two categories of events—DNA sites in the chromosomes that function specifically in chromosome synapsis and crossing-over, and RNA transcripts that function either as

TABLE II

Meiosis-Specific Proteins[a]

Stage	Type	Proposed function
1. G_2–zygotene	L-Protein	Suppression of zygDNA replication and synapsis
2. Leptotene–pachytene	R-Protein	Reassociation of ssDNA; synapsis, recombination
3. Leptotene–pachytene	U-Protein	Unwinding DNA; synapsis, recombination
4. Zygotene–pachytene	Endonuclease	Potential initiating sites for recombination
5. Zygotene–pachytene	m-*rec* Protein	Recombination
6. Zygotene–pachytene	Psn-Protein	Endonuclease accessibility

[a] Each of the proteins is discussed in the text. A full description of the Psn-protein has not yet been published. A functional account is provided by Hotta and Stern (1981). Direct evidence for a synaptic or recombinational function is not available for any of the proteins. Demonstration of recombinogenic activity during meiotic prophase has been possible with partially purified cell extracts.

messages or as agents in modifying chromosome structure for synapsis or crossing-over. It is suggested that chromosomes have DNA sites that function exclusively in meiosis, some of the sites being addressed to synapsis and the other sites being addressed to recombination (see Fig. 1). The limited evidence thus far obtained also suggests that the base sequences within these sites are conserved to a significant degree among a broad variety of species.

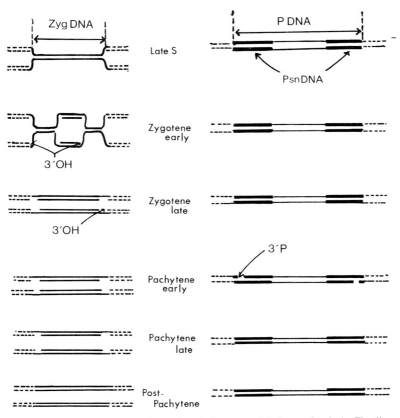

Fig. 1. Events at meiosis-specific DNA sites during sequential phases of meiosis. The diagrams of the single zygDNA and PDNA segments shown are representative of the many segments present in the genome. zygDNA sequences are not replicated at S phase but remain in duplex association. Their replication is initiated at zygotene but small gaps remain at the end of that stage. Replication is completed after pachytene. The precise timing has not been established; it is no earlier than diplotene and no later than meiosis I. The PDNA diagrams are not drawn to scale. As indicated in the text, the flanking PsnDNA segments are between $\frac{1}{5}$ to $\frac{1}{10}$ the length of the internal segments, the latter ranging in size from 800 to 3000 bp. Nicks that are extended to gaps are introduced at early pachytene by meiotic endonuclease. The 3' ends of the strands at the gaps are phosphorylated in contrast with those of the zygotene gaps, which are hydroxylated.

A list of meiosis-specific proteins that appear transiently during meiotic prophase is given in Table II. Some of the proteins are undoubtedly involved in synapsis. However, regulation of synaptic events is difficult to explain even in a preliminary way because of the ambiguity in the observed correlations between reduced synapsis and the reduction in synapsis-related events. The ambiguity lies in our being unable to distinguish between cause and effect with respect to synapsis and the processes analyzed. The reductions in zygRNA and R-protein under achiasmatic conditions might be consequences of an absence of homologous pairing or might be factors in reducing the capacity for pairing. Temporal coordination between synapsis and biochemical events is not the result of a fixed relationship between them. What is clear is that the occurrence of synapsis is directly tied to the occurrence of certain molecular events but not to others, even though both groups of events are normally coordinated with synapsis. One set of

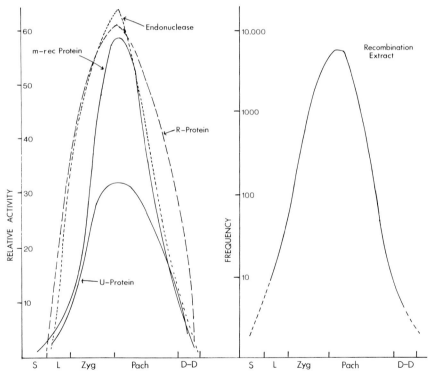

Fig. 2. Meiotic profiles of recombination-related proteins. The ordinates represent relative activities for each of the proteins. Recombination activities of the extracts are plotted on a logarithmic scale. The curves are approximations based on the original data discussed in the text. S, S phase; L, leptotene; Zyg, zygotene; Pach, pachytene; D-D, diplotene/diakinesis.

events, the rendering of PsnDNA regions accessible to endonuclease, regularly follows synapsis but the details of how synapsis regulates the mechanism responsible for effecting accessibility are unknown. If events at PDNA sites serve as a model, it may be concluded that chromosomal site specificity is achieved in meiosis by site-specific alteration of chromatin. This might be the mechanism of choice where activities are transient in nature. If this be so, then it is likely that snRNA molecules are the agents for specifying the sites of chromatin alteration. Given these several considerations, it may well be concluded that no other cellular event in eukaryotes has been analyzed in the same degree of temporal detail as meiosis. The comparison, however, does nothing to reduce the concern that our understanding of the molecular aspects of meiosis is severely limited.

ACKNOWLEDGMENTS

We wish to thank the National Science Foundation, the American Cancer Society, and the National Institute for Child Health and Human Development for support of the various studies on which this article is based.

REFERENCES

Appels, R., Bouchard, R. A., and Stern, H. (1982). cDNA clones from meiotic-specific poly(A)+ RNA in *Lilium:* Homology with sequences in wheat, rye, and maize. *Chromosoma* **85,** 591–602.

Bouchard, R. A., and Stern, H. (1980). DNA synthesized at pachytene in *Lilium:* A non-divergent subclass of moderately repetitive sequences. *Chromosoma* **81,** 349–363.

Callan, H. G. (1972). Replication of DNA in the chromosomes of eukaryotes. *Proc. R. Soc. London, Ser. B* **181,** 19–41.

Carpenter, A. T. C. (1984). Recombination nodules and the mechanism of crossing-over in *Drosophila. Symp. Soc. Exp. Biol.* **38,** 233–244.

Darlington, C. D. (1937). "Recent Advances in Cytology." McGraw-Hill (Blakiston), New York.

Friedman, B. E., Bouchard, R. A., and Stern, H. (1982). DNA sequences repaired at pachytene exhibit strong homology among distantly related higher plants. *Chromosoma* **87,** 409–424.

Holliday, R. (1984). The biological significance of meiosis. *Symp. Soc. Exp. Biol.* **38,** 381–394.

Hotta, Y., and Stern, H. (1971a). Analysis of DNA synthesis during meiotic prophase in *Lilium. J. Mol. Biol.* **55,** 337–355.

Hotta, Y., and Stern, H. (1971b). A DNA-binding protein in meiotic cells of *Lilium. Dev. Biol.* **26,** 87–99.

Hotta, Y., and Stern, H. (1974). DNA scission and repair during pachytene in *Lilium. Chromosoma* **46,** 279–296.

Hotta, Y., and Stern, H. (1975). Zygotene and pachytene-labeled sequences in the meiotic organization of chromosomes. *In* "The Eukaryote Chromosome" (W. J. Peacock and R. D. Brock, eds.), pp. 283–300. Aust. Acad. Sci., Canberra.

Hotta, Y., and Stern, H. (1976). Persistent discontinuities in late replicating DNA during meiosis in *Lilium*. *Chromosoma* **55**, 171–182.

Hotta, Y., and Stern, H. (1978a). Absence of satellite DNA synthesis during meiotic prophase in mouse and human spermatocytes. *Chromosoma* **69**, 323–330.

Hotta, Y., and Stern, H. (1978b). DNA unwinding protein from meiotic cells of *Lilium*. *Biochemistry* **17**, 1872–1880.

Hotta, Y., and Stern, H. (1979). The effect of dephosphorylation on the properties of a helix-destabilizing protein from meiotic cells and its partial reversal by a protein kinase. *Eur. J. Biochem.* **95**, 31–38.

Hotta, Y., and Stern, H. (1981). Small nuclear RNA molecules that regulate nuclease accessibility in specific chromatin regions of meiotic cells. *Cell* **27**, 309–319.

Hotta, Y., and Stern, H. (1984). The organization of DNA segments undergoing repair synthesis during pachytene. *Chromosoma* **89**, 127–137.

Hotta, Y., Bennett, M. D., Toledo, L. A., and Stern, H. (1979). Regulation of R-protein and endonuclease activities in meiocytes by homologous chromosome pairing. *Chromosoma* **72**, 191–201.

Hotta, Y., Tabata, S., and Stern, H. (1984). Replication and nicking of zygotene DNA sequences: Control by a meiosis-specific protein. *Chromosoma* **90**, 243–253.

Hotta, Y., Tabata, S., Stubbs, L., and Stern, H. (1985a). Meiosis-specific transcripts of a DNA component replicated during chromosome pairing: Homology across the phylogenetic spectrum. *Cell* **40**, 785–793.

Hotta, Y., Tabata, S., Bouchard, R. A., Piñon, R., and Stern, H. (1985b). General recombination mechanisms in extracts of meiotic cells. *Chromosoma* **93**, 140–151.

Hotta, Y., de la Peña, A., Tabata, S., and Stern, H. (1985c). Control of enzyme accessibility to specific DNA sequences during meiotic prophase by alterations in chromosome structure. *Cytologia* **50**, 611–620.

Howell, S. H., and Stern, H. (1971). The appearance of DNA breakage and repair activities in the synchronous meiotic cycle of *Lilium*. *J. Mol. Biol.* **55**, 357–378.

Ito, M., and Takegami, M. H. (1982). Commitment of mitotic cells to meiosis during the G2 phase of premeiosis. *Plant Cell Physiol.* **23**, 943–952.

Kurata, N., and Ito, M. (1978). Electron microscope autoradiography of ^{3}H-thymidine incorporation during the zygotene stage in microsporocytes of lily. *Cell Struct. Funct.* **3**, 349–356.

Moses, M. J., Dresser, M. E., and Poorman, P. A. (1984). Composition and role of the synaptonemal complex. *Symp. Soc. Exp. Biol.* **38**, 245–270.

Ninnemann, H., and Epel, B. (1973). Inhibition of cell division by blue light. *Exp. Cell Res.* **79**, 318–326.

Radding, C. M. (1978). Genetic recombination: Strand transfer and mismatch repair. *Annu. Rev. Biochem.* **47**, 847–880.

Roth, T. F., and Ito, M. (1967). DNA-dependent formation of the synaptonemal complex at meiotic prophase. *J. Cell Biol.* **35**, 247–255.

Sakaguchi, K., Hotta, Y., and Stern, H. (1980). Chromatin-associated DNA polymerase activity in meiotic cells of lily and mouse; its stimulation by meiotic helix-destabilizing protein. *Cell Struct. Funct.* **5**, 323–334.

Sega, G. A. (1974). Unscheduled DNA synthesis in the germ cells of male mice exposed *in vivo* to the chemical mutagen, ethylmethanesulfonate. *Proc. Natl. Acad. Sci. U.S.A.* **71**, 4955–4959.

Stern, H., and Hotta, Y. (1969). Biochemistry of meiosis. *In* "Handbook of Molecular Cytology" (A. Lima-de-Faria, ed.), pp. 520–539. North-Holland Publ., Amsterdam.

Stern, H., and Hotta, Y. (1977). Biochemistry of meiosis. *Philos. Trans. R. Soc. London, Ser. B* **277**, 277–293.

10. The Biochemistry of Meiosis

Stern, H., and Hotta, Y. (1978). Regulatory mechanisms in meiotic crossing-over. *Annu. Rev. Plant Physiol.* **29,** 415–436.

Stern, H., and Hotta, Y. (1980). The organization of DNA metabolism during the recombinational phase of meiosis with special reference to humans. *Mol. Cell. Biochem.* **29,** 145–158.

Stern, H., and Hotta, Y. (1984). Chromosome organization in the regulation of meiotic prophase. *Symp. Soc. Exp. Biol.* **38,** 161–175.

Stern, H., and Hotta, Y. (1985). Molecular biology of meiosis: Synapsis-associated phenomena. *In* "Aneuploidy: Etiology and Mechanisms" (V. Dellarco, P. E. Voytek, and A. Hollaender, eds.), pp. 305–316. Plenum, New York.

11

Proteins of the Meiotic Cell Nucleus

MARVIN L. MEISTRICH AND WILLIAM A. BROCK

Department of Experimental Radiotherapy
The University of Texas M. D. Anderson Hospital and
Tumor Institute at Houston
Houston, Texas 77030

I. INTRODUCTION

Meiosis is distinguished from mitosis primarily by the way the DNA and the chromosomes are affected in the process. Homologous chromosomes pair during the long meiotic prophase, but not during the short prophase of mitosis. This pairing is apparently necessary for genetic recombination, which involves a physical interchange of DNA strands between the homologues. Other metabolic events involving DNA also occur during meiotic prophase and include DNA repair and possibly gene conversion. At the first meiotic division the homologous chromosomes move to opposite poles, but the chromatids do not separate; whereas at mitosis one chromatid from each chromosome moves to each pole. The processes of chromosome pairing and genetic recombination are intranuclear; chromosome segregation also involves chromosome elements (centromeres) as well as their interaction with cytoplasmic elements (spindles). To achieve an understanding of the molecular basis of this entire process of meiosis, it is essential to characterize the biochemical events that occur in the cell nucleus.

The biochemical alterations in the DNA have been studied by Stern and co-workers (Hotta *et al.*, 1966; Friedman *et al.*, 1982). Protein is the other major component of the nucleus and plays an all-important role in meiosis. Proteins, such as the core histones, are responsible for the configuration of the DNA into its nucleosomal structure; histone 1 (H1) and the high-mobility-group (HMG) proteins are responsible for higher-order DNA structures. The synaptonemal complex, which is the structure along which the chromosomes align, is largely composed of protein (Comings and Okada, 1972). In addition, enzymes and DNA-binding proteins, involved in the metabolism of DNA essential for chromosomal rearrangements during meiosis, are present at high levels in meiotic cells (Howell and Stern, 1971; Hotta and Stern, 1971a). Furthermore, proteins are the primary constituent of the nuclear matrix (Berezney and Coffey, 1977), which is responsible for the structural order and integrity of the nucleus. Unique proteins or protein variants, for example, particular histones, HMG proteins, proteins of the synaptonemal complex–nuclear matrix structure, and proteins that act on DNA, that are absent from mitotically dividing or interphase somatic cells have been found in the meiotic cells of the testis. Also, inhibition of protein synthesis with cycloheximide immediately blocks meiotic cell development (Hotta *et al.*, 1968). Thus, studies of these newly synthesized, unique proteins are extremely important for gaining an understanding of the distinctive events of meiosis.

II. METHODS FOR STUDYING NUCLEAR PROTEINS IN MEIOSIS

Methods for studying the proteins of meiotic nuclei fall into two categories. These are cytochemical, including immunocytochemistry, and biochemical.

Cytochemical studies have employed three methods: (1) determining the reactivity of classes of macromolecules with specific stains (Moses and Coleman, 1964); (2) evaluating the loss of structures or the loss of reactivity with a stain after treatment of the preparations of cells with agents that selectively remove specific classes of macromolecules (Solari, 1972); or (3) assessing the presence of specific moieties by immunocytochemistry. These studies can be performed on heterogeneous populations of cells because the meiotic cell or nucleus can be identified by its morphological characteristics. Except for immunocytochemistry, these methods are not specific for recognition of different molecules within a broad category. Cytochemical methods do have an advantage over biochemical methods in that very precise stages of meiosis can be identified and molecules can be spatially localized within the nucleus.

Biochemical studies of meiotic nuclei require the preparation of homogeneous populations of cells or nuclei at specific stages of the process. Studies of male germ cells of plants, such as *Lilium,* have been possible because of the synchronous development of germ cells in the anther (Hotta and Stern, 1963). In contrast, the mammalian testis contains germ cells at all stages of development plus a variety of nongerminal cells, so that cell separation methods have been required to obtain reasonably pure populations of meiotic cells for such biochemical studies. Velocity sedimentation, either at unit gravity (Meistrich *et al.,*

11. Proteins of the Meiotic Cell Nucleus

1973) or by centrifugal elutriation (Grabske *et al.*, 1975), and equilibrium density centrifugation (Meistrich *et al.*, 1981) have all been employed to achieve this goal. Sufficient numbers of cells and nuclei have been obtained by these methods to allow quantitation of the proteins of these cells by gel electrophoresis. The advantage of the biochemical over cytochemical approach is that the levels of many different proteins can be quantitatively determined.

The greatest enrichments of mammalian meiotic cells and their nuclei have been achieved using the mouse and the rat. Unit gravity sedimentation of cells prepared from testes with trypsin was shown to yield populations of mid- to late pachytene spermatocytes that were 79% pure in the case of the mouse (Meistrich *et al.*, 1973) and 61% pure in the case of the rat (Platz *et al.*, 1975). A subsequent modification of the cell suspension procedure employing an initial collagenase digestion increased the purity of mouse spermatocytes to 85% (Romrell *et al.*, 1976). Using a similar procedure, Chandley *et al.* (1977) were also able to obtain enrichments to about 30 to 55% for earlier stage spermatocytes.

In order to increase the numbers of cells that could be separated and decrease the separation time, centrifugal elutriation was applied. Pachytene spermatocytes from the mouse or rat could be obtained in 80 or 86% purity, respectively (Meistrich *et al.*, 1977; Meistrich, 1977).

Further enrichment of the mid- to late pachytene spermatocytes from the rat was achieved by a second step of purification, equilibrium density centrifugation (Meistrich *et al.*, 1981); purities of up to 98% were achieved by elutriation followed by equilibrium density centifugation in Percoll gradients. Similarly, unit gravity sedimentation followed by equilibrium density centrifugation in Percoll yielded 95% pure mid- to late pachytene spermatocytes from the mouse (Stern *et al.*, 1983).

Purification of leptotene, zygotene, and early pachytene spermatocytes has proved more difficult than that of the much larger mid- to late pachytene spermatocytes because they are more similar in size to other testicular cell types. To reduce the number of contaminating cell types, researchers have employed immature animals with incomplete spermatogenesis. Separation of cells from 18-day-old mice by unit gravity sedimentation yielded a population composed of 51% leptotene plus zygotene primary spermatocytes (Bellvé *et al.*, 1977). Separation of cells from immature rats (age range: 22–28 days old) by centrifugal elutriation followed by equilibrium density centrifugation yielded populations composed of about 55% zygotene plus early pachytene spermatocytes and about 65% mid-pachytene spermatocytes (Meistrich *et al.*, 1985; Bucci *et al.*, 1986).

Despite the only moderate enrichments of the leptotene plus zygotene and early pachytene spermatocytes, biochemical studies of these cells in separated fractions can be successful provided the nature of the contaminating cell type is known and its contribution to the protein of interest can be quantitated. In this way, it has been shown that preferential synthesis of the testis-specific forms of histone H2A and H2B occurs in the early pachytene stages (Meistrich et al., 1985), whereas the nicking and repair of DNA (Hotta et al., 1977), the increase in levels of the high-mobility-group protein HMG2 (Bucci et al., 1986), and the synthesis of the testis-specific form of H1 occur predominantly in later pachytene stages.

III. HISTONES AND HMG PROTEINS OF MEIOTIC CELLS

A. Studies in the Male Rat

Histones constitute the major class of proteins responsible for the local structure of the DNA within the chromatin. H2A, H2B, H3, and H4 form the nucleosomal cores around which 140 base pairs of DNA are wound in approximately two superhelical turns. The small HMG proteins, HMG14 and HMG17, also bind to the cores and may modify their structure and function. These proteins may have a functional role in processes such as replication, transcription, or repair, in which they may have to be dissociated from or at least slide along the DNA. The linker region of 30 to 60 base pairs of DNA between the adjacent nucleosomes is partially covered by H1 and also by the large HMG proteins. Their location as outer nucleosomal proteins has led to the proposal that interactions between H1 molecules (and possibly HMGs as well) are responsible for the higher-order, superhelical configuration of the nucleosome beads that results in the 30-nm diameter chromatin fibers visible by electron microscopy.

Four of the histone classes, H1, H2A, H2B, and H3, possess nonallelic variants in somatic and germinal cells (Lennox and Cohen, 1984; Zweidler, 1984). The variants of each class of histone differ in primary amino acid sequence and are products of different gene loci. The known variants for each of the somatic histones are indicated in Table I. They can be separated by polyacrylamide gel electrophoresis on sodium dodecyl sulfate (SDS) gels or acetic acid–urea–Triton (AUT) gels (Fig. 1). The genes for each type of histone are present in multiple copies in the mammalian genome. The histones found in somatic cells are conserved in evolution, and with the exception of some of the H1 variants, possess identical electrophoretic mobilities in different species.

Additional variants of H1, H2A, H2B, and H3, designated H1t (Seyedin et al., 1981), TH2A (Trostle-Weige et al., 1982), TH2B (Shires et al., 1975), and TH3 (Trostle-Weige et al., 1984), respectively, are prominently observed in extracts of the rat testis (Fig. 1). These variants are found only in germinal cells, and it has often been proposed that they must have a specific role in meiosis. In addition, the germ cells of the rat testis contain high levels of two other histone variants, H1a (Seyedin and Kistler, 1979; Bucci et al., 1982) and H2A.X (Bran-

TABLE I
Variant Forms of Histones in the Rat

Histone class	Variants in somatic cells	Variants in germinal cells
H1	H1a, H1b, H1c, H1d, H1e, H1°	H1t, H1a
H2A	H2A.1, H2A.2, H2A.X, H2A.Z	TH2A, H2A.X
H2B	H2B.1 (H2B.2 in mouse only)	TH2B
H3	H3.1, H3.2, H3.3	TH3
H4	(no variants)	

son et al., 1975; Trostle-Weige et al., 1984); these are present at low levels in somatic cells but are major histones of the testis, suggesting that they also might exhibit a specific function in meiosis.

The testis is also unique with regard to the HMG proteins, containing higher levels and greater heterogeneity of HMG2 than any other tissue (Bucci et al., 1985). These increased levels are localized to the germ cells, which reveal three electrophoretically distinct bands of HMG2 (Bucci et al., 1986). Spermatogonia and early spermatocytes contain all three bands, whereas late pachytene spermatocytes and round spermatids contain predominantly the fast band and perhaps the middle band too, and somatic tissues only the slowest band. However, there is no conclusive evidence for the existence of a testis-specific or -enriched variant, since no amino acid differences could be observed between the HMG2 proteins extracted from rat testis and somatic tissues (Bucci et al., 1984). Furthermore, in rat spermatogenic calls, radioactive lysine is incorporated mainly into the slow (or in some cases intermediate) form, suggesting that the fast band(s) might correspond to posttranslational modifications of the slow forms (Bucci et al., 1986).

The sequence of replacement of the histones during the process of spermatogenesis is described in Table II and Fig. 2. Three of these proteins, H1a, H2A.X, and TH3, are already at high levels in spermatogonia and remain present throughout meiotic prophase. TH2A and TH2B first appear and achieve high levels during early meiotic prophase. On the other hand, H1t first appears and is present at low levels in the early meiotic stages and only reaches its maximal level after chromosome pairing is complete. No further histone synthesis or change in relative levels of the variants occurs between the first meiotic division and the end of the round spermatid stage. The above sequence suggests possible roles for H1a, H2A.X, and TH3 during spermatogonial proliferation, for TH2A and TH2B during early meiotic prophase at about the time of chromosome pairing, and for H1t later in meiosis; it minimizes their importance in postmeiotic events.

The increase in HMG2 appears to occur concomitantly with that of H1t, in the

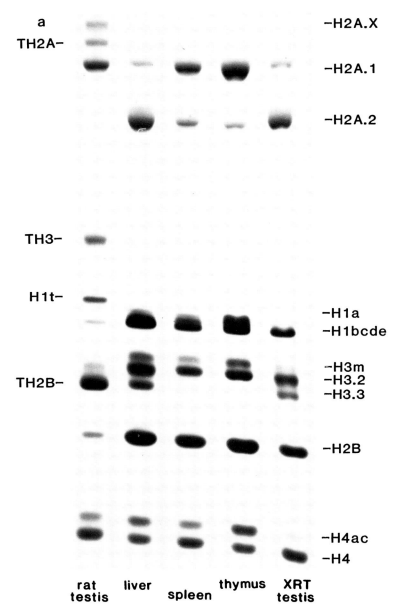

Fig. 1. Separation of somatic and testis histones from rat tissues by electrophoresis in gels containing (a) acetic acid–urea–Triton X-100 or (b) sodium dodecyl sulfate (H1 region only). Testis-specific histones are marked on the left of the gels and testis-enriched and somatic histones on the right. The lane marked XRT testis contains an extract from adult rat testes that had been sterilized by radiation and consisted only of somatic cells.

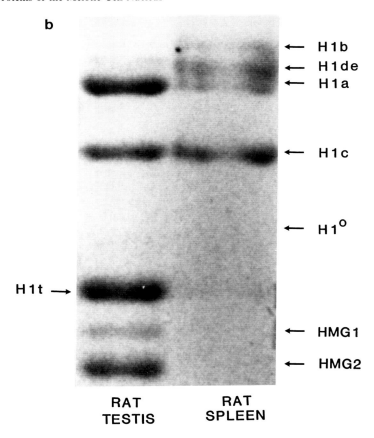

mid- to later part of the pachytene stage. Whereas the total level of H1 protein appears to remain constant during development of pachytene spermatocytes, the levels of HMG proteins increase during this stage (Bucci *et al.*, 1984). The preferential binding of HMG1 and HMG2 to single-stranded, as opposed to double-stranded, DNA (Isackson *et al.*, 1979) would be consistent with a role for HMG2 in the unwinding of local regions of DNA, which must occur during genetic recombination and other events of the meiotic prophase.

It is known that the testis histone variants assume the same intranucleosomal locations as their somatic counterparts (Rao *et al.*, 1982; Bhatnagar and Faulkner, 1983), but the distribution of the testis variants in the chromatin of meiotic cells is not known. The possibility that testis variants only form nucleosomes with each other is ruled out by the different extents of their replacement of the somatic histones and by the observation that they appear at different stages. It is

TABLE II

Levels and Synthesis of Testis-Specific and Testis-Enriched Histone Variants and HMG Proteins in Premeiotic and Meiotic Cells of Rat Testes

Histone class	Variant form	Spermatogonia	Early primary spermatocytes	Late pachytene
H1	H1a	High protein levels	Maximal protein levels	Slight decline in protein
		Actively synthesized	Lower rate of synthesis	Lower rate of synthesis
	H1t	Absent	Low protein levels	Maximal protein levels
		No synthesis	Synthesis begins	Maximal rate of synthesis
H2A	H2A.X	Protein present	Protein level unchanged	Protein level unchanged
		Synthesis	Synthesis	Little or no synthesis[a]
	TH2A	Absent or low levels	Maximal protein levels	Protein level unchanged
		Synthesis at end of stage[a]	Actively synthesized	Continued synthesis
H2B	TH2B	Absent or at low levels	High protein levels	Slight increase in protein
		Synthesis at end of stage[b]	Maximal rates of synthesis	Continued synthesis
H3	TH3	Protein present	Slight increase in protein	Protein level unchanged
		Actively synthesized	No synthesis	No synthesis
HMG2	HMG2.2 HMG2.3	Moderate levels of all three forms (10 μg/mg DNA)		Maximal level of rapidly migrating forms (56 μg/mg DNA)
		Synthesis	Synthesis	Most active synthesis

[a] A marked reduction in radioactivity in this region was observed after extraction of nonhistone proteins with 0.7 M NaCl (Trostle-Weige *et al.*, 1982).

[b] The high specific activity of the protein in the separated population of spermatogonia and preleptotene spermatocytes (Meistrich *et al.*, 1985) indicated that the synthesis occurs at the latest stage of cells present, probably the preleptotene spermatocytes.

Fig. 2. Relative levels of histones and estimates of their rates of synthesis in spermatogonia and spermatocytes (calculated from Meistrich *et al.*, 1985). The panels on the left indicate the percentage of protein of each class of histone contributed by the testis-specific (———), testis-enriched (- - - -), and somatic (- · · -) variants. The panels on the right indicate an estimation of the relative rates of synthesis of the proteins based on the amounts of [^{3}H]arginine incorporated into the different histones corrected by dividing by the percentage of arginine in each protein. Comparison of different cell types is limited by possible differences in uptake of labeled precursor and in pool sizes, which could result in differences in specific activities of the arginine pool. The incorporation into late pachytene spermatocytes from adult rats was corrected for comparison with the other cell types from immature rats by assuming that the population of spermatogonia plus early primary spermatocytes isolated from

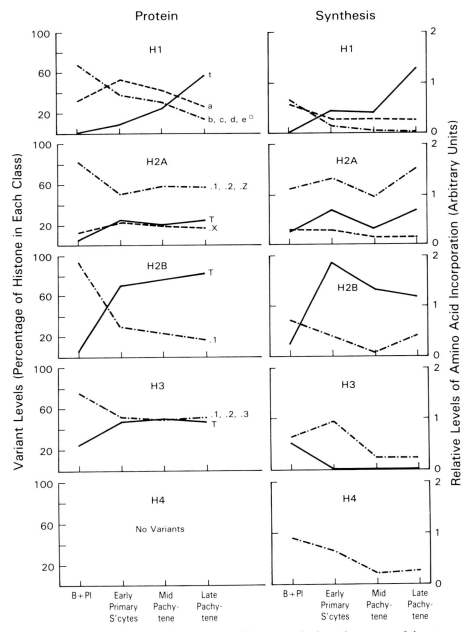

the same adult rats should have the same rate of histone synthesis as the average of these two populations isolated from immature rats.

not known whether there is a preferential association of testis-specific histones with each other. There is an indication from immunocytochemical studies that, in somatic cell and spermatid nuclei, H2A.X is preferentially localized in the nucleolar region (Bhatnagar *et al.*, 1984), but there appears to be no localization within spermatocyte nuclei (Bhatnagar *et al.*, 1985).

It is not known if the replacement of somatic by testis histones alters the properties of the chromatin. One consistent trend is that all six testis histone variants migrate more slowly in acetic acid-urea polyacrylamide gels than do their somatic counterparts. These gels separate proteins primarily on the basis of charge; molecular weight differences cannot be the cause of slower electrophoretic mobilities in acid-urea gels for five of the six testis variants, since they migrate either as fast as or faster than their somatic counterparts upon SDS–gel electrophoresis. The exception is H2A.X, which migrates more slowly than the other H2A forms in SDS gels. Thus, the testis histones, with the possible exception of H2A.X, are less electropositive than the somatic forms and hence should display weaker binding to DNA. Furthermore, H1t differs from the other H1 variants in that it contains a higher proportion of arginine and less lysine. However, the significance of these differences for meiosis is not known.

The possibility also exists that the most significant feature of the testis-specific histones is not their physiocochemical properties but rather the regulation of their synthesis (Brock *et al.*, 1980). The high rate of synthesis of histones (particularly core histones) observed during meiotic prophase is unique among non-S-phase cells (Brock *et al.*, 1983). Subsets of the histones can be synthesized in cells not in S phase, but the overall rate of histone synthesis is much lower than in S (Wu and Bonner, 1981). Although the rates of synthesis shown in Fig. 2 should be regarded only as estimates since they were not corrected for possible variations in the specific activities of the arginine pools, they do indicate the high level of histone synthesis during meiotic prophase. The histones that are newly synthesized during meiotic prophase appear to replace already existing histones, which are presumably degraded. This replacement process may result in exposure of regions of the DNA, which may be required for chromosome pairing, recombination, or repair processes. Another possibility is that histone replacement is a prerequisite for a rearrangement of the chromatin necessary for the shift in pattern of gene expression in spermatocytes and round spermatids, in which most somatic genes are repressed and numerous testis-specific variants of these genes activated (Bellvé, 1979; Kleene *et al.*, 1984).

B. Generalizations to Other Mammalian Species

Since the process of spermatogenesis has many morphological similarities in different mammalian species, we expect that important biochemical processes are also similar. Although testis-specific histones are a general feature of mammalian spermatogenesis, there are qualitative and quantitative differences between species.

Histones have also been characterized in the mouse testis. Both H1a and H1t are present, but mouse H1a displays different electrophoretic mobilities than those of the rat (Lennox and Cohen, 1984), whereas H1t has the same mobility in both species. Levels of H2A.X [also designated H2A.4 by Zweidler (1984) and protein "A" by Bellvé *et al.* (1977)] in mouse testes appear comparable to those in the rat and the electrophoretic mobility is the same. On the other hand, no protein comigrating with rat TH2A is observed in the mouse, and contrary to a statement by Zweidler (1984), we have been unable to resolve any additional testis-specific form of H2A in our electrophoretic systems. The mouse has high levels of a testis-specific variant of H2B, designated H2B.S (Shires *et al.*, 1975; Bhatnagar and Bellvé, 1978; Zweidler, 1984), with electrophoretic mobility in AUT gels slower than that of rat TH2B. A testis-specific variant of H3, called H3.S (Zweidler, 1984), is present in the mouse and has both the same electrophoretic mobility and relative level of abundance as does TH3 in the rat (Trostle-Weige *et al.*, 1984). Synthesis of histones, including some of the testis variants, during meiotic prophase has been shown to occur in the mouse (Goldberg *et al.*, 1977; Bhatnagar *et al.*, 1985).

Other mammals show more divergence from the rat. H1t has been identified in hamster, rabbit, boar, bull, rhesus monkey, chimpanzee, and human testes (Shires *et al.*, 1975; Seyedin *et al.*, 1981; Seyedin and Kistler, 1983; P. K. Trostle-Weige, unpublished results). The relative levels of H1t in some of these species appear to be similar to those in the rat. The situation with H2A is different. We found that the rabbit has a protein that comigrates with rat TH2A; the guinea pig testis shows an extra band migrating just faster than H2A.1; in the dog, bull, rhesus monkey, chimpanzee, and baboon, no extra bands are observed in this region of the gel. However, a band with the electrophoretic mobility of H2A.X is present in testes of all of these species, but elevated levels are observed only in the mouse, guinea pig, dog, and baboon. Electrophoretic analysis has revealed a distinct band corresponding to a testis-specific form of H2B in rodents (mouse, rat), rabbit, and primates (rhesus monkey, chimpanzee, baboon, man) (Shires *et al.*, 1975; Bhatnagar *et al.*, 1983; Wattanaseree and Svasti, 1983; P. K. Trostle-Weige, unpublished observations). The electrophoretic mobility on AUT gels of the testis-specific form of H2B in most of these species differs from that observed in the rat. In the testes of the guinea pig, dog, and bull, the levels of the somatic form of H2B correspond to those of the other core histones, making it unlikely that an electrophoretically distinct H2B variant is present in levels as high as in the rat testis. No species other than the mouse shows a band comigrating with the TH3 band of the rat or for that matter any other band that could be a candidate for a unique H3 variant.

The presence of high levels of HMG2 and the enrichment of a rapidly migrating band are common to the testes of most mammalian species (Bucci et al., 1985). The electrophoretic mobility of HMG2 varies among species, indicating a degree of conservation as in the testis-specific histones rather than in the somatic histones.

Differences in quantitative amounts of histone variants and the apparent absence of some variants in certain species would seem to argue against a requirement for all variant proteins for the completion of meiosis. A more general phenomenon might be the activation of specific histone genes in the testis providing new histones for the turnover that may be required prior to and during meiosis. In some species this tissue-specific histone gene may have evolved, perhaps because of some selective advantage to having a different form of the protein, to encode a variant protein. In other species, this evolution may not have occurred and there might be a testis-specific histone gene that codes for the same protein as the somatic gene from which it must have arisen by gene duplication.

C. Generalizations to Nonmammalian Species

The first observation of a germ-cell-specific histone was actually in the male meiotic cells of *Lilium* (Sheridan and Stern, 1967), in which an extra electrophoretic band with the characteristics of H1 was found. As is the case in mammals, this histone has a slower electrophoretic mobility and higher arginine content than somatic H1 proteins. This histone is undetectable in premeiotic interphase, but increases during leptotene and zygotene achieving maximal levels at pachytene (Strokov et al., 1973). Although no variants of the core histones have yet been found in *Lilium*, the synthesis of the core histones and both the somatic and "meiotic" forms of H1 have been shown to occur during meiotic prophase (Bogdanov et al., 1973; Bogdanov and Strokov, 1976).

Synthesis of histones appears to occur in meiotic cells of the cricket *Gryllus (Acheta) domesticus*. A combination of cytochemical and microspectrophotometric measurements demonstrated an accumulation of histone from a level of 3C (three-fourths of that expected based on DNA content) in leptotene, zygotene, and early pachytene to 4C in late pachytene (Bogdanov et al., 1968). In the newt *Triturus vulgaris*, histone doubling also appears to be delayed beyond the termination of premeiotic DNA synthesis, however, the levels of histone reach 4C in the zygotene stage which is earlier than in the cricket (Bogdanov and Antropova, 1971). Although, as in mammals, histone synthesis occurs during meiotic prophase in the cricket and the newt, there appear to be two essential differences: (1) No stages during meiotic prophase in the rat showing signs of greatly lowered histone : DNA ratio have been isolated, although the leptotene and zygotene stages have not been obtained in sufficient purity for analysis

(Bucci, 1983; Meistrich *et al.*, 1985), and (2) no histone variants have yet been identified in meiotic prophase cells of the cricket (Kaye and McMaster-Kaye, 1974) or the newt.

In another amphibian, the frog *Rana pipiens*, one difference has been observed between somatic and testicular histones. Testis tissue, as well as spermatozoa, shows upon electrophoresis a rapidly migrating form of a protein with some of the properties of H1 (Alder and Gorovsky, 1975), but neither the precise nature of this protein nor its stage of synthesis during spermatogenesis has been determined.

Sperm-specific histone variants have been extensively characterized in the sea urchin, however, the testes have not been analyzed for their histone composition. The sperm-specific histones include variants of H1 and H2A and as many as three H2B variants in some species (Brandt *et al.*, 1979; von Holt *et al.*, 1979). Other echinoderms, such as the starfish, also have a sperm-specific form of H1 (Zalenskaya *et al.*, 1980). These male germ cell-specific H1 variants also have a higher arginine content than do the somatic forms, as is the case in mammals and plants. These variants are most likely first synthesized during spermatogenesis, but no data are available to determine if they are synthesized in premeiotic and meiotic cells, as are the mammalian testis histones, or during spermiogenesis, as is the case with protamines (Ling *et al.*, 1971) and other spermatidal basic proteins (Grimes *et al.*, 1977). In vertebrates, spermatidal basic proteins and protamines replace the histones as the spermatid nucleus condenses.

The occurrence of tissue-specific histone variants as the major basic proteins of the highly condensed sea urchin sperm nucleus raises the question of whether the role of testis histone variants is in meiosis; it could be that they are important for nuclear condensation, which occurs in spermatids. The synthesis of spermatidal basic proteins and protamines at the time of nuclear condensation in mammals, as well as in many other species, may represent an evolutionary step in which the role of the testis-specific histones in condensation was superceded. The testis-specific histones still might be important for the initiation of condensation or for facilitating the replacement of histones by spermatidal basic proteins or protamine.

D. Female Meiotic Cells

Since the genetic aspects of meiosis are identical in males and females, the question of whether the histone variants described above are in oocytes is relevant. The results of such a study could indicate whether these histones are important for processes common to males and females or in other steps of the spermatogenic process. Because of the existence of high levels of specific histone variants in the rat testis, we chose rat oocytes for comparison.

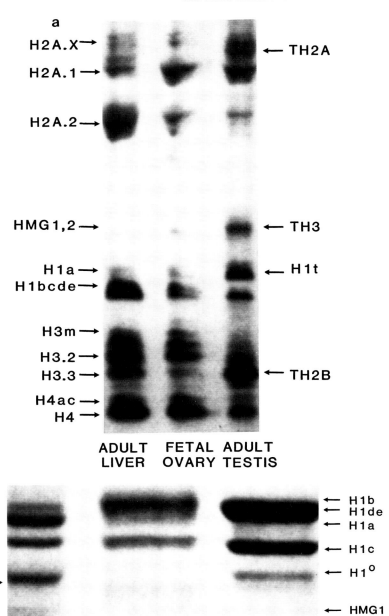

The fraction of cells in the adult ovary that are germ cells is very small because of atresia of these cells. However, ovaries of 18-day postcoitum female embryos contain about 40% meiotic germ cells (Beaumont and Mandl, 1962). We compared histones extracted from adult testis and liver with the fetal ovarian histones (Fig. 3). All samples contained the complete array of somatic histones, and the testis contained the variants described above. Figure 3a shows the separation of histones extracted from rat liver, fetal ovary, and adult testis by electrophoresis in AUT gels. Compared to adult testis, fetal ovary contains little or no detectable TH2A, TH2B, or TH3. The faint protein band migrating in the region of TH3 could represent some HMG1 or HMG2. The protein in fetal ovary migrating with TH2B can be predominantly H3.3. The small amount of protein comigrating with H1t can be shown to be H1a when the histones from these same tissues are separated on an SDS gel (Fig. 3b). Fetal ovary shows no H1t but does contain H1a (the testis-enriched but not testis-specific form of H1). Histological analysis revealed that occytes comprised 35% of the total cells in the fetal ovaries. Our method of dissection resulted in 40% contamination by nonovarian tissue, reducing the total percentage of germ cells to 20%. We conclude from these numbers that the same high levels of histone variants found in the spermatocytes are not present in oocytes, but the presence of these histones in oocytes in reduced amounts cannot be excluded at this point. Thus it appears that the testis-specific histone variants do not play an essential role in those events that are common to both male and female meiosis. They may be involved in chromosome pairing and genetic recombination, but they do not appear to be required for these processes.

Although mature ova can be readily obtained from sea urchins, isolation of large populations of oocytes has required the separation of gonadal cells by velocity sedimentation (Cognetti *et al.*, 1977). The synthesis of H2B in these oocytes, arrested in meiotic prophase, has been demonstrated, but these histones appear to be stored in the cytoplasm for use during the cleavage divisions (Salik *et al.*, 1981), rather than being associated with the nucleus for participation in the meiotic process.

IV. PROTEINS OF STRUCTURAL SUBCOMPONENTS OF THE MEIOTIC CELL NUCLEUS

Isolation of the synaptonemal complex would be an important step for further biochemical investigations of meiosis. In several recent investigations the synap-

---

Fig. 3. Separation of histones from male and female tissues of the rat containing high numbers of meiotic cells (the adult testis and the 18-day postcoitum fetal ovary were used) and the adult liver for comparison. Separations were performed in gels containing (a) acetic acid–urea–Triton X-100 or (b) sodium dodecyl sulfate (H1 region only).

tonemal complex was shown to be part of purified nuclear matrix structures. These preparations, which will be referred to here as synaptonemal complex/nuclear matrix (SC/NM) structures, have been isolated from male meiotic cells in the mouse and rat (Ierardi et al., 1983; Li et al., 1983) and from female meiotic cells of the flour moth (Raveh and Ben-Ze'ev, 1984). There is one report of isolated synaptonemal complex structures, but that was not done with purified meiotic cells (Walmsley and Moses, 1981).

In these isolation methods nuclei are prepared from pachytene cells with Triton X-100. Many of the proteins are removed with NaCl concentrations up to 2 M and the nucleic acids digested (with DNase I and/or micrococcal nuclease and in some preparations with RNase), which facilitates further protein removal. These treatments remove 97% of the RNA, 98–99% of the DNA, and 85–92% of the nuclear proteins, yet many of the characteristics of the synaptonemal complex, including the two lateral and one central elements, a presumptive centromere, attachment plaques to the nuclear envelope, and the ability to be selectively stained with silver nitrate, are preserved.

The SC/NM structures isolated from spermatocytes possess a complex series of proteins when extracts are separated by SDS–gel electrophoresis. This mixture of proteins does show several distinctive features. The absence of the histones in these preparations, which show preservation of synaptonemal complex morphology, is an indication that the histones are not essential for the maintenance of this structure. In addition to the histones, several other more abundant proteins of the pachytene nucleus are absent from the SC/NM structures, and conversely several of the more prominent bands of the SC/NM structure are minor constituents of the nucleus (Li et al., 1983). In addition, the proteins of SC/NM structures are different from those of liver nuclear matrices (Ierardi et al., 1983). A set of three proteins at molecular weights of 61,000 to 65,000, which are the major polypeptides of most somatic nuclear matrices, is absent or only present in low levels in spermatocyte SC/NM structures.

Thus, although the SC/NM structure consists of a well-defined subset of the nuclear proteins, it is still a complex mixture of proteins. The lack of one or a small number of major structural proteins makes further analysis more difficult. The production of monoclonal antibodies may offer a possibility of specific recognition of individual proteins in this mixture and of localization of them within the SC/NM structure.

Another interesting complex has been isolated from nuclei in meiotic prophase following disruption in a phosphate-EDTA solution and equilibrium density gradient centrifugation (Hecht and Stern, 1971). This complex, containing protein, DNA, and lipid, is enriched in a distinctive class of DNA when it is newly synthesized during zygotene. One of the proteins associated with this complex in both plants and mammals (Hotta and Stern, 1971a,b) is designated the R-protein, binds to single-stranded DNA, and facilitates either unwinding or renaturation of

11. Proteins of the Meiotic Cell Nucleus

DNA depending on the ionic conditions (Mather and Hotta, 1977). In addition, DNA polymerase is associated with this complex. The relationship of this complex to the SC/NM structure is not clear although there are certain analogous characteristics. The stability of the complex to high salt or dissociating conditions and Triton X-100 treatment (Hecht and Stern, 1971) and the association of newly replicated DNA with the complex are both similarities with the nuclear matrix (Berezney, 1979). In further studies, the possibility that this complex is a fragment of the SC/NM structure should be considered to combine and reconcile studies of the proteins of both structures.

V. CONCLUSIONS

In parallel with the dramatic morphological changes and unique structures that occur during the meiotic prophase, marked changes in nuclear protein composition occur. Some of the changes in the histones and HMG proteins already exist in the spermatogonia. These may function in the proliferative activity of the spermatogonia, but they may also be involved in events occurring in spermatogonia or premeiotic S phase in preparation for meiosis or possibly even meiotic crossing-over itself (Grell *et al.*, 1972). The appearance of many new proteins during meiotic prophase suggests that these proteins might be involved in the processes of chromosome pairing, genetic recombination, or the reductive divisions. Further characterization of these proteins is important. However, localization of these proteins within identifiable structures, such as the synaptonemal complex, nuclear matrix, or chromatin regions, is essential for developing concepts as to how these proteins might function in meiosis.

ACKNOWLEDGMENTS

We thank Patricia K. Trostle-Weige and Dr. Luke Bucci for their contributions to many of our experiments reviewed here and Sheri Chase for secretarial assistance. Our studies reported here have been supported partly by NIH Grants HD-16843 and CA-17364.

REFERENCES

Alder, D., and Gorovsky, M. A. (1975). Electrophoretic analysis of liver and testis histones of the frog *Rana pipiens*. *J. Cell Biol.* **64,** 389–397.
Bellvé, A. R. (1979). The molecular biology of mammalian spermatogenesis. *In* "Oxford Reviews of Reproductive Biology" (C. A. Finn, ed.), Vol. 1, pp. 159–261. Oxford Univ. Press (Clarendon), London and New York.

Bellvé, A. R., Cavicchia, J. C., Millette, C. F., O'Brien, D. A., Bhatnagar, Y. M., and Dym, M. (1977). Spermatogenic cells of the prepubertal mouse; isolation and morphological characterization. *J. Cell Biol.* **74**, 68–85.

Beaumont, H. M., and Mandl, A. M. (1962). A quantitative and cytological study of oogonia and oocytes in the foetal and neonatal rat. *Proc. R. Soc. London, Ser. B* **155**, 557–579.

Berezney, R. (1979). Effect of protease inhibitors on matrix proteins and the association of replicating DNA. *Exp. Cell Res.* **123**, 411–414.

Berezney, R., and Coffey, D. S. (1977). Nuclear matrix. Isolation and characterization of a framework structure from rat liver nuclei. *J. Cell Biol.* **73**, 616–637.

Bhatnagar, Y. M. (1985). Immunocytochemical localization of an H2A variant in the spermatogenic cells of the mouse. *Biol. Reprod.* **32**, 599–609.

Bhatnagar, Y. M., and Bellvé, A. R. (1978). Two-dimensional electrophoretic analysis of major histone species and their variants from somatic and germ-line tissue. *Anal. Biochem.* **86**, 754–760.

Bhatnagar, Y. M., and Faulkner, R. D. (1983). Germ cell nucleosomes contain remodeled core protein complex. *Biochim. Biophys. Acta* **739**, 132–136.

Bhatnagar, Y. M., McCullar, K., Faulkner, R. D., and Ghai, R. D. (1983). Biochemical and immunological characterization of a histone variant associated with spermatogenesis in the mouse. *Biochim. Biophys. Acta* **760**, 25–33.

Bhatnagar, Y. M., McCullar, M. K., and Chronister, R. B. (1984). Immunocytochemical localization of a histone H2A variant in the mammalian nucleolar chromatin. *Cell Biol. Int. Rep.* **8**, 971–979.

Bhatnagar, Y. M., Romrell, L. J., and Bellvé, A. R. (1985). Biosynthesis of specific histones during meiotic prophase of mouse spermatogenesis. *Biol. Reprod.* **32**, 599–609.

Bogdanov, Y. F., and Antropova, E. N. (1971). Delayed termination of nuclear histone doubling after premeiotic DNA synthesis in *Triturus vulgaris* male meiosis. *Chromosoma* **35**, 353–373.

Bogdanov, Y. F., and Strokov, A. A. (1976). Peculiarities of histone synthesis in meiosis and its relation to chromosome structure and behaviour. *Biol. Zentralbl.* **95**, 335–342.

Bogdanov, Y. F., Liapunova, N. A., Sherudilo, A. A., and Antropova, E. N. (1968). Uncoupling of DNA and histone synthesis prior to prophase I of meiosis in the cricket *Grillus (Acheta) domesticus* L. *Exp. Cell Res.* **52**, 59–70.

Bogdanov, Y. F., Strokov, A. A., and Reznickova, S. A. (1973). Histone synthesis during meiotic prophase in *Lilium*. *Chromosoma* **43**, 237–245.

Branson, R. E., Grimes, S. R., Yonushot, G., and Irvin, J. L. (1975). The histones of rat testis. *Arch. Biochem. Biophys.* **16**, 400–412.

Brandt, W. F., Strickland, W. N., Strickland, M., Carlisle, L., Woods, D., and von Holt, C. (1979). A histone programme during the life cycle of sea urchin. *Eur. J. Biochem.* **94**, 1–10.

Brock, W. A., Trostle, P. K., and Meistrich, M. L. (1980). Meiotic synthesis of testis histones in the rat. *Proc. Natl. Acad. Sci. U.S.A.* **77**, 371–375.

Brock, W. A., Trostle-Weige, P. K., Williams, M., and Meistrich, M. L. (1983). Histone and DNA synthesis in differentiating and rapidly proliferating cells *in vivo* and *in vitro*. *Cell Differ.* **12**, 47–55.

Bucci, L. R. (1983). Characterization of high mobility group and histone H1 protein levels and synthesis during spermatogenesis in the rat. Ph.D. Thesis, Graduate School of Biomedical Sciences, University of Texas Health Science Center at Houston.

Bucci, L. R., Brock, W. A., and Meistrich, M. L. (1982). Distribution and synthesis of histone 1 subfractions during spermatogenesis in the rat. *Exp. Cell Res.* **140**, 111–118.

Bucci, L. R., Brock, W. A., Goldknopf, I. L., and Meistrich, M. L. (1984). Characterization of high mobility group protein levels during spermatogenesis in the rat. *J. Biol. Chem.* **259**, 8840–8846.

11. Proteins of the Meiotic Cell Nucleus

Bucci, L. R., Brock, W. A., and Meistrich, M. L. (1985). Heterogeneity of high-mobility-group protein 2. Enrichment of a rapidly migrating form in the testis. *Biochem. J.* **229**, 233–240.
Bucci, L. R., Brock, W. A., Johnson, T. S., and Meistrich, M. L. (1986). Isolation and biochemical characterization of enriched populations of spermatogonia and early primary spermatocytes from rat testes. *Biol. Reprod.* **34**, 195–206.
Chandley, A. C., Hotta, Y., and Stern, H. (1977). Biochemical analysis of meiosis in the male mouse. I. Separation and DNA labelling of specific spermatogenic stages. *Chromosoma* **62**, 243–253.
Cognetti, G., Platz, R. D., Meistrich, M. L., and diLiegro, I. (1977). *Cell Differ.* **5**, 283–291.
Comings, D. E., and Okada, T. A. (1972). Architecture of meiotic cells and mechanisms of chromosome pairing. *Adv. Cell Mol. Biol.* **2**, 309–383.
Friedman, B. E., Bouchard, R. A., and Stern, H. (1982). DNA sequences repaired at pachytene exhibit strong homology among distantly related higher plants. *Chromosoma* **87**, 409–424.
Goldberg, R. B., Geremia, R., and Bruce, W. R. (1977). Histone synthesis and replacement during spermatogenesis in the mouse. *Differentiation* **7**, 167–180.
Grabske, R. J., Lake, S., Gledhill, B., and Meistrich, M. L. (1975). Centrifugal elutriation: Separation of spermatogenic cells on the basis of sedimentation velocity. *J. Cell. Physiol.* **86**, 177–190.
Grell, R. F., Bank, H., and Gassner, G. (1972). Meiotic exchange without the synaptonemal complex. *Nature (London), New Biol.* **240**, 155–157.
Grimes, S. R., Meistrich, M. L., Platz, R. D., and Hnilica, L. S. (1977). Nuclear protein transitions in rat testis spermatids. *Exp. Cell Res.* **110**, 31–39.
Hecht, N. B., and Stern, H. (1971). A late replicating DNA protein complex from cells in meiotic prophase. *Exp. Cell Res.* **69**, 1–10.
Hotta, Y., and Stern, H. (1963). Inhibition of protein synthesis during meiosis and its bearing on intracellular regulation. *J. Cell Biol.* **16**, 259–279.
Hotta, Y., and Stern, H. (1971a). A DNA-binding protein in meiotic cells of *Lilium*. *Dev. Biol.* **26**, 87–99.
Hotta, Y., and Stern, H. (197b). Meiotic protein in spermatocytes of mammals. *Nature (London) New Biol.* **234**, 83–86.
Hotta, Y., Ito, M., and Stern, H. (1966). Synthesis of DNA during meiosis. *Proc. Natl. Acad. Sci. U.S.A.* **56**, 1184–1191.
Hotta, Y., Parchman, L. G., and Stern, H. (1968). Protein synthesis during meiosis. *Proc. Natl. Acad. Sci. U.S.A.* **60**, 575–582.
Hotta, Y., Chandley, A. C., and Stern, H. (1977). Biochemical analysis of meiosis in the male mouse. II. DNA metabolism at pachytene. *Chromosoma* **62**, 255–268.
Howell, S. H., and Stern, H. (1971). The appearance of DNA breakage and repair activities in the synchronous meiotic cycle of *Lilium*. *J. Mol. Biol.* **55**, 357–378.
Ierardi, L. A., Moss, S. B., and Bellvé, A. R. (1983). Synaptonemal complexes are integral components of the isolated mouse spermatocyte nuclear matrix. *J. Cell Biol.* **96**, 1717–1726.
Isackson, P. J., Fishback, J. L., Bidney, D. L., and Reeck, G. R. (1979). Preferential affinity of high molecular weight high mobility group non-histone proteins for single-stranded DNA. *J. Biol. Chem.* **254**, 5569–5572.
Kaye, J. H., and McMaster-Kaye, R. (1974). Histones of spermatogenic cells of the house cricket. *Chromosoma* **46**, 379–419.
Kleene, K. C., Distel, R. J., and Hecht, N. B. (1984). Translational regulation and deadenylation of a protamine mRNA during spermatogenesis in the mouse. *Dev. Biol.* **105**, 71–79.
Lennox, R. W., and Cohen, L. H., (1984). The alterations in H1 histone complement during mouse spermatogenesis and their significance for H1 subtype function. *Dev. Biol.* **163**, 80–84.
Li, S., Meistrich, M. L., Brock, W. A., Hsu, T. C., and Kuo, M. T. (1983). Isolation and

preliminary characterization of the synaptonemal complex from rat pachytene spermatocytes. *Exp. Cell Res.* **144,** 63–72.
Ling, V., Jergil, B., and Dixon, G. H. (1971). The biosynthesis of protamine in trout testis. III. Characterization of protamine components and their synthesis during testis development. *J. Biol. Chem.* **246,** 1168–1176.
Mather, J., and Hotta, Y. (1977). A phosphorylatable DNA-binding protein associated with a lipoprotein fraction from rat spermatocyte nuclei. *Exp. Cell Res.* **109,** 181–189.
Meistrich, M. L. (1977). Separation of spermatogenic cells and nuclei from rodent testes. *Methods Cell Biol.* **15,** 15–52.
Meistrich, M. L., Bruce, W. R., and Clermont, Y. (1973). Cellular composition of fractions of mouse testis cells following velocity sedimentation separation. *Exp. Cell Res.* **79,** 213–227.
Meistrich, M. L., Trostle, P. K., Frapart, M., and Erickson, R. P. (1977). Biosynthesis and localization of lactate dehydrogenase X in pachytene spermatocytes and spermatids of mouse testes. *Dev. Biol.* **60,** 428–441.
Meistrich, M. L., Longtin, J., Brock, W. A., Grimes, S. R., and Mace, M. L. (1981). Purification of rat spermatogenic cells and preliminary biochemical analysis of these cells. *Biol. Reprod.* **25,** 1065–1077.
Meistrich, M. L., Bucci, L. R., Trostle-Weige, P. K., and Brock, W. A. (1985). Histone variants in rat spermatogonia and primary spermatocytes. *Dev. Biol.* **112,** 230–240.
Moses, M. J., and Coleman, J. R. (1964). Structural patterns and the functional organization of chromosomes. *In* "The Role of Chromosomes in Development" (M. Locke, ed.), pp. 11–49. Academic Press, New York.
Platz, R. D., Grimes, S. R., Meistrich, M. L., and Hnilica, L. S. (1975). Changes in nuclear proteins in rat testis cells separated by velocity sedimentation. *J. Biol. Chem.* **250,** 5791–5800.
Rao, M. R. S., Rao, B. J., and Ganguly, J. (1982). Localization of testis-variant histones in rat testis chromatin. *Biochem. J.* 205, 15–21.
Raveh, D., and Ben-Ze'ev, A. (1984). The synaptonemal complex as a part of the nuclear matrix of the flour moth, *Ephestia kuehniella*. *Exp. Cell Res.* **153,** 99–108.
Romrell, L. J., Bellvé, A. R., and Fawcett, D. W. (1976). Separation of mouse spermatogenic cells by sedimentation velocity. *Dev. Biol.* **49,** 119–131.
Salik, J., Herlands, L., Hoffmann, H. P., and Poccia, D. (1981). Electrophoretic analysis of the stored histone pool in unfertilized sea urchin eggs: Quantification and identification by antibody binding. *J. Cell Biol.* **90,** 385–395.
Seyedin, S. M., and Kistler, W. S. (1979). H1 histone subfractions of mammalian testes. 1. Organ specificity in the rat. *Biochemistry* **18,** 1371–1375.
Seyedin, S. M., and Kistler, W. S. (1983). H1 histones from mammalian testes. H1t is associated with spermatogenesis in humans. *Exp. Cell Res.* **143,** 451–454.
Seyedin, S. M., Cole, R. D., and Kistler, W. S. (1981). H1 histone subfractions of mammalian testes. The widespread occurrence of H1t. *Exp. Cell Res.* **136,** 399–405.
Sheridan, W. F., and Stern, H. (1967). Histones of meiosis. *Exp. Cell Res.* **45,** 323–335.
Shires, A., Carpenter, M. F., and Chalkley, R. (1975). New histones found in mature mammalian testes. *Proc. Natl. Acad. Sci. U.S.A.* **72,** 2714–2718.
Solari, A. J. (1972). Ultrastructure and composition of the synaptonemal complex in spread and negatively stained spermatocytes of the golden hamster and albino rat. *Chromosoma* **39,** 237–263.
Stern, L., Gold, B., and Hecht, N. B. (1983). Gene expression during mammalian spermatogenesis. I. Evidence for stage-specific synthesis of polypeptides in vivo. *Biol. Reprod.* **28,** 483–496.
Strokov, A. A., Bogdanov, Y. F., and Reznickova, S. A. (1973). A quantitative study of histones of meiocytes. II. Polyacrylamide gel electrophoresis of isolated histones from *Lilium* microsporocytes. *Chromosoma* **43,** 247–260.

Trostle-Weige, P. K., Meistrich, M. L., Brock, W. A., Nishioka, K., and Bremer, J. W. (1982). Isolation and characterization of TH2A, a germ cell-specific variant of histone 2A in rat testis. *J. Biol. Chem.* **257,** 5560–5567.

Trostle-Weige, P. K., Meistrich, M. L., Brock, W. A., and Nishioka, K. (1984). Isolation and characterization of TH3, a germ cell-specific variant of histone 3 in rat testis. *J. Biol. Chem.* **259,** 8769–8776.

von Holt, C., Strickland, W. N., Brandt, W. F., and Strickland, M. S. (1979). More histone structures. *FEBS Lett.* **100,** 201–218.

Walmsley, M., and Moses, M. J. (1984). Isolation of synaptonemal complexes from hamster spermatocytes. *Exp. Cell Res.* **133,** 405–411.

Wattanaseree, J., and Svasti, J. (1983). Human testis-specific histone TH2B: Fractionation and peptide mapping. *Arch. Biochem. Biophys.* **225,** 892–897.

Wu, R. S., and Bonner, W. M. (1981). Separation of basal histone synthesis from S-phase synthesis in dividing cells. *Cell* **27,** 321–330.

Zalenskaya, I. A., Zalenskaya, E. O., and Zalensky, A. O. (1980). Basic chromosomal proteins of marine invertebrates. II. Starfish and *Holothuria*. *Comp. Biochem. Physiol.* **65B,** 375–378.

Zweidler, A. (1984). Core histone variants in the mouse: Primary structure and differential expression. *In* "Histone Genes: Structure, Organization and Regulation" (G. S. Stein, J. L. Stein, and W. F. Marzluff, eds.), pp. 339–371. Wiley, New York.

12

Transcription during Meiosis

P. T. MAGEE

Department of Microbiology and Public Health
Michigan State University
East Lansing, Michigan 48824-1101

I. INTRODUCTION

Although meiosis is a fundamental part of the process of sexual reproduction, and its genetic consequences are thus of paramount importance, it may also be viewed as a developmental pathway and studied as such. When it is considered in this light, a number of questions can be posed. These have to do with the mechanisms by which the expression of the necessary genes is ordered temporally. Specifically, one can ask whether there are genes which are unique to meiotic as distinct from mitotic nuclear divisions. If so, how many? Is there a single interdependent pathway (or series of pathways) or are there several pathways, each regulated independently of the others, which converge to achieve the final process? At what level (transcriptional, posttranscriptional, translational, or posttranslational) are the genes regulated? Is regulation sequential (i.e., the product of A is necessary to turn on gene B) or are there groups of genes subject to a common regulatory signal? Determining the answers to all these questions depends on a genetic and molecular biological analysis of the meiotic process, an analysis which has been hard to achieve.

The problems in studying the genetics of meiosis are, of course, related to its fundamental role in the life cycle of the cells in which it occurs and to its occurrence being limited to diploid cells, so that recessive mutations affecting the process must be homozygous to be expressed. These two considerations limit rather narrowly the organisms in which genetic studies on meiosis can be carried out, since any mutations must be conditional in order to be useful for genetic analysis. Microorganisms have been the traditional system in which conditional mutations have been selected, but there are only a few in which meiosis occurs under conditions which are experimentally manipulatable, which have a genetic system sufficiently advanced to allow exploitation of conditional mutants, and which can easily be made homozygous to allow expression of the mutations. The organism which fits all these criteria most closely and which has indeed been most extensively exploited to study the genetics of meiosis is the yeast *Saccharomyces cerevisiae*. Over the last 15 years a large number of studies on the genetics, biochemistry, physiology, and molecular biology of this organism have appeared, and the fraction of this body of literature that bears on transcription will be the major subject of this chapter. However, there are a few observations from mammalian and plant meiocytes which are also pertinent to the subject of transcription during meiosis and these will also be reviewed. Since the latter studies are more general and the systems are for the most part less well defined, I will begin with this work and then discuss the results in *S. cerevisiae*, pointing out the similarities and differences between the results obtained in the various sorts of organisms. I will not attempt to review the voluminous literature on transcription in amphibian meiocytes.

II. TRANSCRIPTION IN MAMMALIAN MEIOCYTES

A number of studies have examined RNA synthesis in mammalian meiocytes, specifically with a view to trying to determine whether chromosomes of the "lampbrush" type found in amphibians occur during meiosis in these organisms. Baker and Franchi (1967) found ultrastructural evidence for such structures in human primordial oocytes, which are arrested at early diplotene. However, other studies indicate that transcription is rather low in primordial oocytes. Although pachytene–early diplotene cells incorporate [^{3}H]uridine (Bakken and McClanahan, 1978), transcription units visualized in the electron microscope are sparsely spaced (Bakken and Hamkalo, 1978). Since in mammalian oocytes there is a period between the prophase of the first meiotic division and its completion, called the dictyate stage, during which the occytes grow and accumulate gene products, one might expect this to be the most active time for lampbrush-type chromosomes. However, Bachvarova (1981) has shown that even though this period is a very active one for transcription in the mouse oocyte, it is still about 40 times lower than the rate of transcription in *Xenopus laevis*, where lampbrush chromosomes have been most extensively studied. Light and electron microscopic studies on the chromosomes of growing oocytes have demonstrated that although these structures superficially resemble lampbrush chromosomes, the lateral projections, rather than being single fibers covered with a

12. Transcription during Meiosis

ribonucleoprotein matrix, are in fact bundles of chromatin fibers, possessing on the average only one or two nascent ribonucleoprotein fibrils (Bachvarova et al., 1982).

There is no doubt, however, that oocytes not only incorporate RNA precursors but also synthesize all three classes of RNA, ribosomal, transfer, and messenger (Brower et al., 1981; Bachvarova, 1974; Jahn et al., 1976). The ratio of these classes remains constant throughout oocyte development, at least until the oocytes reach full size (Brower et al., 1981). The newly synthesized RNA is quite stable: the half-life of the bulk of the radioactive products is about 28 days in growing oocytes; if the oocytes are viable but not growing, the half-life is about 4.5 days. Poly(A)-containing RNA has a half-life of about 8 days under growth conditions. De Leon et al. (1983) separated poly(A)$^+$ RNA into two classes: 20% is associated with polysomes, with a half-life of about 6 days, and the remaining material was completely stable under their labeling conditions. They assign the discrepancy to the fact that they included a chase of cold uridine in their experiments, while Brower et al. did not. Fourcroy (1982) also found evidence for complete stability of cytoplasmic RNA, using in vivo labeling and radioautography. Of course, his experiments did not involve a chase, and he could not distinguish between rRNA and mRNA. The bulk of the evidence, then, supports the idea that there is significant transcription in oocytes, even though there is no evidence for lampbrush chromosomes, and that a significant portion of the transcripts remain stable for the life of the cell. These may be maternal messages important for the development of the embryo. In any case, no one has shown any specific gene to be transcribed in oocytes, except of course the rRNA genes.

In mammalian spermatocytes, there also appears to be significant transcription, but in these cells there is somewhat less ribosomal RNA transcribed. The fraction of poly(A)-containing RNA in oocytes is about 20%, while it ranges from 19% in pachytene spermatocytes labeled for 1 hr in vivo to 29% in secondary spermatocytes and diploid spermatids. There appears to be more transcription in the haploid cell products of spermatogenesis than there is in the meiotic cells themselves. Kleene et al. (1983) made a cDNA library from RNA from mouse testis, and they found 17 clones which hybridized preferentially to labeled cDNA from round spermatids as compared to cDNA from pachytene spermatocytes. The difference in hybridization levels was 10-fold. These plasmids did not hybridize to cDNA from a variety of other tissues. They thus indicate that there are gene products specific to spermatids. Erickson's group (Erickson et al., 1980; Gold et al., 1983) has shown that two of these gene products are a protaminelike histone and an isozyme of phosphoglycerate kinase. Once again, no specific genes have been shown to be preferentially transcribed during meiosis in spermatocytes.

III. TRANSCRIPTION IN MEIOCYTES OF *LILIUM*

Work at about the same level of detail has been done on RNA metabolism in *Lilium* microsporocytes, even though the system has been extensively studied from the point of view of DNA synthesis (Stern and Hotta, 1977). Radioautographic studies indicate that [^{3}H]uridine is incorporated at a relatively high rate in cells in the leptotene and early and mid-zygotene stages, while in the late zygotene and early pachytene incorporation falls off dramatically. At late pachytene, incorporation begins to rise again and continues at a relatively high rate at least until the tetrad stage. The first stage of RNA synthesis appears to occur in the intrachromosomal regions of the nucleus. There was not very much label in the cytoplasm, indicating that much of the incorporation might be heterogeneous nuclear RNA which is not exported to the cytoplasm. Synthesis also occurs over the nucleolus, with activity being greater during the first synthetic period than during the second (Porter *et al.*, 1982). The level of poly(A)$^+$ RNA isolatable from the cytoplasm follows kinetics different from that of uridine incorporation. The level of this RNA begins to decline during leptotene and falls to a minimum in pachytene. It remains more or less constant during metaphase, anaphase, and telophase, then rises in the dyads and tetrads (Porter *et al.*, 1983). This pattern is consistent with previous observations of a general clearing of all kinds of RNA during meiotic prophase in these cells (Dickinson and Heslop-Harrison, 1977). The transcription observed may be due to a general relaxation of transcriptional control during meiosis (see below); alternatively, the RNA may be a message for paternal transcripts which pass through meiosis and enter the gametocyte cytoplasm, for example, pollen-wall pattern determinants (Godwin, 1968; Dickinson and Heslop-Harrison, 1970; Sheldon and Dickinson, 1983). However there is no unequivocal identification in these cells of a gene which is transcribed during meiosis.

IV. TRANSCRIPTION IN MEIOCYTES OF *SACCHAROMYCES CEREVISIAE*

A. Physiological and Biochemical Features of Meiosis and Sporulation in *Saccharomyces cerevisiae*

In *Saccharomyces cerevisiae*, meiosis is part of the life cycle leading to ascospore formation. This process occurs in cultures of the appropriate mating

12. Transcription during Meiosis

type, which can be induced to sporulate by transfer to medium lacking a usuable nitrogen source and containing a fermentable carbon source, such as acetate (Miller and Hoffman-Ostenhof, 1964). Ammonia and glucose both repress the process. Only cells which contain both an active MAT*a* and an active MATα allele will undergo meiosis and sporulate (Roman and Sands, 1953). The level of ploidy is not important for the process, although cells of uneven ploidy yield a high proportion of inviable progeny. One is therefore able to use isogenic or nearly isogenic strains which differ only in the MAT configuration to investigate the physiology of meiosis and sporulation in this organism, since cultures of MAT*a*/MATα and MATα/MATα cells can be subjected to identical culture conditions, whereupon the former will undergo meiosis while the latter will not. Several groups have taken advantage of this to examine the physiology and biochemistry of sporulating cells, and these studies have led to a detailed picture of the events which are specific to cells undergoing meiosis (Esposito *et al.*, 1969; Roth and Lusnak, 1970; Hopper *et al.*, 1974; Colonna and Magee, 1978). It is clear that a significant amount of transcription takes place during adjustment to the culture conditions, during premeiotic DNA synthesis, and during prophase. This is similar to the results discussed earlier in mammalian meiocytes, except that no one has looked at the poly(A)$^+$ fraction specifically. The question of how much transcription occurs as the cells enter meiosis I and II and form ascospores is much harder to answer, due to the difficulty of getting labeled precursors into cells at this stage of the developmental process (Mills, 1972). However, addition of cycloheximide to cells at virtually any stage of sporulation causes a block within an hour of any further development (Esposito *et al.*, 1969; Magee and Hopper, 1974), indicating that at least new translation takes place late in the process. It is possible that the new products which have been found to appear at late times [spore wall proteins (Becket *et al.*, 1973), sporulation amyloglucosidase (Colonna and Magee, 1978)] are made from stored mRNA, although there is no evidence for this. Ascospores, however, do contain stored mRNA. Harper *et al.* (1980) showed that spores contain polysomes and that the poly(A)$^+$ RNA isolated from them is capped and resembles in most properties the mRNA from vegetative cells. The major difference between the two kinds of cells is the ribosomal profile. Gradients prepared from ascospores show a large peak of monosomes and polysomes of five monomers or less. Vegetative cells show a small 80 S peak and polysomes up to at least seven. Studies on germinating ascospores indicate that spores are poised to begin protein synthesis immediately, and that protein synthesis may precede RNA synthesis (Rousseau and Halvorson, 1969; Armstrong *et al.*, 1984). Although these studies are complicated by problems of the entry of labeled precursors, they nevertheless suggest that some transcription of genes required for spore germination must take place during the sporulation process.

B. Evidence for Meiosis-Specific Gene Products

1. Genetic Evidence

An approach which can give an indication not only of the occurrence of transcription but also of some of the genes which are likely to be transcribed is the identification through mutation of functions which are required for the process of meiosis. Two major genetic approaches have been used to dissect sporulation. The first is the isolation of mutants conditionally unable to sporulate and the second is to examine the sporulation capability of cells mutant for some known function. Both methods have yielded significant information about genes involved in sporulation. A comprehensive catalogue of mutations affecting sporulation is given in Esposito and Klapholz (1983).

The most extensively characterized group of mutations which affect sporulation comprises the *spo* mutants by Esposito and Esposito (1969, 1974). These were isolated on the basis of their inability to complete ascus formation at 34°C while being sporulation-competent at 20° to 23°C. Fourteen separate complementation groups have been identified among these mutants, and from the number of repeat isolations of the same gene Esposito and Esposito calculated that there were 48 ± 27 genes required exclusively for meiosis and capable of mutating to thermosensitivity. The *SPO* genes fall into four general classes: those that fail to complete premeiotic DNA synthesis (*SPO7, 8,* and *9*); those that undergo premeiotic S and recombine at normal or near-normal levels but fail to complete either meiosis or spore formation (*SPO1–5*); one that fails to commit to recombination although it does replicate its DNA (*SPO11*); and two that skip the reductional meiotic division and produce two diploid or near-diploid spores (*SPO12* and *13*).

The genes which appear to have dual mitotic–meiotic functions include all the cell division cycle (*CDC*) genes which are not involved in budding or cytokinesis. Thus, all DNA synthesis mutations, all nuclear division mutations, and all "start" (G_1-arrest) mutations seem to have an effect on meiosis. In addition, mutations which affect the production of two of the three main proteinases of the yeast cell block sporulation (Zubenko *et al.*, 1979; Zubenko and Jones, 1981; Betz, 1979). Mutations affecting recombination, such as the rad loci and the rec loci, also sometimes block meiosis, although not always (Dowling *et al.*, 1985). Finally, there are mutations which affect functions apparently unrelated to sporulation, such as thymidine monophosphate uptake (*tup*1) and killer exclusion (*kex*1), which also block ascus formation.

One large group of genes which appears to be transcribed during meiosis includes the rRNA and ribosomal protein genes. Emanuel and Magee (1981) showed that cycloheximide resistance, a recessive character, is expressed in ascospores without any phenotypic lag. This implies some mechanism for segregation of ribosomes to spores in such a way that the genotypically Cyh^R cells get phenotypically resistant ribosomes. The simplest mechanism for this would be for each spore to contain a complement of ribosomes synthesized after its nucleus had become functionally haploid. The extensive breakdown of RNA during sporulation must include ribosomes, since it comprises 50 to 70% of preexisting RNA (Croes, 1967; Esposito et al., 1969; Chaffin et al., 1974; Hopper et al., 1974), therefore, any newly synthesized ribosomes would constitute a large fraction of the ascospore complement. Emanuel and Magee showed that rRNA was in fact synthesized at about the time of tetranucleate cell appearance, and that 90% of the label incorporated during this time survived in ascospores compared with 50% of vegetative label that survived; a considerable portion of the vegetative label is of course reutilized during sporulation. Pearson and Haber (1980) showed that ribosomal protein synthesis slows during the first 4 hr of protein synthesis to about 20% that of vegetative cells; this new rate seems to persist for the next 8 hr at least. It may be that this lower rate is sufficient to provide the ribosomal proteins needed for ascospores. This synthesis occurred even in a diploid homozygous for the temperature-sensitive mutation $rna2$ when shifted to restrictive temperature (34°C). Since $rna2$ has since been shown to affect the splicing of intron-containing mRNA for ribosomal proteins, this result might mean either that sporulating cells have a different splicing function or that this synthesis occurs from preexisting processed mRNA. In the latter case, of course, ribosomal proteins would not be among those transcribed during sporulation.

2. Biochemical Evidence

Two kinds of enzyme activities have been postulated to be apparently restricted to sporulating yeast cells. These are the α-1,4-amyloglucosidase (SAG) first observed by Colonna and Magee (1978) and a β-1,3-glucanase observed by del Rey et al. (1980). SAG is synthesized only in cells undergoing sporulation; its function is apparently to degrade the glycogen which is accumulated during the early hours in sporulation medium. Cells homozygous at the mating-type locus accumulate glycogen at a rate similar to MATa/MATα cells, but only the latter break down the storage carbohydrate relatively rapidly beginning at about the time of meiosis I. Enzyme activity is undetectable until about 7 hr, after which it increases roughly in parallel to the rate of ascospore formation. Whether the activity is one enzyme activity or two is not resolved. Colonna and Magee reported partial resolution of an α-1,4- and an α-1,6-glucosidase activity, but M. J. Clancy and P. T. Magee (unpublished data) purified the activity to near homogeneity and found no evidence for two entities; the purified enzyme was as active on glycogen as on starch, which contains very few α-1,6 linkages. Immunoprecipitation studies on pulse-labeled cells with antibodies prepared against the nearly homogeneous enzyme show that the major protein brought down (MW = 68,000) is synthesized *de novo* during sporulation; this band, which appears to be glycosylated, is not found in vegetative cells nor in cells homozygous at the MAT locus (M. J. Clancy and P. T. Magee, unpublished data). An extensive series of studies was carried out on the appearance of the enzyme activity in cells genetically blocked at various stages of meiosis and sporulation (Clancy *et al.*, 1982). SAG was not found in cells blocked at or before pachytene in meiotic prophase, but it did appear in cells arrested at spindle pole body duplication (*spo 1* mutants) or at later stages. Its appearance was not dependent on successful completion of recombination. The sporulation-specific character of the enzyme was further supported by the demonstration that it was not found in cells coming out of stationary phase, a phase of growth when glycogen is extensively degraded. The gene encoding this enzyme has now been cloned and will be discussed in Section IV,E.

The case for the β-1,3-glucanase is controversial. Yeast cells make six such glucanases, four *endo-* and two *exo*-β-glucanases; these activities are chromatographically separable (Hien and Fleet, 1983a). The specific activity of total β-1,3-glucanase was shown by del Rey *et al.* (1980) to rise slowly until about the time of the midpoint of DNA synthesis, then dramatically during the late stages of sporulation. The steep rise does not occur in α/α cells, nor is it found if the cells are treated with hydroxyurea (HU) at 0 hr; if HU is added at a time midway through DNA synthesis the increase in β-glucanase activity is considerably reduced. Chromatographic resolution of the activities reveals that the early increase in activity is the result of the two vegetative enzymes; the later, rapid rise is due to the appearance of a new activity which resembles the endoglucanase from vegetative cells in its substrate specificity but is chromatographically distinct (del Rey *et al.*, 1980). Hien and Fleet (1983b) found that total β-glucanase activity is constant from the time the cells are placed in sporulation medium until the appearance of asci, where there was a two- to threefold increase. The increase they found was due to endoglucanase activity, specifically endoglucanase V, which makes up less than 6% of the activity in vegetative cells but accounts for 70 to 75% of the sporulation activity (Hien and Fleet, 1983b). The reason for the discrepancy between the two sets of results is not clear. It may be the difference in the strains used; del Rey *et al.* (1980) used AP1, which completes sporulation in 16 to 20 hr, while Hien and Fleet used S90, which requires 96 hr. It is obviously very important to decide whether these two groups are observing the same activity.

A chromatographically distinct peak of RNA polymerase activity has been found in MAT*a*/MATα cells but not in α/α cells in sporulation medium. This peak appears at about the time of meiosis I and is transient in that it is not found in extracts of asci. It has sensitivity to the fungal toxin α-amanitin intermediate between that of RNA polymerase I and RNA polymerase II (Magee, 1974).

12. Transcription during Meiosis

Whether it is in fact a functional sporulation-specific change (resulting from the synthesis of a new gene product or the modification of an old one) or a product of nonspecific proteolysis (Klar *et al.*, 1976) is not resolved.

One can therefore suppose *a priori* that a number of genes may be transcribed during meiosis and sporulation in *Saccharomyces*. These include the genes predicted by the Espositos' calculation and the mitotic genes whose function is essential for meiosis. Those gene products which have been shown to be specifically expresed during meiosis (β-glucanase, spore wall proteins) ought also to be transcribed. (Until mutants are isolated, one cannot decide whether they fall into either of the two previous groups.) Other genes which might be expressed include "housekeeping genes" and genes under strong negative control whose regulation is lost during the developmental process. The last two groups might be expected to be transcribed in cells of any mating-type configuration which are subjected to the conditions which induce sporulation in competent cells.

However, attempts to demonstrate large classes of meiosis-specific gene products in sporulating *S. cerevisiae* have not in general been successful. Several groups have attempted to identify proteins synthesized in MAT*a*/MATα cells under sporulation conditions but not present in MATα/MATα or MAT*a*/MAT*a* cells using two-dimensional gel electrophoresis (O'Farrell, 1975). Although a few new proteins were seen in the sporulation-competent cells, these were also found in the MAT*a*/*a* or MATα/α cells (Kraig and Haber, 1980; Hopper *et al.*, 1974; Peterson *et al.*, 1979; Trew *et al.*, 1979; Wright and Dawes, 1979; Wright *et al.*, 1981). In general, only quantitative differences were found in these experiments, indicating changes in the amount of transcription rather than the presence of new transcripts. The one demonstration of qualitative differences in the pattern of proteins occurred when the cells were prelabeled in vegetative medium and then switched to sporulation medium (Wright *et al.*, 1981). In this case, it is not clear whether the new spots were due to functional modifications of preexisting proteins and were related to sporulation or were artifacts caused by the heightened level of proteolysis in sporulating cells.

3. *Evidence from in Vitro Translation of RNA from Meiotic Cells*

Two groups have shown that *in vitro* translation of RNA from sporulating cells yields a number of products which are not found when analogous experiments are carried out with RNA from control cells (Weir-Thompson and Dawes, 1984; Kurtz and Lindquist, 1984). In 1980, Mills reported at the 10th International Congress of Yeast Genetics and Molecular Biology at Louvaine, Belgium, that approximately 5–7% of the poly(A)$^+$ RNA isolated from sporulating cells is not found in vegetative cells. Mills conducted RNA–cDNA kinetic hybridization experiments and found that while vegetative poly(A)$^+$ RNA protected its homologous cDNA to greater than 90% from S_1 nuclease digestion, cDNA from 12-hr sporulating cells was protected only to about 80%. Sporulation poly(A)$^+$ RNA protected its own cDNA completely, within the limits of the error of the experiment. Twelve-hour poly(A)$^+$ RNA from MATα/MATα cells was similar to the vegetative material in its protection of the sporulation cDNA. These findings led the two groups mentioned above to look for differences in the coding properties of the poly(A)$^+$ RNA from sporulating cells.

Weir-Thompson and Dawes isolated total cellular RNA from a MAT*a*/MATα strain and from both MAT*a*/MAT*a* and MATα/MATα as controls. Using a rabbit reticulocyte system, they found that the total availability of translatable RNA first rose and then declined to about 50% of the original value in the sporulation-competent strain, while in the two control strains the decline was much less. This is in agreement with the observations of Hopper *et al.* (1974) on RNA degradation in such near-isogenic strains in sporulation medium. Two-dimensional gels were used to resolve the *in vitro* translation products. A total of 80 changes were found to occur during the developmental period. Forty-three of these were specific to MAT*a*/MATα cells, 36 occurred in all the cells, and 1 was found only in α/α cells. Samples were analyzed at 2, 4, 6, 8, 12, and 16–24 hr; in their system, meiotic prophase begins at about 2 hr and commitment to meiosis occurs at about 5 hr. Ascus formation is observable at 12 hr and is 80% complete by 24 hr. Only four totally new translation products were found; these appeared between 4 and 8 hr. The rest of the sporulation-specific changes were quantitative ones: 20 species of translatable RNA appeared to undergo permanent increases in level (i.e., were still elevated in the 16- to 24-hr samples), while another 8 showed a transient increase. Eleven showed a decline or disappearance during the developmental process.

The four polypeptides that appeared *de novo* included two of 46,000 to 47,000 MW, with isoelectric points of 5.6 and 5.2. The extent of increase of these two peptides is so great that they can be detected on a one-dimensional electrophoretic gel. They, along with a polypeptide of 30,500 MW and a pI of 6.1, appeared first in the 6-hr samples; the capacity for translation of the larger polypeptides continued to increase up to 12 hr, after which they remained a constant (albeit major) fraction of the translatable RNA. The 30,500 MW peptide did not appear to increase after its first appearance. The other *de novo* product was a polypeptide of 31,500 Da and a pI of 6.8 which first appeared as a translation product from RNA isolated at 8 hr and remained stable throughout the rest of the experiment. In accordance with the results of the pulse-labeling studies referred to above, there were 36 changes which were common to all cells incubated in sporulation medium, whatever the mating-type configuration. These included 5 new polypeptide spots, 17 translation products that showed a permanent increase, 3 that showed a transient increase, and 11 that declined in concentration or disappeared.

Of the new or increased species of mRNA which were specific to **a**/α cells, all but three appeared or increased after 6 hr in sporulation medium. This provides further evidence that they encode functions that are involved in meiosis, rather than adjustment to the medium. Which specific functions are involved is of course impossible to say.

Kurtz and Lindquist (1984) have also examined the translatable mRNA species in *S. cerevisiae* undergoing meiosis and sporulation. Although they did not present quantitative data analogous to those of Weir-Thompson and Dawes, it seems clear from their one-dimensional polyacrylamide gels that they found a similar time course of appearance and decline of translatable mRNA species. The major experimental differences between the two reports are that Weir-Thompson and Dawes used ^{35}S]methionine as a label and a rabbit reticulocyte system for translation, while Kurtz and Lindquist labeled with [^{3}H]leucine and translated the RNA in a wheat germ system. The results reported, on the other hand, differ significantly. Where Weir-Thompson and Dawes found 4 new polypeptides, Kurtz and Lindquist observed 12 new polypeptides in their *in vitro* translation system. Two heat shock proteins (of 26 and 83 kDa) appeared before 6 hr. After 6 hr, mRNA for peptides of 17, 20, 14, 31, 38, 50, 65, and 68.5 kDa appeared. Finally, after 16 hr, RNA coding for two polypeptides of MW 21,500 and 34,000 appeared. Of all these peptides, only one and possibly two were among those observed by Weir-Thompson and Dawes. The 31-kD peptide is probably the 30,500 translation product mentioned earlier, and since Kurtz and Lindquist used only one-dimensional gels, the 31,000-kDa protein which appears at 8 hr may well have escaped detection in their system. The major differences are the absence in Kurtz's experiments of the two 46- to 47-kDa proteins observed by Weir-Thompson to be a major portion of the translation products, and the large number of new polypeptides observed by Kurtz. There are two explanations for some of these discrepancies; either or both may apply, depending on the case. The first is that since Weir-Thompson and Dawes used methionine, they may have failed to label some polypeptides sufficiently to detect them. The second explanation is the one proposed by Kurtz and Lindquist: these sequences code for spore wall proteins, whose messages may be expected to have leader sequences. Such mRNAs translate well in wheat germ, which does not require microsomes to recognize them, but poorly in a reticulocyte system unless it is supplemented with microsomes. Neither argument explains why Kurtz and Lindquist missed the 46- to 47-kDa polypeptides. An intriguing possibility is that there are translational controls operating during meiosis and sporulation, and that these can be demonstrated by using heterologous *in vitro* translation systems. This is not a possibility envisaged by Kurtz and Lindquist, who imply that the appearance of translatable RNA implies *de novo* synthesis. In such a complex developmental pathway it is important to be extremely conservative in evaluating results; their failure to consider the possibility of posttranscriptional and translational controls is a drawback to their arguments.

C. Identification and Isolation of Genes Transcribed during Meiosis in Yeast

With the advent of sophisticated techniques for gene isolation, great progress has been made in the identification and characterization of genes which are transcribed during meiosis and sporulation in yeast. Three general approaches have been used. The first to be tried was to use differential hybridization to a genomic library to identify sequences which occur in the mRNA in sporulating cells but not in vegetative cells nor in cells which are genetically unable to sporulate. This approach was used by Clancy et al. (1983), Percival-Smith and Segall (1984), and Ninfa et al. (1984). A second approach is to ask whether cloned genes whose vegetative function is known are differentially transcribed during meiosis. Several genes have been analyzed in this way by Kaback and Feldberg (1985). Finally, R. E. Esposito and R. Elder and their co-workers have cloned three of the *SPO* genes and analyzed their transcription patterns. Each of these approaches has yielded interesting information about transcriptional regulation during meiosis in *Saccharomyces cerevisiae*.

The very small number of new gene products found by either two-dimensional gel electrophoresis of proteins from pulse-labeled sporulating cells or identified physiologically led Clancy *et al.* (1983) to the view that a more rapid way to identify important functions might be to search directly for the genes involved. The availability of near-isogenic strains of yeast which are genetically unable to undergo premeiotic DNA synthesis and thus are most likely blocked at an early step in the process provided an ideal control for the differential plaque hybridization technique used by Benton and Davis (1977) and St. John and Davis (1979). Clancy and her co-workers therefore prepared poly(A)$^+$ RNA from both MATa/MATα and MATα/MATα cells in sporulation medium. Samples taken at 2-hr intervals from 7 to 13 hr were pooled and labeled cDNA was made. cDNA from each kind of cell was then used to probe a library of 3500λ-*S. cerevisiae* chimeric bacteriophage. The library was prepared from an Mbo I partial digest of yeast DNA. Positive clones (those which hybridized with a/α cDNA but not α/α cDNA) were rescreened; finally, 46 separate recombinant phage were isolated. These were separated into groups containing the same sequence by cross-hybridization experiments on Southern blots (Southern, 1975) of restriction digests of the phage DNA. For those which gave equivocal results in these experiments, further resolution was accomplished by using end-labeled poly(A)$^+$ RNA to identify the restriction fragments which are differentially transcribed in meiotic cells. The 46 original clones were grouped on the basis of these experiments into 26 separate classes, corresponding (to a first approximation) to 26 separate genes (Holaway *et al.*, 1985). These genes were named *SPR* (SPorulation-Regulated) genes and numbered 1–26. In order to get some idea of their possible functions, the time at which each was turned on was determined.

12. Transcription during Meiosis

Dot blots of RNA isolated at various times from sporulating cells were probed with labeled representatives of most of the *SPR* families, the resulting radioautographs were scanned with a densitometer, and the amount of homologous RNA present at any time was determined. Figure 1 shows representative data. It is evident that the *SPR* genes fall into three main groups. One, the early group, begins to show an increase in transcription at the time the cells are shifted to sporulation medium and transcripts continue to accumulate throughout the developmental process. Another, the middle group, shows no accumulation of transcripts until about the middle of premeiotic S phase. Transcripts then begin to rise sharply in abundance until about meiosis I, at which time their rate of increase declines. A third class, the late *SPR* genes, show no change in the level of transcription until about the time of meiosis I, when their transcripts begin to accumulate dramatically, eventually reaching a level which is 7- to 10-fold higher than at the time of shift. Their period of increase culminates at about the time of spore formation. Table I, from Holaway *et al.* (1985), summarizes the general characteristics of the *SPR* genes.

The identification and characterization of the transcriptional program of the *SPR* sequences is a first step toward understanding the role of transcription during meiosis. However, one needs to know the products of the genes, their role in meiosis, and the signals which regulate their transcription. Fortunately, the

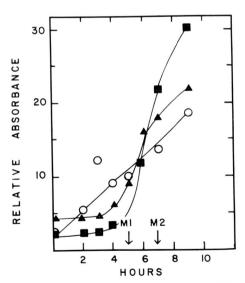

Fig. 1. Examples of transcript accumulation from early (○), middle (▲), and late (■) *SPR* genes. RNA from cells underoing meiosis and sporulation was isolated and measured by the dot-blot technique, using nick-translated λ-*spr* DNA or probe.

TABLE I

General Characteristics of *SPR* Genes[a]

SPR gene	Number of members	Size of diagnostic *Eco*R1 fragments[b]	Classification[c]
1	5	2.85, 5.2	L
2	6	2.65	L
3	2	1.65, 2.4 (*Bam*H1)	L
4	3	1.2	M
5	1	2.7, 1.8	E
6	1	6.6	L
7	5	5.9	L
8	2	1.44	E
9	1	1.45, 0.9	E
10	3	1.51[d]	M
11	1	3.65, 2.65	E
12	1	3.35, 0.75	M
13	1	2.6	L
14	2	2.8	M
15	1	2.8[d], 0.4[d]	M
16	1	5.1, 1.25, 0.8	M
17	1	5.1[d]	EL
18	1	n.d.[e]	L
19	1	0.95	E
20	1	>10 (λ arm)	L
21	1	Arm	M
22	1	Arm 2.4[d]	L
23	1	n.d.	L
24	1	n.d.	n.d.
25	1	n.d.	n.d.
26	1	n.d.	n.d.

[a] From Holaway *et al.* (1985).

[b] The DNAs were digested with *Eco*R1 or other enzymes as indicated, blotted to nitrocellulose, and hybridized as described in the legend to Fig. 3. In cases where *Eco*R1 did not yield a satisfactory pattern, other enzymes were used (these are noted in parentheses).

[c] E, early; M, middle; L, late.

[d] Relatively small quantitative increase in intensity.

[e] n.d., not determined.

genetic system of *S. cerevisiae* facilitates the answering of these questions. It is possible to supply a cell with several copies of a particular gene by cloning it into a multicopy plasmid; this allows the effect of gene dosage to be determined. Gene disruption (Shortle *et al.*, 1982) or gene transplacement (Rothstein, 1983) permits the substitution of a mutated copy of any cloned gene for the intact wild-type sequence. The phenotype of the mutant can then be determined. Fusions of yeast sequences with the *Escherichia coli lacZ* gene can be constructed. These produce hybrid proteins which can be used in several ways: the β-galactosidase activity can be used to monitor gene expression in a way much more convenient than measuring RNA or (usually) the wild-type gene product. The hybrid protein can be purified and antiserum prepared against it; this antiserum usually reacts with the yeast part of the protein as well as the bacterial part and can be used to purify the wild-type protein. Finally, for genes which are more or less tightly regulated, mutations which affect the regulation can be obtained by isolating cells which express the fusion protein where the yeast gene is normally turned off. All these approaches are being used with *SPR* genes, although they are not yet far along. Two *SPR* genes, *SPR1* and *SPR3*, are being studied in some detail. Since these experiments are illustrative and may be representative, I will discuss them in some detail.

SPR1 is a late gene whose expression is particularly strong. It was found on an 11-kb insert. Surprisingly, Northern blots (Thomas, 1983) indicate that *SPR1* is the only transcribed sequence on this insert, at least in vegetative cells grown on acetate as a carbon source and in sporulating cells. The single transcript is 1500 kb in length; a restriction map, based on the work of D. Primerano (unpublished data), is shown in Fig. 2a. The *Xho*I/*Xho*I fragment, containing the intact *SPR1*, has been cloned into two multicopy plasmids; one (YRp12) contains an autonomously replicating sequence (*ARS*) and the other (YEp16) contains the replication origin of the yeast 2μ circle. Each of these plasmids has been transformed into a sporulation-proficient yeast strain to see whether the increased copy number of the *SPR1* gene will stimulate or inhibit sporulation. Surprisingly, the plasmids appear to have no effect on growth or sporulation (C. Garcia, unpublished data). Disruption experiments have been carried out; the homozygous null mutation has no effect on sporulation (D. Primerano, unpublished data).

SPR3 is a late gene which encodes a transcript of 2 kb. A map of this gene, constructed by M. Clancy, is shown in Fig. 2b. M. J. Clancy (unpublished data) has carried out an extensive series of experiments with this gene. *SPR3* has been inserted into each of two plasmids, one containing an *ARS* and one containing an *ARS* and a centromere, and the effect of each on sporulation determined. Interestingly, *SPR3* as a plasmid seems to cause cells growing on acetate to go into stationary phase at a cell density $\frac{1}{10}$ that of the plasmid-free strain, but such strains sporulate normally. Two *lac* fusions have been constructed; one is at the *Bgl*II site and the other at the *Bam*HI site, about 100 bp away from the first. Neither is expressed strongly, even during sporulation, when carried on a plasmid. A possible gene disruption has been carried out by insertion of *LEU2* gene from yeast at the *Bgl*II site. A strain homozygous for this construction sporulates normally. The direction of transcription is not known for sure, but it is most likely from right to left. It is possible, therefore, that the insert does not affect an important coding region, since it may lie rather close to the 3' end of the gene.

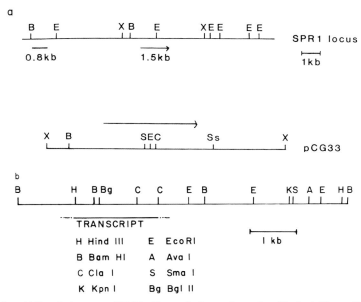

Fig. 2. (a) Restriction map of *SPR1*. Above, the largest insert found in the λ Charon 28 library. Below, the subclone used for S1 nuclease mapping. The 5' end of the transcript entends to within 300–400 bp of the *Bam*HI site. The 3' end is not mapped, but it does not extend past the *Sst*I site. Restriction sites: E, *Eco*RI; B, *Bam*HI; X, *Xho*I; S, *Sal*I; C, *Cla*I; Ss, *Sst*I. (b) Restriction map of *SPR3*. Transcription is probably from right to left.

Several other *SPR* genes (*SPR7*, *SPR9*, and *SPR2*) are in the process of being analyzed, but no definitive data have been forthcoming from that analysis.

A search identical to that of Clancy *et al.* (1983) for sporulation-specific genes was carried out by Percival-Smith and Segall (1984), except that their library was constructed in the plasmid pBR322, using a *Sau*3A partial digest sized to between 4 to 8 kb. RNA was extracted from cells at 3, 7, 10, and 15 hr, but in this case the material from the different time points was not pooled. Two rounds of differential colony hybridization and one of differential dot-blot hybridization, using cDNA from *a*/α and α/α cells, yielded 15 colonies which appeared to contain genes expressed only during sporulation. Twenty-three other colonies appeared to contain genes expressed in all cell types during vegetative growth but expressed at a qualitatively greater rate during sporulation in *a*/α cells. These 38 clones were classified into groups thought to contain the same gene based on the size of restriction fragments generated with *Hin*fI and the size of the sporulation-specific transcripts. Fourteen groups emerged: of these, one (with five independent isolates) appeared to contain a gene whose transcript (950 bp long) is expressed at a qualitatively different rate during sporulation; the others all con-

12. Transcription during Meiosis

tained a gene or at least one transcript which appeared only in sporulating cells. These transcripts ranged in size from 860 to 2800 bases. Interestingly, only one of the groups showed changes in transcript level at any time after 7 hr; this was Group 11, whose single isolate hybridized strongly with 10-hr cDNA but not with that from 7 or 15 hr.

The actual number of separate genes included in the plasmids isolated by Segall's laboratory is rather hard to determine. The method of classifying, by size of restriction fragments and of transcript, can be deceiving since common size does not necessarily indicate identity. Furthermore, in the restriction digests illustrated, there is ample evidence of partial digestion. However, it is safe to say that they have at least 14 and quite possibly several more separate sporulation-specific genes. M. J. Clancy, P. T. Magee, and J. Segall (unpublished data) have exchanged clones and carried out cross-hybridization to determine which genes have been independently isolated in the two searches. Surprisingly, there is evidence for only one duplicate isolation: *SPR9* cross-hybridizes with Group 8 of Percival-Smith and Segall. Since the isolation techniques were extremely similar, down to the use of related strains as the source of the mRNA, it is hard to understand the lack of overlap. However, the clear implication is that a number of other sporulation-regulated genes remain to be isolated by this technique.

A. Garber and J. Segall (personal communication, 1984) have used hybrid-selected *in vitro* translation to examine the gene products of plasmid 27, a representative of Percival-Smith and Segall's Group 2. This plasmid contains a gene which codes for a protein of 44,000 MW which is a predominant sporulation product. This most likely corresponds to one of the 46,000- to 47,000-Da proteins of Weir-Thompson and Dawes (1984). Gene disruption experiments are in progress to attempt to determine the role of this protein in meiosis.

A. Percival-Smith and J. Segall (personal communication) have shown that plasmid 84 of Group 6 contains three sporulation-specific transcripts of 1.9, 1.6, and 1.75 kb. The gene encoding the 1.6-kb transcript has been disrupted by integration of an internal fragment in the 5' region, leading to a 0.8-kb deletion and the insertion of the *URA3* gene. MAT**a**/MATα strains homozygous for this disruption do not form asci; the stage at which they are blocked is not yet determined (A. Percival-Smith and J. Segall, unpublished data). A fusion of the first 403 bp of this gene with the *E. coli lacZ* gene has been constructed; this fusion product expresses the *lac* gene (as determined by S_1 mapping) during sporulation but enzyme activity is not detectable.

Ninfa *et al.* (1984) have also isolated DNA sequences encoding meiosis-specific transcripts using differential plaque hybridization. In this study, 2- to 4-kb inserts in λ607 were used. In a screen of 5500 plaques, four sporulation-specific sequences were found, each of which hybridized to a different transcript. These transcripts all appeared during premeiotic S, reached a maximum at about the time of ascus formation, and declined over the next 8 hr. These transcripts were present at about 20–25 copies per cell at their maximum levels. Two of these genes have been disrupted, one by insertion of a selectable marker into the coding region and one by deleting a portion of the gene and then inserting a new sequence. In neither case was the homozygote blocked in meiosis or sporulation.

To summarize the work on genes isolated by the use of the property of being transcribed only in meiotically competent cells, approximately 40 different sequences have been identified. In one case the gene product has been identified electrophoretically, but its function is not known. The timing of transcription of such sequences varies, but the most comprehensive study suggests that there are three main groups: early, beginning at the time the cells are transferred to sporulation medium; middle, beginning during premeiotic S; and late, beginning at meiosis I. There is no evidence for extensive clustering of sequences transcribed during meiosis, but at least one cluster has been identified. The phenotype of *a*/α diploids homozygotic for apparent disruptions of several of these genes has been determined, and in one case the cells were ascoporogenous, while in four others there was no apparent phenotype. It is important to note that such gene disruption experiments must be viewed extremely critically; unless the transcription map is known in great detail, it is quite easy to make a chromosome insertion which does not disrupt the gene but only makes a partial duplication (A. Percival-Smith, personal communication; D. Primerano, personal communication). Therefore, one must always show that the transcript is missing or altered before one can be certain that the gene has been disrupted. (It is of course possible that even a truncated transcript might code for a partially active gene product and thus yield a pseudowild phenotype.) Furthermore, the phenotype of a disruption might be extremely subtle. Effects on recombination, chromosome disjunction, or spore viability will not be detected if the only assay is the appearance of asci. At the time of this writing, only Ninfa *et al.* (1984) have fulfilled all the criteria necessary to show that disruption of a sporulation-specific gene does not change the phenotype of the homozygote, and they have done it in only one of two disruptions. While it is therefore clear that some fraction of the sporulation-specific sequences will be dispensable for meiosis and sporulation, the size of that fraction cannot be determined as yet.

The existence of these sequences and their similarity in kinetics of appearance renders them interesting regardless of the nature of their role in sporulation. Like the heat shock genes (Ingolia *et al.*, 1980), these sequences seem to share a common regulatory pattern, and elucidation of the mechanism of their regulation may provide insight into general questions about the control of expression of gene families, regardless of whether they are required for meiosis.

D. Transcriptional Analysis of Cloned *SPO* Genes

R. Esposito and R. Elder and their co-workers (R. Esposito, personal communication) have cloned three of the *SPO* genes, *SPO11*, *SPO12*, and *SPO13*, and

12. Transcription during Meiosis

analyzed their transcription patterns. *SPO11* was cloned by complementation; the phenotype of mutants in this gene is lack of recombination during meiosis, leading to chromosome nondisjunction and aneuploidy of the haploid products. The selection, therefore, was for cells which yielded viable spores expressing recessive markers (cycloheximide or canavanine resistance). *SPO12* and *13* presented more difficult cloning problems; mutations in these genes lead to a suppression of the reductional meiotic division and the development of two-spored asci containing viable diploid spores. These two genes were cloned by selecting for markers known to be genetically closely linked, *ARG4* for *SPO13* and *SUF8* for *SPO12*, and using complementation to demonstrate the presence of the *SPO* gene. Each of the genes was localized to a specific chromosomal fragment and transcription from these fragments was analyzed by the S_1 nuclease technique, using single-stranded probes generated in M13 vectors, as well as by Northern blots.

The results of the *SPO13* analysis showed that there is a major 1.1-kb and a minor 1.5-kb transcript which are rapidly induced when the cells enter sporulation mediums, appearing at 1.5 hr. These transcripts differ at the 5' end but have similar or identical 3' end. They are induced some 75-fold, apparently greater than any of the *SPR* genes of Holoway *et al.* (1985). However, this ratio depends greatly on the baseline in vegetative cells, and the S_1 technique gives much better precision than the dot-blot one. The transcripts reach a maximum at 4.5 hr and persist until well after asci begin to appear, but they are greatly reduced after 30 hr in sporulation medium. Both transcripts are transiently induced in *a*/*a* or α/α cells. The 1.5-kb transcript is initiated at the 5' site of the 1.0-kb one; its function is not known. Complementation analysis and gene disruption using the cloned fragments demonstrated that the 1.0- (major) kb transcript encodes the *SPO13* function, since the fragment that contains the start site for the 1.4-kb transcript was not needed.

The *SPO11* fragment encodes two transcripts, one of 1.66 and one of 1.5 kb. These arise from the same strand of DNA and are not overlapping. The 1.5-kb transcript is induced beginning almost immediately after the shift to sporulation medium and reaches a maximum by 6 hr, when it has increased about 65-fold. The 1.66-kb transcript, which terminates 50 bp upstream from the start site for the 1.5-kb one, is turned on in vegetative cells but is undetectable in sporulating cells. Disruption of the region encoding the 1.5-kb transcript gives a *SPO11* phenotype.

The transcription pattern of the *SPO12* clone is rather more complicated. The 6-kb insert encodes two transcripts, a 1.2-kb RNA molecule which is induced 60-fold during sporulation and a 0.7-kb molecule which is unregulated. These transcripts are convergent and overlapping; the 3' end of the 1.4-kb transcript falls within 100 bp of the 5' end of the 0.7-kb RNA molecule (Fig. 3). The 1.2-kb transcript, like the 1.1-kb one from *SPO13*, shows a transient induction in MAT*a*/MAT*a* and MATα/MATα cells. The *SPO12-11* mutation has been

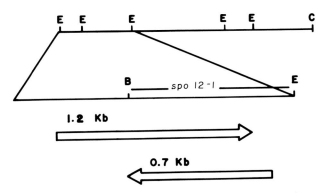

Fig. 3. Restriction map of the *SPO12* region. These are convergent transcripts, one of which (1.4 kb) is found at low levels in both vegetative cells and cells in sporulation medium, and the other (0.7 kb) occurs at moderate levels in vegetative and high levels in sporulating cells. The 0.7-kb transcript has been found to encode the *SPO12* function; the 1.4-kb transcript is involved in sporulation; deletion of most of this gene reduces sporulation 4- to 10-fold in a homozygous diploid.

mapped to the region in which the two transcripts overlap. Deletion of almost all of the 0.7-kb transcript sequence indicates that this sequence encodes the *SPO12* function. Deletion of the nonoverlapping part of the gene encoding the 1.2-kbp transcript also gives a spo⁻ phenotype, suggesting that this gene is essential for sporulation. This gene has been designated *SPO16*.

These results on known *SPO* genes suggest some rather interesting ideas about the transcriptional program during meiosis and sporulation in *S. cerevisiae*. One common feature is the early onset of transcription of these genes, in contrast to the *SPR* genes, many of which seem to be turned on at the time at which the *SPO* gene transcripts are reaching their maximum or declining. These observations suggest that the *meiosis*-specific genes may be among the very early sporulation-specific transcripts. The early genes of Holaway *et al.* (1985) show kinetics not dissimilar to those of the regulated transcripts of *SPO12* and SPO13, although they do not seem to be so strongly induced. One might imagine, then, that the early *SPR* genes will tend to be meiosis-specific, the middle ones to be important in spore formation, and the late ones to be involved in spore maturation. Although this is doubtless an oversimplification, it may be useful as a guide in designing experiments to determine the phenotype of a disruptant.

The prospect of identifying the products of such fascinating genes as *SPO12* and *SPP13*, as well as *SPO11*, is enticing. Until now no proteins directly involved in the mechanics of chromosome disjunction have been identified, but the *SPO12* and *13* gene products are clearly very strong candidates. The *SPO11* gene product seems to be involved in the same recombination pathway as *RAD52* (Esposito and Klapholz, 1981), and its gene product might fulfill any of a number of functions; the fact that *SPO11* has no mitotic phenotype might indicate that it is involved in meiotic synapsis of chromosomes.

E. Transcriptional Analysis of the Sporulation Amyloglucosidase Gene

Yamashita and Fukui (1985) have cloned the sporulation amloglucosidase gene by using its homology to the *STA* gene which encodes an enzyme necessary for starch degradation in *Saccharomyces diastaticus* (Yamashita *et al.*, 1985). They showed that the gene was regulated at the level of mRNA. Its 2.0-kb transcript was very evident at 8 hr after cells entered sporulation medium and increased in intensity until 12 hr, after which its level declined. Since Northern blots only measure RNA abundance, this implies but does not prove transcriptional regulation. The transcript did not appear in MAT*a* or MATα cells under any conditions nor was it found in MAT*a*/MATα cells on vegetative medium. Its appearance was dependent on *MATa1* and *MATα2*. Disruption of the gene by integrating an internal fragment caused loss of the enzyme activity and accumulation of glycogen rather than breakdown of the polysaccharide in a homozygote, but there was no effect on sporulation.

F. Transcription of Mitotic Genes during Meiosis

Kaback and Feldberg (1985) have examined the transcription during meiosis and sporulation of a number of genes whose function appears to be mitotic, not meiotic, as well as genes which either are or might be required in both parts of the life cycle. Among the former are *CDC10*, a gene whose function appears to be in cytokinesis, and *GAL10*, *TRP1*, and *HO*. The genes expected to function in both mitosis and meiosis included the *H2A* and *H2B* histone genes, *URA3*, and *TDH* (the gene for triosephosphate dehydrogenase, an enzyme required for glycolysis and gluconeogenesis). Except for *HO*, all these genes were transcribed at some time during meiosis and sporulation. *URA3* and *TDH* were turned on immediately after transfer to sporulation medium; by the middle of premeiotic S, transcripts declined to a level equal to or below the initial one. For both genes, the level of mRNA rose late in sporulation. The *H2A* and *H2B* genes behaved as expected; their transcripts began to rise at the beginning of premeiotic S, reached a maximum late in S, and declined. In contrast to the situation in vegetative cells (Osley and Hereford, 1981), the transcripts remained at an intermediate level throughout the rest of the developmental cycle. Histone mRNA was in fact isolatable from polysomes from fresh ascospores, although the amount of mRNA in ascospores declined over four days to an undetectable level.

GAL10, surprisingly, behaved very similarly to the histone genes except that its transcripts began to increase about an hour later than those of the *H2a* and *H2b* genes, declined to a lower level, and then rose late in sporulation. *CDC10*, another gene expected to be required only during mitosis, also showed sporulation-medium-enhanced transcription, reaching a peak in transcript accumulation at about the time of spore formation, declining over the next 4–6 hr, and disappearing by 24 hr. In both *GAL10* and the *CDC10* cases, cells genetically incapable of meiosis and sporulation either did not accumulate the specific transcript or accumulated it at a much reduced level. The haploid-specific gene, *HO*, was not expressed by cells of **a**/α mating-type configuration in sporulation medium.

Kaback and Feldberg's results demonstrate that regulation of gene expression during meiosis and sporulation in yeast is not as tightly ordered as one might expect. At least two genes, *CDC10* and *GAL10*, which are demonstrably nonessential for sporulation (homozygotes for mutations in either gene sporulate normally) are transcribed differentially in sporulation medium, and the levels that their transcripts reach are significantly higher than those occurring in vegetative medium. There are a number of possible explanations for this behavior. One attractive one is that the general proteolysis which occurs in sporulating cells inactivates a number of negative controlling elements, leading to synthesis of a variety of negatively controlled genes. In the case of galactose metabolism, we know that the *GAL80* gene product plays such a role (Torchia *et al.*, 1984) albeit one more complicated than that of the *lac* repressor. Since nothing is known about the regulation of the *CDC10* gene, we can form no hypotheses about the reason for its increased expression.

The kinetics with which the *GAL10* and *CDC10* transcripts appear and disappear differ from those of any of the genes isolated by Clancy *et al.* (1983) and Holoway *et al.* (1985). The transcripts of the nonessential genes increase abruptly, over 4 hr, whereas the *SPR* transcripts remain elevated for at least 4–8 hr. This behavior might be expected if the transcription of the mitosis-specific genes were the result of a transient decline in the concentration of a regulatory molecule.

V. SUMMARY AND FUTURE PROSPECTS

As the work in this review shows, progress is beginning to be made in the area of identifying the genes expressed during meiosis in a variety of cells. Some inferences about the need for specific gene products can be drawn from the experiments showing that mutations causing a variety of phenotypes in mitotic cells may affect meiosis. (These studies, of course, are only possible in organ-

12. Transcription during Meiosis

isms with a well-developed genetic system.) However, since the physiology of the meiotic process is still not well understood, interpretation of many of these observations is somewhat problematic. It seems quite possible that many gene products may persist from premeiotic cells, and these genes, while they may be necessary for the process, would not be considered to be expressed during meiosis. This is not a trivial point, since one of the objects of studying meiosis is to identify the way in which gene expression is regulated, and essential gene products which are synthesized before meiosis and persist constitute a class different from those which are synthesized during the process. The observation by several laboratories, especially Bachvarova and her co-workers, that mammalian meiocytes show active transcription and contain significant amounts of long-lived poly(A)$^+$ RNA is also interesting but represents the earliest stages of the work required to look at specific gene expression; the same is true of the *Lilium* studies.

The identification of specific genes which are activated during meiosis is a major step forward. However, there are several caveats which must be observed in evaluating these results. The first point to be made is that meiosis may be accompanied by a loosening of the regulatory controls over the expression of a variety of genes. This is not only plausible but likely in yeast, where extensive proteolysis takes place in meiotic cells (Hopper *et al.*, 1974), and it may be the case in the meiocytes of higher organisms, although perhaps to a lesser extent than in fungi. A significant fraction of the transcription observed, then, could be noise rather than signal. The results of Kaback and Feldberg (1985) certainly suggest that there is a significant amount of noise in sporulating yeast.

Even if the transcription observed is meiosis-specific, in that the genes are responding to signals which occur only in meiocytes, the transcripts may not be vital to the meiotic process. Evidence is accumulating of functional redundancy in developmental pathways, Evidence includes the apparent dispensability of the *SAG* gene in *S. cerevisiae* (Yamashita and Fukui, 1985) and a condidiation-specific tubulin which can be replaced by mitotic tubulin in *Aspergillus*, but only if there is a null mutation in the condiation tubulin gene (Weatherbee, *et al.*, 1985). Genes whose products provide auxiliary functions, such as helping the meiotic products to survive, may be transcribed during meiosis. The transcripts or the gene products may be stored, since the function may be necessary soon after meiosis is complete. Analysis of such genes may not tell us much about the molecular events of meiosis per se, but their regulation is likely to be quite interesting. They may respond to a signal which also turns on meiotic genes, or they may be regulated by a simultaneous, but parallel set of signals. In any case, their analysis is likely to be worthwhile.

If we define meiotic genes, as mentioned above, as those whose products play a vital role in the classical stages of meiosis, their number is likely to be rather small and the difficulty of identifying them among the other types mentioned, short of having a test for function, will be great. However, the recent demonstrations that genes whose products are ubiquitous, such as actin (Ng and Abelson, 1980), tubulin (Neff et al., 1983), or ras (Powers et al., 1984), are highly conserved makes it likely that as such genes are identified in one system, like yeast, they will be able to be identified in a variety of organisms by cross-hybridization. Genes functioning in kinetochore organization, synaptonemal complex formation, and recombination might be identifiable in this way.

The advantage of using S. cerevisiae as a model system for examining transcription during meiosis relates of course to the possibility of carrying out genetic studies on the genes transcribed. Most importantly, specific functions can in principle be assigned to such genes by examining the phenotype of mutations constructed in vivo and then replaced in the organism. It is somewhat disappointing that of the four genes studied only the one disrupted by Percival-Smith and Segall has been known to lead to an asporogenesis phenotype, and since the only parameter determined was ascus formation, meiosis may be unaffected, as it is, for example, in the spo 3 mutation studied by Moens et al. (1974). Whether studies on the rest of these genes will yield a similar ratio of essential to nonessential genes remains to be seen. It is difficult to imagine, despite the observation of activation of vegetative genes by Kaback and Feldberg (1985), that these tightly and coordinately regulated groups of genes do not have specific functions in meiosis. It is therefore important to examine the phenotype of any authentic gene disruptants in detail to avoid overlooking subtle phenotypes.

The molecular examination of the genes whose spo^- phenotype is known, by R. Esposito, R. Elder, and their co-workers, has yielded some very interesting results. Among these are the observation that transcription of these genes begins very early. Clancy, Magee, and Segall all chose to use probes isolated when the cells were undergoing premeiotic DNA synthesis, on the view that most of the earlier gene activity would be common to all cells adjusting to the starvation conditions of sporulation medium, as indicated by the work of Hopper et al. (1974). While this assumption is probably correct, early-expressed genes that are meiosis-specific appear to be very important, as shown by the fact that SPO11, 12, and 13 all appear to be transcribed beginning very early after the shift to sporulation medium, and all affect functions which occur 5–7 hr later at the earliest. The possible mode of regulation suggested by the SPO12 results, convergent transcription, is extremely exciting; confirmation and further elucidation of this phenomenon will be of great general interest in addition to casting light on the developmental regulation of meiosis.

As the details of the regulation of transcription of the SPO and SPR genes in yeast are worked out, the data obtained will be useful in heterologous systems. The identification of poly(A)$^+$ transcripts in mammalian meiocytes will in principle allow the isolation of meiosis-regulated genes from these organisms. The complexity of the mammalian genome increases the logistical problems, but

12. Transcription during Meiosis

there is precedent for overcoming these problems in a variety of systems. Indeed, the use of interspecific homology may circumvent them altogether. It thus appears that in the next few years we may see rapid advances in our understanding of the molecular bases for the complex developmental phenomena associated with meiosis.

ACKNOWLEDGMENTS

I would like to thank Dr. Donald Primerano for reading the manuscript and making helpful suggestions. The work from my laboratory reported here was supported by NIGMS Grant GM 33305.

REFERENCES

Armstrong, R. L., West, T. P., and Magee, P. T. (1984). Protein synthesis in germinating *Saccharomyces cerevisiae* ascospores. *Can. J. Microbiol.* **30,** 345–352.

Bachvarova, R. (1974). Incorporation of tritiated adenosine into mouse ovum RNA. *Dev. Biol.* **40,** 52–58.

Bachvarova, R. (1981). Synthesis, turnover, and stability of heterogeneous RNA in growing mouse oocytes. *Dev. Biol.* **86,** 384–392.

Bachvarova, R., Burns, J. P., Spiegelman, I., Choy, J., and Chaganti, R. S. K. (1982). Morphology and transcriptional activity of mouse oocyte chromosomes. *Chromosoma* **86,** 191–196.

Baker, T. G., and Franchi, L. L. (1967). The structure of the chromosomes in human primordial oocytes. *Chromosoma* **22,** 358–377.

Bakken, A. H., and Hamkalo, B. (1978). Techniques for visualizing genetic material. *In* "Principles of Techniques of Electron Microscopy" (M. Hayat, ed.), pp. 84–106. Reinhardt, New York.

Bakken, A. H., and McClanahan, M. (1978). Patterns of RNA synthesis in early prophase oocytes from fetal mouse ovaries. *Chromosoma* **67,** 21–40.

Becket, A., Illingworth, R. F., and Rose, A.H. (1973). Ascospore wall development in *Saccharomyces cerevisiae. J. Bacteriol.* **115,** 1054–1057.

Benton, D., and Davis, R. W. (1977). Screening λgt recombinant clones by hybridization to single plaques *in situ. Science* **196,** 180–182.

Betz, H. (1979). Loss of sporulation ability in a yeast mutant with low proteinase A levels. *FEBS Lett.* **100,** 171–174.

Brower, P. T., Gizang, E., Boreen, S. M., and Schultz, R. M. (1981). Biochemical studies of mammalian oogenesis: Synthesis and stability of various classes of RNA during growth of the mouse oocyte *in vitro. Dev. Biol.* **86,** 373–383.

Chaffin, W. L., Sogin, S. J., and Halvorson, H. O. (1974). Nature of ribonucleic and synthesis during early sporulation in *Saccharomyces cerevisiae. J. Bacteriol.* **120,** 872–879.

Clancy, M. J., Smith, L. M., and Magee, P. T. (1982). Developmental regulation of a sporulation-specific enzyme activity in *Saccharomyces cerevisiae. Mol. Cell. Biol.* **2,** 171–178.

Clancy, M. J., Buten-Magee, B. D., Straight, D., Kennedy, A. L., Partridge, R. M., and Magee, P. T. (1983). Isolation of genes expressed preferentially during sporulation in the yeast, *Saccharomyces cerevisiae. Proc. Natl. Acad. Sci. U.S.A.* **80,** 3000–3004.

Colonna, W., and Magee, P. T. (1978). Glycogenolytic enzymes in sporulating yeast. *J. Bacteriol.* **134**, 844–853.

Croes, A. F. (1967). Induction of meiosis in yeast. II. Metabolic factors leading to meiosis. *Planta* **76**, 227–237.

De Leon, V., Johnson, A., and Bachvarova, R. (1983). Half-lives and relative amounts of stored and polysomal ribosomes and poly(A)+ RNA in mouse oocytes. *Dev. Biol.* **98**, 400–408.

del Rey, F., Sautos, T., Garcia-Acha, I., and Nombela, C. (1980). Synthesis of β-glucanases during sporulation in *Saccharomyces cerevisiae;* formation of a new sporulation-specific 1,3 β-glucanase. *J. Bacteriol.* **143**, 621–627.

Dickinson, H. J., and Heslop-Harrison, J. (1970). The ribosome cycle, nucleoli, and cytoplasmic nucleotides in the meiocytes of *Lilium*. *Protoplasma* **69**, 187–200.

Dickinson, H. G., and Heslop-Harrison, J. (1977). Ribosomes, membranes, and organelles in meiosis. *Philos. Trans. R. Soc. London Ser. B* **277**, 327–342.

Dowling, E. L., Maloney, D. H., and Fogel, S. (1985). Meiotic recombination and sporulation in repair-deficient strains of yeast. *Genetics* **109**, 283–302.

Emanuel, J., and Magee, P. T. (1981). Timing of ribosome synthesis during ascosporogenesis of yeast cells: Evidence for early function of haploid daughter genomes. *J. Bacteriol.* **145**, 1342–1350.

Erickson, R. P., Erickson, J. M., Betlach, C. J., and Meistrich, M. L. (1980). Further evidence for haploid gene expression during spermatogenesis: Heterogeneous poly(A)-containing RNA is synthesized post-meiotically. *J. Exp. Zool.* **214**, 13–19.

Esposito, M. S., and Esposito, R. E. (1969). The genetic control of sporulation in *Saccharomyces.* I. The isolation of temperature-sensitive sporulation-deficient mutants. *Genetics* **61**, 79–89.

Esposito, M. S., and Esposito, R. E. (1974). Genes controlling meiosis and spore formation in yeast. *Genetics* **78**, 215–225.

Esposito, M. S., Esposito, R. E., Arnaud, M., and Halvorson, H. O. (1969). Acetate utilization and macromolecular synthesis during sporulation of yeast. *J. Bacteriol.* **100**, 180–186.

Esposito, R. E., and Klapholz, S. (1981). Meiosis and ascospore development. *In* "The Molecular Biology of the Yeast *Saccharomyces:* Life Cycle and Inheritance" (J. Strathern, E. Jones, and J. Broach, eds.), pp. 211–287. Cold Spring Harbor Lab., Cold Spring Harbor, New York.

Fourcroy, J. (1982). RNA synthesis in immature mouse oocyte development. *J. Exp. Zool.* **219**, 257–266.

Godwin, H. (1968). The origin of the exine. *New Phytol.* **67**, 667–676.

Gold, B., Fujimoto, H., Kramer, J. M., Erickson, R. P., and Hecht, N. B. (1983). Haploid accumulation and translational control of phosphoglycerate kinase-2 messenger RNA during mouse spermatogenesis. *Dev. Biol.* **98**, 392–399.

Harper, J., Clancy, M. J., and Magee, P. T. (1980). Properties of polyadenylate-associated ribonucleic acid from *Saccharomyces cerevisiae* ascospores. *J. Bacteriol.* **143**, 958–965.

Hien, N. H., and Fleet, G. H. (1983a). Separation and characterization of six (1-3) β-glucanases from *Saccharomyces cerevisiae*. *J. Bacteriol.* **156**, 1204–1213.

Hien, N. H., and Fleet, G. H. (1983b). Variation of (1-3) β-glucanases in *Saccharomyces cerevisiae* during vegetative growth, conjugation, and sporulation. *J. Bacteriol.* **156**, 1214–1220.

Holoway, B. L., Lehman, D. J., Primerano, D. A., Magee, P. T., and Clancy, M. J. (1985). Sporulation-regulated genes of *Saccharomyces cerevisiae*. *Curr. Genet.* **10**, 163–169.

Hopper, A. K., Magee, P. T., Welch, S. M., Friedman, M., and Hall, B. D. (1974). Macromolecule synthesis and breakdown in relation to sporulation of meiosis in yeast. *J. Bacteriol.* **119**, 619–629.

Ingolia, T. D., Slater, M. R., and Craig, E. A. (1980). *Saccharomyces cerevisiae* contains a complex multigene family related to the major heat shock-inducible gene of *Drosophila*. *Mol. Cell. Biol.* **2**, 1388–1398.

12. Transcription during Meiosis

Jahn, C. L., Baran, M. M., and Bachvarova, R. (1976). Stability of RNA synthesized by the mouse oocyte during its major growth phase. *J. Exp. Zool.* **197**, 161–172.

Kaback, D. B., and Feldberg, L. R. (1985). Yeast cells exhibit a sporulation-specific temporal program of transcript accumulation. *Mol. Cell Biol.* **5**, 751–761.

Klar, A. J. S., Cohen, A., and Halvorson, H. O. (1976). Control of enzyme synthesis and stability during sporulation of *Saccharomyces cerevisiae*. *Biochimie* **58**, 219–224.

Kleene, K. C., Distel, R. J., and Hecht, N. B. (1983). cDNA clones encoding cytoplasmic poly(A)$^+$ RNAs which first appear at detectable levels in haploid phases of spermatogenesis in the mouse. *Dev. Biol.* **98**, 455–464.

Kraig, E., and Haber, J. E. (1980). Messenger ribonucleic acid and protein metabolism during sporulation in *Saccharomyces cerevisiae*. *J. Bacteriol.* **144**, 1098–1112.

Kurtz, S., and Lindquist, S. (1984). Changing patterns of gene expression during sporulation in yeast. *Proc. Natl. Acad. Sci. U.S.A.* **83**, 7323–7327.

Magee, P. T. (1974). Changes in DNA-dependent RNA polymerase in sporulating yeast. *Mol. Biol. Rep.* **1**, 275–281.

Magee, P. T., and Hopper, A. K. (1974). Protein synthesis in relation to sporulation and meiosis in yeast. *J. Bacteriol.* **119**, 952–960.

Miller, J. J., and Hoffman-Ostenhof, O. (1964). Spore formation and germination in *Saccharomyces cerevisiae*. *Z. Allg. Mikrobiol.* **4**, 273–294.

Mills, D. (1972). Effect of pH on adenine and amino acid uptake during sporulation in *Saccharomyces cerevisiae*. *J. Bacteriol.* **112**, 519–526.

Mills, D. (1980). Quantitative and qualitative analysis of polyadenylated RNA sequences during meiosis of *Saccharomyces cerevisiae*. *Abstr. Int. Conf. Yeast Genet. Mol. Biol., 10th, 1980*, p. 55.

Moens, P. B., Esposito, R. E., and Esposito, M. S. (1974). Aberrant nuclear behavior at meiosis and anucleate spore formation by sporulation-deficient (spo) mutants of *Saccharomyces cerevisiae*. *Exp. Cell Res.* **83**, 166–174.

Neff, N. F., Thomas, J. H., Grisafi, P., and Botstein, D. (1983). Isolation of the β tubulin gene from yeast and demonstration of its essential function *in vivo*. *Cell* **33**, 211–219.

Ng, R., and Abelson, J. (1980). Isolation and sequence of the gene for actin in *Saccharomyces cerevisiae*. *Proc. Natl. Acad. Sci. U.S.A.* **77**, 3912–3916.

Ninfa, E. G., Feldberg, L., and Kaback, D. B. (1984). DNA sequences complementary to sporulation-specific transcripts are not necessarily essential for differentiation. *Abstr. Int. Conf. Yeast Genet. Mol. Biol., 12th, 1984*, p. 265.

O'Farrell, P. H. (1975). High resolution two-dimensional resolution of proteins. *J. Biol. Chem.* **250**, 4007–4021.

Osley, M. A., and Hereford, C. M. (1981). Yeast histone genes show dosage compensation. *Cell* **24**, 377–384.

Pearson, N. J., and Haber, J. E. (1980). Changes in the regulation of ribosomal protein synthesis during vegetative growth and sporulation in yeast. *J. Bacteriol.* **143**, 1411–1419.

Percival-Smith, A., and Segall, J. (1984). Isolation of DNA sequences preferentially expressed during sporulation in *Saccharomyces cerevisiae*. *Mol. Cell. Biol.* **4**, 142–150.

Peterson, J. G., Kielland-Brandt, M. C., and Nilsson-Tillgren, T. (1979). Protein patterns of yeast during sporulation. *Carlsberg Res. Commun.* **44**, 149–162.

Porter, E. K., Bird, J. M., and Dickinson, H. G. (1982). Nucleic acid synthesis in microsporocytes of *Lilium* cv. cinnabar: Events in the nucleus. *J. Cell Sci.* **57**, 229–246.

Porter, E. K., Parry, D., and Dickinson, H. G. (1983). Changes in poly(A)$^+$ RNA during male meiosis in *Lilium*. *J. Cell Sci.* **62**, 177–186.

Powers, S., Kataoka, T., Fasaro, O., Goldfarb, M., Strathern, J., Broach, J., and Wiglur, M. (1984). Genes in *S. cerevisiae* encoding proteins with domains homologous to the mammalian *ras* proteins. *Cell* **36**, 607–612.

Roman, H., and Sands, S. M. (1953). Ieterogeneity of clones of *Saccharomyces* derived from haploid ascospores. *Proc. Natl. Acad. Sci. U.S.A.* **39**, 171–179.

Roth, R., and Lusnak, K. (1970). DNA synthesis during yeast sporulation: Genetic control of an early developmental event. *Science* **168**, 493–494.

Rothstein, R. (1983). One-step gene disruption in yeast. *In* "Methods in Enzymology" (R. W. L. Grossman and K. Moldave, eds.), Vol. 101, pp. 202–211. Academic Press, New York.

Rousseau, P., and Halvorson, H. O. (1973). Macromolecular synthesis during the germination of *Saccharomyces cerevisiae*. *J. Bacteriol.* **113**, 1289–1295.

St. John, T. P., and Davis, R. W. (1979). Isolation of galactose-inducible DNA sequences from *Saccharomyces cerevisiae* by differential plaque filter hybridization. *Cell* **16**, 443–452.

Sheldon, J., and Dickinson, H. G. (1983). The determination of patterning in the pollen wall of *Lilium henryi*. *J. Cell Sci.* **62**, 250–257.

Shortle, D., Haber, J. G., and Botstein, D. (1982). Lethal disruption of the yeast actin gene by integrative DNA transformation. *Science* **217**, 317–373.

Southern, E. M. (1975). Detection of specific sequences among DNA fragments separated by gel electrophoresis. *J. Mol. Biol.* **98**, 503–517.

Stern, H., and Hotta, Y. (1977). Biochemistry of meiosis. *Philos. Trans. R. Soc. London, Ser. B* **277**, 277–293.

Thomas, P. S. (1983). Hybridization of denatured RNA transferred or dotted to nitrocellulose paper. *In* "Methods in Enzymology" (R. Wu, L. Grossman, and K. Moldave, eds.), Vol. 100, 255–266. Academic Press, New York.

Torchia, T. E., Hamilton, R. W., Cano, C. L., and Hopper, J. E. (1984). Disruption of regulatory gene GAL80 in *Saccharomyces cerevisiae:* Effects on carbon-controlled regulation of the galactose/melibiose path genes. *Mol. Cell. Biol.* **4**, 1521–1527.

Trew, B. J., Friesen, J. D., and Moens, P. B. (1979). Two-dimensional protein patterns during growth and sporulation in *Saccharomyces cerevisiae*. *J. Bacteriol.* **138**, 60–69.

Weatherbee, J. A., May, G. S., Gambino, J., and Morris, N. R. (1985). Involvement of a particular species of beta-tublin (beta 3) in condial development in *Aspergillus nidulans*. *J. Cell Biol.* **101**, 706–711.

Weir-Thompson, E., and Dawes, I. W. (1984). Developmental changes in translatable RNA species associated with meiosis and spore formation in *Saccharomyces cerevisiae*. *Mol. Cell. Biol.* **4**, 695–702.

Wright, J. F., and Dawes, I. W. (1979). Sporulation-specific protein changes in yeast. *FEBS Lett.* **104**, 183–186.

Wright, J. F., Ajam, N., and Dawes, I. W. (1981). Nature and timing of some sporulation-specific protein changes in *Saccharomyces cerevisiae*. *Mol. Cell. Biol.* **1**, 910–918.

Yamashita, I., and Fukui, S. (1985). Transcriptional control of the sporulation-specific glucoamylase gene in the yeast *Saccharomyces cerevisae*. *Mol. Cell Biol.* **5**, 3069–3073.

Yamashita, I., Suzuki, C. K., and Fukui, S. (1985). Nucleotide sequence of the extracellular glucoamylose gene STA1 in the yeast *Saccharomyces diastaticus*. *J. Bacteriol.* **161**, 567–573.

Zubenko, G. S., and Jones, E. W. (1981). Protein degradation, meiosis, and sporulation in proteinase-deficient mutants of *Saccharomyces cerevisiae*. *Genetics* **97**, 45–64.

Zubenko, G. S., Mitchell, A. P., and Jones, E. W. (1979). Septum formation, cell division, and sporulation in mutants of yeast deficient in proteinase B. *Proc. Natl. Acad. Sci. U.S.A.* **76**, 2395–2399.

Index

A

Acheta domesticus, 247, 344
Achiasmatic nucleus, 313, 316, 319, 328
Actin, synaptonemal complex, 261
Adenylate cyclase, 166
Adventitious embryony, 80
Agamospermy, 80
Alchemilla, 83
   apomictic hybridization, 51
Alchemilla gelida, 51
Alchemilla glacilis, 51
Alchemilla pentaphyllea, 51
Allium cepa, 223
   colchicine effect, 293–294
   telomere, 277–279
Allium fistulosum, 223, 227, 285
   crossover, 17
Ambystoma tremblayi, 71
   diplotene spread, 96
Ameiotic restitution, 69
α-1,4-Amyloglucosidase, 361
Amyloglucosidase gene, 359
   transcriptional analysis, 375
Anennaria diplospory, 81–82
Apis mellificia male haploidy, 84–86
Apogamety, 81
   hybrids, 49–53
Apomixis, 47–51
   Alchemilla, 51
   fish, 69–70
   forms, 80–81
   Hieracim, 49–50
   Poa, 55
   Taraxacum, 52
Apospory, 80, 82–83
Apotettix, parthenogenesis, 60–63
Apotettix convexus, 61
Apotettix eurycephalus, 61
Apterona helix meiosis, 88, 93
Arrhenotoky, 84–86
Artemia, 56
   cytology, 89, 93
Artemia salina, 56
   meiotic restoration, 90–92, 94
Ascaris, 215, 280
Ascobolus, 151–152
   parental genotype restoration, 123–127
Ascobolus immersus, 148
   recombination in, 108, 114, 115–116, 120
Aspergillus, tubulin, 377
Aspergillus nidulans, 108
Atraphaxis, 83
Autoimmune disease, mild, 259
Autotriploid, origin, 53
Aviemore model, 153
   crossover position control, 148–151
   heteroduplex, 140–142

B

Baboon, spermatogenesis histones, 343
Bacillus rossii, 93
Bacterophage, 22, 179
Batrachoseps attenuatus, 213
Biochemistry
   approach, 304–305
   DNA-unwinding protein, 319–320
   endonuclease, 319
   meiosis-specific products, 361–363
   nuclear proteins, 333–353
   prezygotene phase, 305–308
   reassociation protein, 320–321
   recombination catalyzing protein, 321–324

Biochemistry (*cont.*)
  recombination proteins, 318–324
  Saccharomyces cerevisiae meiosis and sporulation, 358–359
Bivalent
  acrocentric, 227
  telocentric, 227
Black fly, 56
Bombyx, 215, 249, 250
Bombyx mori, 288
Bothriopiana semperi, 89
Brine shrimp, 56
Brookesia spectrum, 77
Bull, spermatogenesis histones, 343

C

C-Banding, 286, 293
  interphase, 276–279
  leptotene, 287
Caledia, 227
Centromere, 227
  differentiation, 248–249
  leptotene, 287–291
  localization, 259
  mitosis
    interphase, 276–279
    prophase, 279–282
  premeiotic mitosis, 284–287
Chara crinita, 47
Chiasmata
  control, 225–229
    genetic, 230
    mechanism, 232–238
    models, 231–232
  distribution, 15–17
  formation mechanism, 217–220
  frequency, 236
  hidden, 219
  historical, 213–216
  interchiasma distance, 228
  sequential, 237
  terminal bivalent, 221–223
  terminalization, 220–221
  variation, 223–224
Chiasmatype theory, 216–217
Chicken, oocyte, 289
Chimpanzee, spermatogenesis histones, 343

Chloealtis abdominalis
  meiotic behavior, 1–5
  translocation, 15–16
Chloealtis conspersa, 227, 233
  recombination, 13–17
  synaptonemal complex, 12–13
Chorthippus, 223
Chorthippus brunneus, 224
  chiasmata, 227–229
Chromosome
  meiotic behavior, 1–5
  lampbrush, 356
  pairing, 9–13
    presynaptic, 285
  processing, 167–168
Chyletus eruditus, 93
Clarkia xanthia, 26
Cleavage nuclei, fusion, 93–96
Clitumnus extradentatus, 94
Cnemidophorus, 69
Cnemidophorus cozumela, 76
Cnemidophorus dixoni, 76
Cnemidophorus exsanguis, 76
Cnemidophorus flagellicauda, 76
Cnemidophorus laredoensis, 76
Cnemidophorus lemniscatus, 76
Cnemidophorus neomexicanus, 76
Cnemidophorus opatae, 76
Cnemidophorus rodecki, 76
Cnemidophorus sonorae, 76
Cnemidophorus tesselatus, 76
Cnemidophorus uniparens, 31, 71, 76
Cnemidophorus velox, 76
Cnephia mutata, 56
Colchicine, 313
  synapsis effect, 292–294
Colchicine-binding protein, 293
Coprinus, 167, 171, 179, 215
Crepis, 83
Crepis capillaris, 230, 281
  chiasmata, 224, 226
CREST syndrome, 259
Cricket, 346
  cave, 56
Cross-hybridization, 311
Crossing over, 311, *see also* Recombination
Crossover position, 125–127
  control, 148–151
  interference, 108
    intragenic, 111

Index

Culex, 227
Cytokinesis, gene, 375

D

Danthonia spicata, selfing propensity, 26–27
Dense body, 255–259
  double, 225
Diabrotica, 223
Diplospory, 81–82
Diplotene, 309
DNA
  HMG protein binding, 339
  metabolism, meiotic, 177–178
  polymerase, meiosis, 167, 318
  processing, 164
  ratio, meiotic, 344
  replication
    pachytene, 313–318
    prezygotene phase, 305–306
    suppression, 312
    zygotene, 307, 309–310
  transcription
    meiotic, 355–382
    zygotene, 310–311
DNA-unwinding protein, 319–320
DNase II, 316
Dog, spermatogenesis histones, 343
Double-strand break repair model, 145–146
Double-strand gap filling, 132
Drosophila, 216, 231
  meiosis-defective mutant, 176–177
  parthenogenesis, 47
Drosophila mangabeirai, 32, 95
Drosophila melanogaster, 227, 230, 250, 251, 287–288
  chiasmata, 234–235
  polytene chromosome, 282–283
Drosophila mercatorum, 32
Drosophila parthenogenetica, 32
Drosophila virilis, 282
Duiprion polytomium, 94

E

Echinostelium, 255
Egg, diploid production, 71
Endoduplication, 71
Endomitosis, 71
Endonuclease, 319, 326, 328

Ephestia kuehniella synaptonemal complex, 10–11
Ernst's hypothesis, 50
Escherichia coli, 159, 314
  RecA protein, 144–145, 151
Euhadenoecus insolitus, 56
Euthystira brachyptera, 227, 234
Excision correction, mismatch, 124
Excision repair, 130–131
  alternatives to, 131–135

F

Fish, parthenogenesis, 68–79
Fitness, genotype covariance, 22–25

G

G_1 phase, 277
Gap-filling, 132
Gecko, parthenogenetic, 75–78
Gehyra variegata, 78
Gehyra variegaga ogasawarasimae, 78
Gene
  amplification, 163
  conversion, 162–163
  meiosis-specific, 179–182
  products, meiosis-specific, 360–365
Genetic imbalance hypothesis, 48
Genotype, fitness covariance, 22–25
Giant chromosome, 10
β-1,3-Glucanase, 361–363
α-Glucosidase, 361
Guinea pig, spermatogenesis histones, 343
Gynogenesis, 68

H

Hadenoecini, 56
Hadenoecus cumberlandicus, 56
Hamster
  chiasmata, 219
  Chinese, 249
    zygotene, 289
  spermatogenesis histones, 343
Haplopappus gracilis, 281
HeLa S-3 cells, 283
Hemidactylus granoti, 75, 77
Hemidactylus karenorum, 75, 77
Hemidactylus vietnamensis, parthenogenesis, 71–74

Heterochromin, 235
  block, 285
  mass, 286
Heteroduplex conversion, *see also*
  Recombination
  co-correction, 122–123
  hypothesis status, 130–131
  mappability, 129–130
  parental restoration, 123–129
  tetrad ratio, aberrant, 121–122
Heteroduplex formation, 320
  Aviemore model, 143–144
  conversion and crossing-over separation, 151–153
  direction of travel, 139–140
  double-strand break repair model, 145–146
  Holliday model, 140–142
  position control, 148–151
  RecA model, 144–145
  RecBC model, 146–147
Heteronotia binoei, 77
  parthenogenesis, 71–74
Heterozygote, translocation, 15
Hieracium, 83
  parthenogenesis, 48, 49
Hieracium aurantiacum, 49–50
Hieracium auricula, 50
Hieracium flagellare, apospory, 84
Histone
  classes, meiotic, 336–347, 336
  meiotic
    male rat, 336–342
    mammalian, other than rat, 342–344
    nonmammalian, 344–345
  oocyte, 345–347
  separation, 338
  sporulation, 375–376
  synthesis
    meiotic, 340–342, 344
    rate, 342
  testis-specific, 342
  variant, 336–337, 339
    percentage, 341
HMG protein, 337–347
Holliday model, heteroduplex, 140–142
Homo sapiens, 288
Honey bee, male haploidy, 84–86
Human, spermatogenesis histones, 343
Hyalophora columbia, synaptonemal complex, 10–12

Hybridization, parthenogenesis
  hypothesis, 47–55
  insect, 58–66
  vertebrate, 66–79
Hypochaeric radicata, 230, 236–237

I

Insect, hybridization parthenogenesis, 58–66
Interference, 235
Interphase, mitosis, Rabl Orientation, 276–279
Interstitial differentiation, 249
Ionizing radiation, 276

K

Karogamy, 291–292
Kentropyx borckionus, 77
Kinetochore, 248
  synaptonemal complex, 259–261

L

L-protein, 321, 326–328
  leptotene, 306–308
  zygotene, 312–313
Larcerta armeniaca, 77
Larcerta dahli, 77
Larcerta unisexualis, 77
Lateral element, 248
  length change, 250
  thickness change, 249–250
Lecanium hemisphaericum, 94
Leiolepis triploida, 74, 77
Lepidodactylus lugubris, 77–79
Leposoma percarinatum, 77
Leptotene, 326–328
  DNA synthesis, 307
  Rabl Orientation, 287–291
Lespedeze cuneata, 26
Lilium, 179
  centromere, 285
  colchicine effect, 292–293
  meiocyte transcription, 358
  meiotic histones, 344
  recombination processing, 172–174
Lilium longiflorum, 5, 6, 237, 287, 249
Lizard
  parthenogenesis, 68–79

Index

parthenogenetic chromosome number, 76–77
Locusta, 249
Locusta migratoria, 219, 237
   chiasmata, 214, 222
   chromosome pairing, 10
Lolium, 223
Lonchoptera dubia, 32, 95
Lumbricellus lineatus, 94, 95
Lundstroemia parthogenetica, 90
Lycosperiscon esculentum, 230
Lymnophyes biverticillatus, 90
Lymnophyes virgo, 90

M

Maize, 249
   leptotene, 288
Male haploidy, 84–86
Mapping
   flanking marker combinations, 110–111
   heteroduplex conversion, 129–130
   intragenic, 170
Meiocyte differentiation, 304–305
Meiosis-specific transcript, 371–372, 374
Meiotic commitment, 5–9
Meiotic restitution
   apospory, 83–84
   arrhenotoky, 84–86
   diplospory, 81–82
   parthenogenetic, 79–98
   thelytoky, 86–96
Melanoplus differentialis, 248
Mesostoma ehrenbergii ehrenbergii, 232
Mitosis
   premeiotic, 284–287
   Rabl Orientation, 276–284
   reversion, 306
Mitotic genes, meiotic transcription, 375–376
Mitotic reversion, 7
Model
   parthenogenesis evolution, 33–37
   selfing evolution, 27–31
Monkey, rhesus, spermatogenesis histones, 343
Monoclonal antibody, synaptonemal complex, 265–268
Mouse, 250, 277
   chiasmata, 219
   rec protein, 321, 323
   spermatogenesis histones, 343

Multiple complex, 247
Mutation
   frameshift, 114, 123–124
   isolation and characterization, 175–182
   meiosis-defective, 176
   meiotic, 166, 230
      preconditioned, 231
   recombination-deficient, 176
   substitution, 114
Myosin, synaptonemal complex, 261–262
Myrmeleotettix, 223

N

Natural selection, single multiallelic locus, 23–25
Neotiella, 215
Neurospora, 152, 130, 215
   crossover control, 149–151
   recombination, 169
Neurospora crassa, 230
   karogamy, 291
   prototrophic frequency, 108–111
   tetrad ratio, 111–114
Newt, 346
Nicking, pachytene, 315–317
Nicotina rustica, 26
Nondisjunction, 168
Nuclear lamins, 264–265
Nuclear protein, study methods, 334–336
Nuclease α, 174
Nuclease control, 194–196
Nucleolar protein C23, 257
Nucleoli, synaptonemal complex, 255–259
Nucleosomal core histone, 336

O

Ochthiphila polystigma, 56
Oenothera, 230
   translocation heterozygote, 15
Omocestus, 223
Oocyte
   meiotic histones, 345–347
   transcription, 357
Ornithogalum virens, 281, 285
Otiorrhincus scaber, 57
Otiorrhincus singularis, 57

P

Pachytene, 291, 322, 325–328
  biochemistry summary, 324
  DNA
    replication, chromosomal site, 313–315
    synthesis, 309
    nicking, 315–317
    nuclear isolation, 348
Paeonea, 227
Pales, 215
Parthenogenesis, 9, 15, *see also* Apomixis
  animal, 55–58
  apospory, 82–83
  Apotettix, 60–63
  arrhenotoky, 84–86
  cost associated, 32–33
  diplospory, 81
  evolution, 43–104
  evolutionary model, 33–37
  genetic variation, 31–32
  hybridization
    hypothesis, 47–55
    insect, 58–66
  meiosis and, 44–47
  meiotic restitution, 71
    mechanism, 80–98
  mutation source, 55–58
  Tephrosia, 59–60
  thelytoky, 86–96
  vertebrate origin, 66–79
Phaedranassa, 249
Phyllodactylus marmoratus, 224
Pisum sativum, 230
Plasmid, zygDNA insert, 312
Poa, apomictic, 55
Poa caespitosa, 55
Poa nevadensis, 55
Poa pratensis, 55
Poa scabrellae, 55
Podospora anserina, 230
Poecilia formosa, 68–69
Polar nuclei, fusion, 92–93
Pollen mother cell, 292
  chaismata, 222, 224
Polycelis nigra, 89
Polycomplex, 247
Polydrosus mollis, 57
Polyploidy, 47
  origin, 53–55

Polytene chromosomes, Rabl Orientation, 282–283
Potentilla, 83
Prealignment, 285, 293
Precocity Theory, 306
Preleptotene, 326
Prezygotene phase, DNA replication, 305–306
Pristophora pallipes, 94
Processing factors, recombination, 172–175
Prometaphase, 281–282
Prophase
  early meiotic, Rabl Orientation, 287–291
  meiotic, 293
    histones, 337
  mitosis, 279–282
Protein, nuclear protein, structural, 347–349
Psn-protein, 326
Pulvinaria hydrangae, 94

R

R-protein, 307, 320, 324, 326–328
RAD genes, biochemical relation, 187–189
RAD52 gene
  meiosis role, 189–196
  nuclease control, 194–196
  phenotypes, 186–187
Rabbit, spermatogenesis histones, 343
Rabl Orientation
  cochicine and temperature effect, 292–294
  mitosis
    interphase, 276–279
    model, 283–284
    prophase, 279–282
  polytene chromosomes, 282–283
  premeiotic mitosis, 284–287
  prophase, early meiotic, 287–291
  zygotic meiosis, 291–292
Rana esculunta, 71
Rana pipiens, 345
Rat
  oocyte histone, 345–347
  spermatogenesis histones, 336–342
Reassociation protein, 320–321
Rec protein, 321–322
RecA
  model, heteroduplex, 144–145, 151
  protein, 173
RecA-like protein, 321–322

Index

RecBC
  enzyme, 169
  model, heteroduplex, 146–147
Recombination, 13–17, *see also* Heteroduplex conversion
  additivity, 118–119
  biochemical events, yeast, 182–196
  catalyzing protein, 321–324
  cell extract activity, 322–324, 328
  conversion and crossing-over separation, 151–153
  conversion spectrum variation, 114–115
  crossing over, 107–108
  crossover location, 116–117
  general description, 119–120
  genetically identifiable factors, 163–175
    chromosomal, 169–172
    indirect, 166–169
    processing, 172–175
  induction, 193
  initiation, 314
  intragenic, 108–111
    interference, 111
  loci variation, 117–119
  mapping, 110–111, 117–118
  meiosis relation, 159–161
  molecular changes, 190–192
  mutation isolation and characterization, 175–182
  nodule, 16, 154, 251
  polarity, 115–116, 119
  symmetry, 116
  tetrad ratio, 111–114
  types and mechanisms, 161–163
Recombination proteins, 318–324
Regulation, meiotic events, 325–329
Repair mutant, categories, 187–188
Repair replication, 313, 315–316
Resolvase, 174
Restoration, measurement, 128
Rheo spatheceae, 290
  translocation heterozygote, 15
Ribonucleoprotein, small nuclear, 263–264
RNA
  in vitro translation, meiotic, 363–365
  meiotic-specific, 179–182
  poly (A$^+$), 311, 325, 359, 363–364, 366
  polymerase, 362
  sporulation, 361

Rubi eubati, 54
Rye, pollen mother cell, 222

S

S phase
  premeiotic, 322
  prezygotene, 305–306
Saccharomyces cerevisiae, 115, 123
  double-strand gap filling, 132
  meiocyte transcription, 358–376
  meiosis regulation, 325
  meiosis-specific gene products, 360–365
  meiotic commitment, 5–9
  meiotic genes, 366–372
  rec protein, 323
  recombination, 108–109, 159, 162, 166, 182–196
Scepticus insularis, 57
Schistocerce gregaria, 215, 227
  chiasmata, 235–236
Schizophyllum commune, 230, 287
Schizosaccharomyces pombe, 115
  recombination, 163, 166, 169
Sea urchin, histone, 345, 347
Secale, 222, 237
Selfing propensity
  cost associated, 26–27
  evolutionary models, 27–31
  genetic variation, 25–26
Sex bivalent, mammalian, 222
Solenboia, 32
Solenboia fenicella, 58
Solenobia lichenella, 58
  meiosis, 88, 91
Solenobia triguetrella, 56, 58, 95
  meiosis, 88
Sordaria, 215
Sordaria brevicollis, 148–150, 152
Sordaria fimicola, 148
  karogamy, 291–292
Sordaria humana, 215
Sperm, diploid production, 71
Spermatocyte
  purification, 335
  synaptonemal complex/nuclear matrix, 348
  transcription, 357
Spermatogenesis
  histone replacement, 337, 342

Spermatogenesis (cont.)
  proteins
    male rat, 336–342
    nonmammalian, 344–345
    other mammalian male, 342–344
Sporulation
  genes, cloned transcriptional analysis, 372–375
  RNA, 361
Sporulation-regulated genes, transcription, 366–372
Sporulation-specific genes, 370–372
Stethophyma, 249
Stethophyma grossum, 218, 227, 232–233
Strophosomus capitatutus, 57
Strophosomus melanogrammus, 57
Streptococcus, transformation, 133–134
Stick insect, 93
Stonewort, 47
Synapsis, 290–291, 308–309, 326
  colchicine and temperature effect, 292–294
Synaptic adjustment, 251
Synaptic association, change, 250–251
Synaptonemal complex, 10–13, 235, 309
  biochemical enrichment, 254
  definition, 245–246
  dynamic nature, 248–252
  extra-synaptonemal complex components, 247–248
  immunocytochemistry, 252–255
  kinetochore, 259–261
  monoclonal antibody, 265–268
  nucleolar component association, 255–259
  protein, 261–265, 333–334
    structural, 347–349
Synaptonemal complex/nuclear matrix structure, 348

T

Taraxacum
  diplospory, 81, 83
  parthenogenesis, 48, 49, 51, 52
Telomere, 248
  colchicine effect, 292–294
  differentiation, 248–249
  leptotene, 287–291
  mitosis
    interphase, 276–279
    prophase, 279–282
  premeiotic, 286–287

Telophase
  mitosis, 277
  premeiotic, 284–285
Temperature
  recA protein, 321
  synapsis effect, 292–294
Tephrosia bistortata, 59–60
Tephrosia crepuscularia, 59–60
Tetrad, crossover classification, 126
Tetrahymena thermophilia, 179
Thelytoky, 31
  evolution, 33
  meiotic restitution, 86–96
Topisomerase, 318
  I, 173
  II, role, 173
Transcription
  Lilium meiocyte, 358
  meiosis-specific, 325–329
  meiotic genes, 366–372
  meiotic, mammalian meiocyte, 356–357
  mitotic transcription during meiosis, 375–376
  Saccharomyces cerevisiae meiocyte, 358–376
Transcriptional analysis
  amyloglucosidase gene, 375
  sporulation genes, 372–375
Transformation, 133–134
Translocation
  heterozygote, 15
  reciprocal, chiasmata, 236
Trialeurodes vapororiorum, 93
Trichoniscus, 56
Trichoniscus elisabethae, 56
Trillium erectum, 229
Triticum, 223, 230
Tricitum aestivum, colchicine effect, 292–293
Triturus vulgaris, histones, 344
Tubulin, 377
  synaptonemal complex, 262–263
Tulip, 224
Tychoparthenogenesis, 46

U

U-protein, 319–320, 324, 326–328
Ultraviolet light, repair, 188
Urodeles, 277
Ustilago maydis, 134, 151
  recombination processing, 173–174

Index

V

Vicia faba, 227

W

Warramba picta, 63
Warramba virgo, 32
　parthenogenesis, 63–66
Weevil, parthenogenetic, 55–56
White fly, 93

X

Xenopus laevis, 356

Y

Yeast
　cytokinesis gene, 375
　meiocyte transcription, 358–376
　meiosis
　　genes transcribed, 366–372
　　mutant, 177
　　RAD52 pathway, 189–196
　　repair mutant, 185–189
　mitotic transcription, 375–376
　recombination
　　biochemical/genetic events, 184–185
　　meiotic events, 182–184
　restoration, 128–129
　sporulation gene, transcriptional analysis, 372–275

Z

Zea mays, 230, 249
Zygotene, 323, 326–328
　DNA
　　synthesis, 307, 309–310
　　transcription, 310–311
　L-protein activity, 312–313
　endonuclease, 319
Zygotic meiosis, Rabl Orientation, 291–292